Renault 14 Owners Workshop Manual

by J H Haynes
Member of the Guild of Motoring Writers
and Ian Coomber

Models covered
Renault 14 Base, TL, GTL, TS, LS, Safrane and Regency;
1218 cc and 1360 cc

ISBN 1 85010 192 2

© Haynes Publishing Group 1979, 1981, 1982, 1984, 1986, 1987, 1988

All rights reserved. No part of this book may be reproduced or transmitted in any form or by any means, electronic or mechanical, including photocopying, recording or by any information storage or retrieval system, without permission in writing from the copyright holder.

Printed in England *(362-4P6)*

ABC

Haynes Publishing Group
Sparkford Nr Yeovil
Somerset BA22 7JJ England

Haynes Publications, Inc
861 Lawrence Drive
Newbury Park
California 91320 USA

British Library Cataloguing in Publication Data

Coomber, Ian
 Renault 14 owners workshop manual. – 3rd ed.
 1. Renault automobile
 I. Title
 629.28'722   TL215.R4
 ISBN 1-85010-192-2

Acknowledgements

No book is the work of one person. This Owners Workshop Manual is the work of many and due credit should go to each and every one of them.

Thanks are due to Régie Renault – in particular Renault Limited (UK), for their assistance with technical information and the provision of certain illustrations. Duckhams Oils provided lubrication data, and the Champion Sparking Plug Company supplied the illustrations showing the various spark plug conditions.

Lastly, special thanks to all of those people at Sparkford who helped in the production of this manual.

About this manual

Its aim

The aim of this manual is to help you get the best value from your vehicle. It can do so in several ways. It can help you decide what work must be done (even should you choose to get it done by a garage), provide information on routine maintenance and servicing, and give a logical course of action and diagnosis when random faults occur. However, it is hoped that you will use the manual by tackling the work yourself. On simpler jobs it may even be quicker than booking the car into a garage and going there twice, to leave and collect it. Perhaps most important, a lot of money can be saved by avoiding the costs a garage must charge to cover its labour and overheads.

The manual has drawings and descriptions to show the function of the various components so that their layout can be understood. Then the tasks are described and photographed in a step-by-step sequence so that even a novice can do the work.

Its arrangement

The manual is divided into twelve Chapters, each covering a logical sub-division of the vehicle. The Chapters are each divided into Sections, numbered with single figures, eg 5; and the Sections into paragraphs (or sub-sections), with decimal numbers following on from the Section they are in, eg 5.1, 5.2, 5.3 etc.

It is freely illustrated, especially in those parts where there is a detailed sequence of operations to be carried out. There are two forms of illustration: figures and photographs. The figures are numbered in sequence with decimal numbers, according to their position in the Chapter – eg Fig. 6.4 is the fourth drawing/illustration in Chapter 6. Photographs carry the same number (either individually or in related groups) as the Section or sub-section to which they relate.

There is an alphabetical index at the back of the manual as well as a contents list at the front. Each Chapter is also preceded by its own individual contents list.

References to the 'left' or 'right' of the vehicle are in the sense of a person in the driver's seat facing forwards.

Unless otherwise stated, nuts and bolts are removed by turning anti-clockwise, and tightened by turning clockwise.

Vehicle manufacturers continually make changes to specifications and recommendations, and these, when notified, are incorporated into our manuals at the earliest opportunity.

Whilst every care is taken to ensure that the information in this manual is correct, no liability can be accepted by the authors or publishers for loss, damage or injury caused by any errors in, or omissions from, the information given.

Introduction to the Renault 14

Although first announced in May of 1976, the Renault 14 did not become available in the UK until March of 1977. Although basically a new concept in Renault ideas, it retains some features of the other models within their range whilst at the same time having its own individual character and incorporating some novel features.

The monocoque bodyshell is of a modern medium size, four-door, hatchback saloon type that is quite different in appearance from any other model within the Renault range. It can comfortably accommodate up to five people when required but, as has become fairly common these days, can be converted to a two-seat utility vehicle with the rear seat folded.

The most individual feature of the car is the layout of the engine and transmission units. Although there is nothing new about a transversely-mounted engine and transmission driving the front wheels, the manner in which Renault has mated the power unit to the transmission, does stray from convention.

The engine has been inclined to the rear with the cylinder head adjacent to the bulkhead. Whilst this reduces the height of the power unit, it also means that the normally simple process of renewing a cylinder head gasket requires the engine to be removed! The transmission assembly is mounted directly underneath the crankcase and incorporates the differential housing. The gearbox also serves as the sump, since the engine and transmission share the same lubricant. Both the engine and transmission housings are manufactured from an aluminium alloy.

In typical Renault fashion removable wet cylinder liners are used. The aluminium cylinder head is located by dowels and is secured by through-bolts which also secure the overhead camshaft bearing pedestals. The bolts pass through the head and the top half of the cylinder block; the securing nuts are located in channels formed in the side of the cylinder block. The camshaft is chain-driven whereas the oil pump is gear-driven; both drives are taken from the nose of the crankshaft.

The distributor is mounted onto the cylinder head at the flywheel end of the engine, driven directly from the camshaft.

Drive from the engine is via a diaphragm clutch and transfer gears which are encased separately on the outside of the clutch housing.

The gearbox has synchromesh on all forward gears and drive to the differential is direct, the crownwheel being a helical spur gear which is driven by the mainshaft pinion gear. The differential unit runs in shell bearings with thrust washers taking up the end play as opposed to the more conventional ball or roller bearing system.

The suspension is fully independent with MacPherson struts at the front, and trailing arm and torsion bar at the rear. Double-acting hydraulic shock absorbers are employed front and rear to soak up the road shocks, and this they do in typical Renault fashion – efficiently.

The Renault 14 has proved to be a popular member of the Renault family, and given the right treatment it will undoubtedly prove as reliable and practical as its relatives. However it must be said that it does not readily endear itself to the DIY mechanic, due to its previously-mentioned unconventional layout. Although most of the basic routine maintenance tasks can be easily undertaken, more serious problems will require careful thought and in some cases special tools. For this reason it is advisable to read through the Sections concerning the job at hand before starting to dismantle in order to assess the special requirements and points to watch out for.

Contents

	Page
Acknowledgements	2
About this manual	2
Introduction to the Renault 14	2
Buying spare parts and vehicle identification numbers	5
Tools and working facilities	6
Jacking and towing	8
Recommended lubricants and fluids	10
Routine maintenance	11
Chapter 1 Engine	14
Chapter 2 Cooling system	45
Chapter 3 Fuel and exhaust systems	53
Chapter 4 Ignition system	62
Chapter 5 Clutch	70
Chapter 6 Transmission	74
Chapter 7 Driveshafts, hubs, wheels and tyres	88
Chapter 8 Braking system	93
Chapter 9 Electrical system	105
Chapter 10 Suspension and steering	122
Chapter 11 Bodywork and fittings	136
Chapter 12 Supplement: Revisions and information on later models	**149**
Safety first!	196
Fault diagnosis	197
General repair procedures	200
Conversion factors	201
Index	202

Renault 14

Buying spare parts and vehicle identification numbers

Buying spare parts

Spare parts are available from many sources, for example: Renault garages, other garages and accessory shops, and motor factors. Our advice regarding spare parts is as follows:

Officially appointed Renault garage – This is the best source of parts which are peculiar to your car and otherwise not generally available (eg; complete cylinder heads, internal gearbox components, badges, interior trim etc). It is also the only place at which you should buy parts if your car is still under warranty; non-Renault components may invalidate the warranty. To be sure of obtaining the correct parts it will always be necessary to give the storeman your car's engine and chassis number, and if possible, to take the old part along for positive identification. Remember that many parts are available on a factory exchange scheme – any parts returned should always be clean! It obviously makes good sense to go straight to the specialists on your car for this type of part for they are best equipped to supply you.

Other garages and accessory shops – These are often very good places to buy material and components needed for the maintenance of your car (eg; oil filters, spark plugs, bulbs, drive belts, oils and grease, touch-up paint, filler paste etc). They also sell general accessories, usually have convenient opening hours, charge lower prices and can often be found not far from home.

Motor factors – Good factors will stock all of the more important components which wear out relatively quickly (eg; clutch components, pistons, valves, exhaust systems, brake cylinders/pipes/hoses/seals/shoes and pads, etc). Motor factors will often provide new or reconditioned components on a part exchange basis – this can save a considerable amount of money.

Vehicle identification numbers

Modifications are a continuing and unpublished process in vehicle manufacture, quite apart from major model changes. Spare parts manuals and lists are compiled upon a numerical basis, the individual vehicle numbers being essential for correct identification of the component required.

Some replacement items for the models covered by this manual are not interchangeable and you must therefore quote the correct Renault model number when describing your car to the storeman.

The storeman will need to know the *oval plate number* under all circumstances. For engine and transmission parts he will need to know the *engine number* and the *gearbox number* respectively in addition. For certain chassis and body parts he may need to know the *diamond plate number*. The *paint code* may be required if the shade of car is not easily described. All these numbers are located in readily visible places. (See the diagrams).

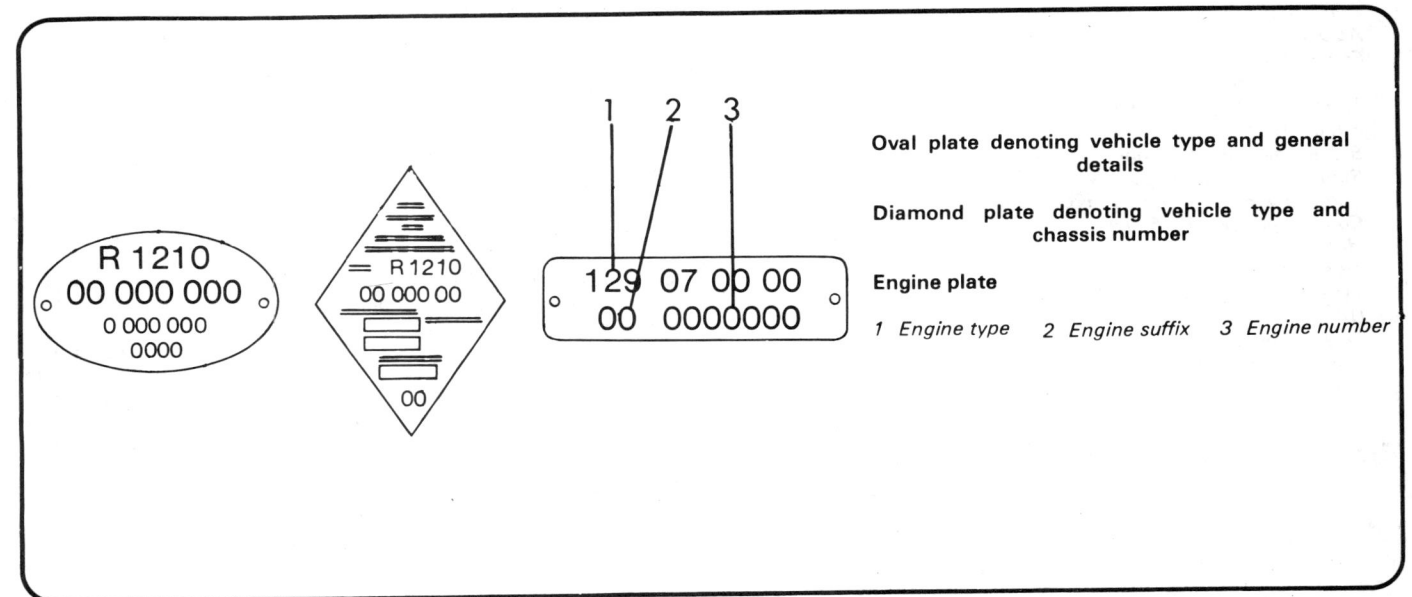

Oval plate denoting vehicle type and general details

Diamond plate denoting vehicle type and chassis number

Engine plate

1 Engine type   2 Engine suffix   3 Engine number

Tools and working facilities

Introduction

A selection of good tools is a fundamental requirement for anyone contemplating the maintenance and repair of a motor vehicle. For the owner who does not possess any, their purchase will prove a considerable expense, offsetting some of the savings made by doing-it-yourself. However, provided that the tools purchased are of good quality, they will last for many years and prove an extremely worthwhile investment.

To help the average owner to decide which tools are needed to carry out the various tasks detailed in this manual, we have compiled three lists of tools under the following headings: *Maintenance and minor repair*, *Repair and overhaul*, and *Special*. The newcomer to practical mechanics should start off with the *Maintenance and minor repair* tool kit and confine himself to the simpler jobs around the vehicle. Then, as his confidence and experience grows, he can undertake more difficult tasks, buying extra tools as, and when, they are needed. In this way, a *Maintenance and minor repair* tool kit can be built-up into a *Repair and overhaul* tool kit over a considerable period of time without any major cash outlays. The experienced do-it-yourselfer will have a tool kit good enough for most repair and overhaul procedures and will add tools from the *Special* category when he feels the expense is justified by the amount of use to which these tools will be put.

It is obviously not possible to cover the subject of tools fully here. For those who wish to learn more about tools and their use there is a book entitled *How to Choose and Use Car Tools* available from the publishers of this manual.

Maintenance and minor repair tool kit

The tools given in this list should be considered as a minimum requirement if routine maintenance, servicing and minor repair operations are to be undertaken. We recommend the purchase of combination spanners (ring one end, open-ended the other); although more expensive than open-ended ones, they do give the advantages of both types of spanner.

Combination spanners – 10, 11, 12, 13, 14, 17 mm
Adjustable spanner - 9 inch
Engine sump/gearbox drain plug key
Spark plug spanner (with rubber insert)
Spark plug gap adjustment tool
Set of feeler gauges
Brake bleed nipple spanner
Screwdriver – 4 in long x $\frac{1}{4}$ in dia (flat blade)
Screwdriver – 4 in long x $\frac{1}{4}$ in dia (cross blade)
Combination pliers – 6 inch
Hacksaw, junior
Tyre pump
Tyre pressure gauge
Oil can
Fine emery cloth (1 sheet)
Wire brush (small)
Funnel (medium size)

Repair and overhaul tool kit

These tools are virtually essential for anyone undertaking any major repairs to a motor vehicle, and are additional to those given in the *Maintenance and minor repair* list. Included in this list is a comprehensive set of sockets. Although these are expensive they will be found invaluable as they are so versatile – particularly if various drives are included in the set. We recommend the $\frac{1}{2}$ in square-drive type, as this can be used with most proprietary torque wrenches. If you cannot afford a socket set, even bought piecemeal, then inexpensive tubular box spanners are a useful alternative.

The tools in this list will occasionally need to be supplemented by tools from the *Special* list.

Sockets (or box spanners) to cover range in previous list
Reversible ratchet drive (for use with sockets)
Extension piece, 10 inch (for use with sockets)
Universal joint (for use with sockets)
Torque wrench (for use with sockets)
'Mole' wrench – 8 inch
Ball pein hammer
Soft-faced hammer, plastic or rubber
Screwdriver – 6 in long x $\frac{5}{16}$ in dia (flat blade)
Screwdriver – 2 in long x $\frac{5}{16}$ in square (flat blade)
Screwdriver – 1$\frac{1}{2}$ in long x $\frac{1}{4}$ in dia (cross blade)
Screwdriver – 3 in long x $\frac{1}{8}$ in dia (electricians)
Pliers – electricians side cutters
Pliers – needle nosed
Pliers – circlip (internal and external)
Cold chisel – $\frac{1}{2}$ inch
Scriber (this can be made by grinding the end of a broken hacksaw blade)
Scraper (this can be made by flattening and sharpening one end of a piece of copper pipe)
Centre punch
Pin punch
Hacksaw
Valve grinding tool
Steel rule/straight edge
Allen keys
Selection of files
Wire brush (large)
Axle-stands
Jack (strong scissor or hydraulic type)

Special tools

The tools in this list are those which are not used regularly, are expensive to buy, or which need to be used in accordance with their manufacturers' instructions. Unless relatively difficult mechanical jobs are undertaken frequently, it will not be economic to buy many of these tools. Where this is the case, you could consider clubbing together with friends (or a motorists' club) to make a joint purchase, or borrowing the tools against a deposit from a local garage or tool hire specialist.

The following list contains only those tools and instruments freely available to the public, and not those special tools produced by the vehicle manufacturer specifically for its dealer network. You will find occasional references to these manufacturers' special tools in the text of this manual. Generally, an alternative method of doing the job without the vehicle manufacturer's special tool is given. However, sometimes, there is no alternative to using them. Where this is the

Tools and working facilities

case and the relevant tool cannot be bought or borrowed you will have to entrust the work to a franchised garage.

Valve spring compressor
Piston ring compressor
Balljoint separator
Universal hub/bearing puller
Impact screwdriver
Micrometer and/or vernier gauge
Dial gauge
Stroboscopic timing light
Dwell angle meter/tachometer
Universal electrical multi-meter
Cylinder compression gauge
Lifting tackle
Trolley jack
Light with extension lead

Buying tools

For practically all tools, a tool factor is the best source since he will have a very comprehensive range compared with the average garage or accessory shop. Having said that, accessory shops often offer excellent quality tools at discount prices, so it pays to shop around.

Remember, you don't have to buy the most expensive items on the shelf, but it is always advisable to steer clear of the very cheap tools. There are plenty of good tools around at reasonable prices, so ask the proprietor or manager of the shop for advice before making a purchase.

Care and maintenance of tools

Having purchased a reasonable tool kit, it is necessary to keep the tools in a clean serviceable condition. After use, always wipe off any dirt, grease and metal particles using a clean, dry cloth, before putting the tools away. Never leave them lying around after they have been used. A simple tool rack on the garage or workshop wall, for items such as screwdrivers and pliers is a good idea. Store all normal spanners and sockets in a metal box. Any measuring instruments, gauges, meters, etc, must be carefully stored where they cannot be damaged or become rusty.

Take a little care when tools are used. Hammer heads inevitably become marked and screwdrivers lose the keen edge on their blades from time to time. A little timely attention with emery cloth or a file will soon restore items like this to a good serviceable finish.

Working facilities

Not to be forgotten when discussing tools, is the workshop itself. If anything more than routine maintenance is to be carried out, some form of suitable working area becomes essential.

It is appreciated that many an owner mechanic is forced by circumstances to remove an engine or similar item, without the benefit of a garage or workshop. Having done this, any repairs should always be done under the cover of a roof.

Wherever possible, any dismantling should be done on a clean flat workbench or table at a suitable working height.

Any workbench needs a vice: one with a jaw opening of 4 in (100 mm) is suitable for most jobs. As mentioned previously, some clean dry storage space is also required for tools, as well as the lubricants, cleaning fluids, touch-up paints and so on which become necessary.

Another item which may be required, and which has a much more general usage, is an electric drill with a chuck capacity of at least $\frac{5}{16}$ in (8 mm). This, together with a good range of twist drills, is virtually essential for fitting accessories such as wing mirrors and reversing lights.

Last, but not least, always keep a supply of old newspapers and clean, lint-free rags available, and try to keep any working area as clean as possible.

Spanner jaw gap comparison table

Jaw gap (in)	Spanner size
0.250	$\frac{1}{4}$ in AF
0.276	7 mm
0.313	$\frac{5}{16}$ in AF
0.315	8 mm
0.344	$\frac{11}{32}$ in AF; $\frac{1}{8}$ in Whitworth
0.354	9 mm
0.375	$\frac{3}{8}$ in AF
0.394	10 mm
0.433	11 mm
0.438	$\frac{7}{16}$ in AF
0.445	$\frac{3}{16}$ in Whitworth; $\frac{1}{4}$ in BSF
0.472	12 mm
0.500	$\frac{1}{2}$ in AF
0.512	13 mm
0.525	$\frac{1}{4}$ in Whitworth; $\frac{5}{16}$ in BSF
0.551	14 mm
0.563	$\frac{9}{16}$ in AF
0.591	15 mm
0.600	$\frac{5}{16}$ in Whitworth; $\frac{3}{8}$ in BSF
0.625	$\frac{5}{8}$ in AF
0.630	16 mm
0.669	17 mm
0.686	$\frac{11}{16}$ in AF
0.709	18 mm
0.710	$\frac{3}{8}$ in Whitworth; $\frac{7}{16}$ in BSF
0.748	19 mm
0.750	$\frac{3}{4}$ in AF
0.813	$\frac{13}{16}$ in AF
0.820	$\frac{7}{16}$ in Whitworth; $\frac{1}{2}$ in BSF
0.866	22 mm
0.875	$\frac{7}{8}$ in AF
0.920	$\frac{1}{2}$ in Whitworth; $\frac{9}{16}$ in BSF
0.938	$\frac{15}{16}$ in AF
0.945	24 mm
1.000	1 in AF
1.010	$\frac{9}{16}$ in Whitworth; $\frac{5}{8}$ in BSF
1.024	26 mm
1.063	$1\frac{1}{16}$ in AF; 27 mm
1.100	$\frac{5}{8}$ in Whitworth; $\frac{11}{16}$ in BSF
1.125	$1\frac{1}{8}$ in AF
1.181	30 mm
1.200	$\frac{11}{16}$ in Whitworth; $\frac{3}{4}$ in BSF
1.250	$1\frac{1}{4}$ in AF
1.260	32 mm
1.300	$\frac{3}{4}$ in Whitworth; $\frac{7}{8}$ in BSF
1.313	$1\frac{5}{16}$ in AF
1.390	$\frac{13}{16}$ in Whitworth; $\frac{15}{16}$ in BSF
1.417	36 mm
1.438	$1\frac{7}{16}$ in AF
1.480	$\frac{7}{8}$ in Whitworth; 1 in BSF
1.500	$1\frac{1}{2}$ in AF
1.575	40 mm; $\frac{15}{16}$ in Whitworth
1.614	41 mm
1.625	$1\frac{5}{8}$ in AF
1.670	1 in Whitworth; $1\frac{1}{8}$ in BSF
1.688	$1\frac{11}{16}$ in AF
1.811	46 mm
1.813	$1\frac{13}{16}$ in AF
1.860	$1\frac{1}{8}$ in Whitworth; $1\frac{1}{4}$ in BSF
1.875	$1\frac{7}{8}$ in AF
1.969	50 mm
2.000	2 in AF
2.050	$1\frac{1}{4}$ in Whitworth; $1\frac{3}{8}$ in BSF
2.165	55 mm
2.362	60 mm

Jacking and towing

The pantograph type jack supplied with the car should only be used at the positions provided below the body sills. *Never* go beneath the vehicle when only the pantograph jack is being used; provide additional support by positioning axle-stands beneath the sills, or below the front or rear subframe side-members. Always chock a roadwheel on the side of the car opposite to the jack.

Other types of jack should be used in conjunction with a suitable piece of wood, between the subframe side members, at the front in line with the suspension arm centres, and at the rear ahead of the fuel tank. For side jacking the piece of wood should have a groove to engage in the flange below the front door sill.

If the vehicle breaks down or becomes immobile, front and rear towing eyes are provided for the attachment of a tow-rope. **Caution** *The gearbox is pressure lubricated from the engine oil system and the front wheels must be lifted off the ground for towing. In exceptional circumstances the car may be towed with the front wheels on the ground provided a speed of 18 mph (30 km/hr) and a distance of 30 miles (50 km) are not exceeded.*

Locate jack in sill slot ...

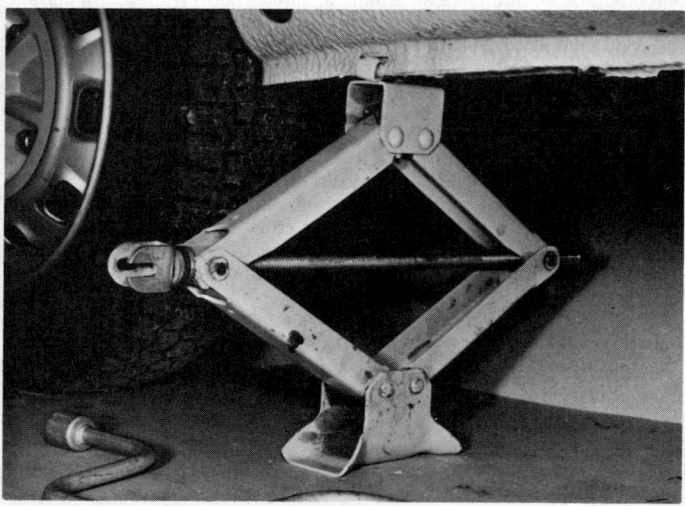

... then raise on firm, level ground

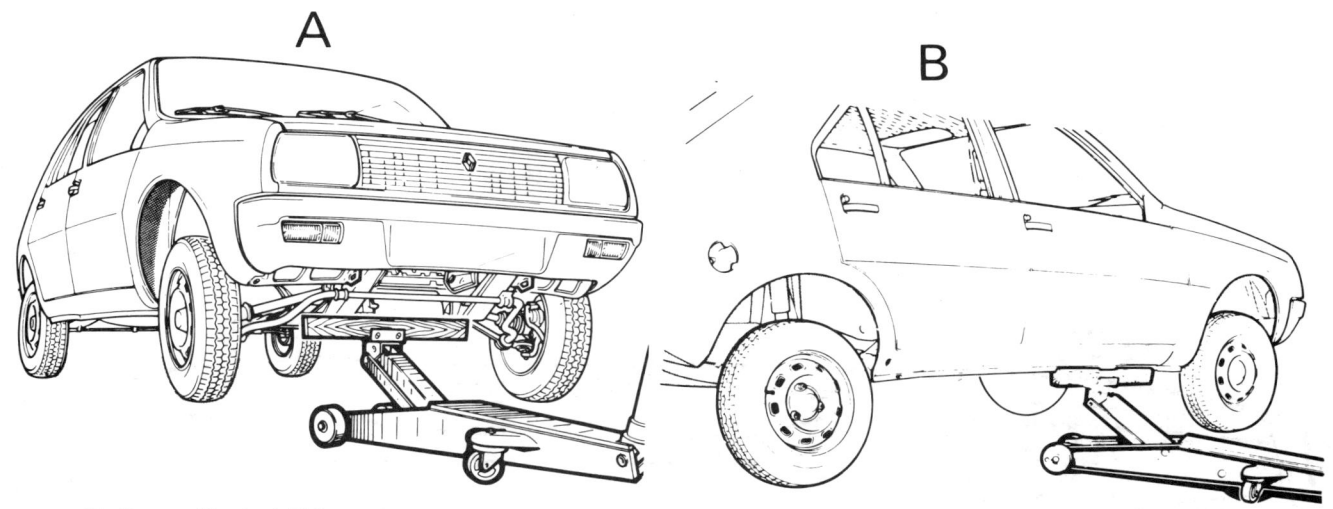

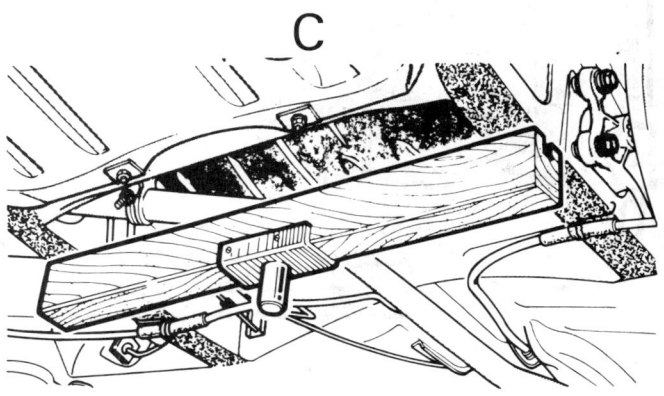

Trolley or pillar jack lifting points

(A) Front   (B) Centre   (C) Rear

Note use of wooden block to prevent underbody damage/distortion, and to spread the load

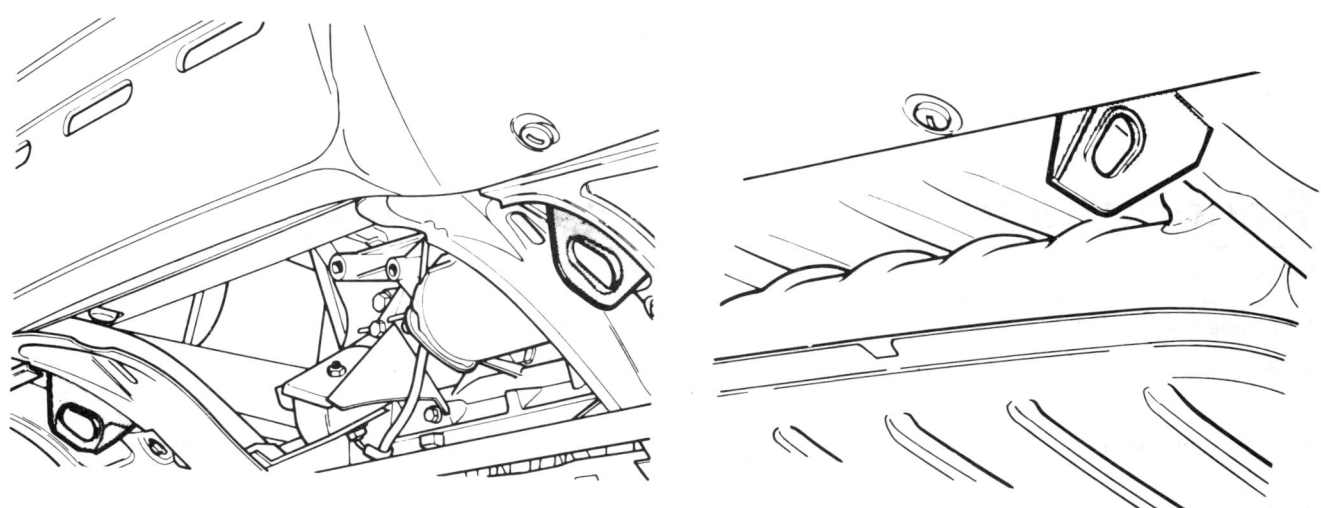

The front and rear tow hooks

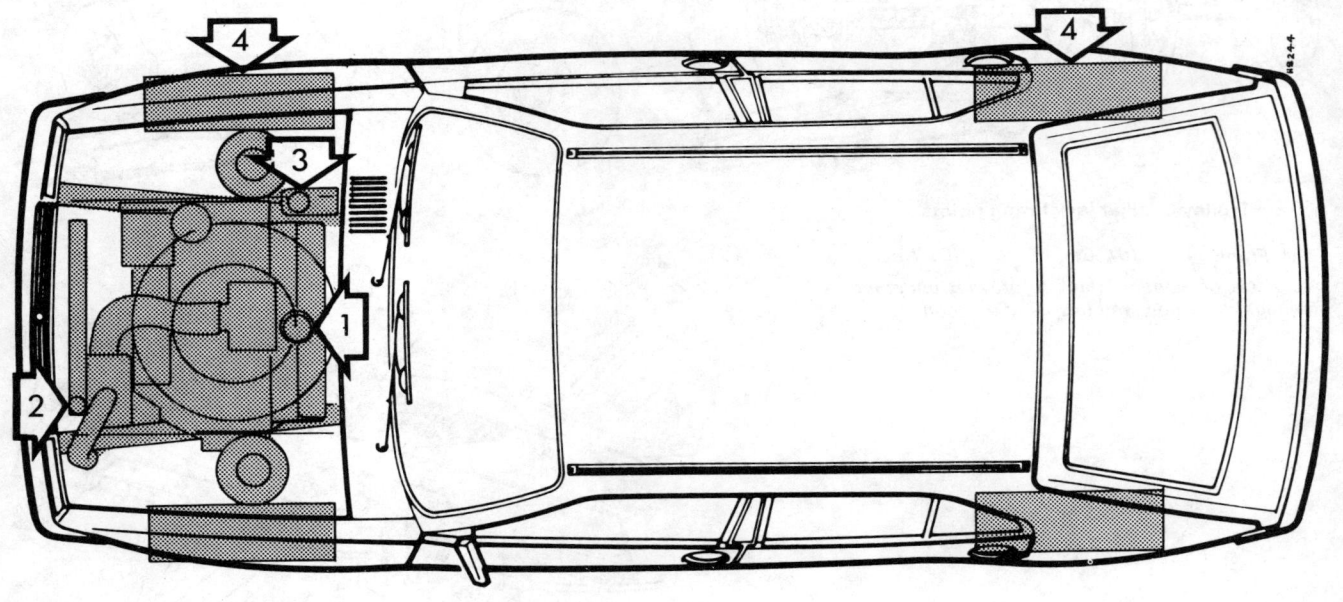

Recommended lubricants and fluids

Component or system	Lubricant type/specification	Duckhams recommendation
1 Engine/transmission	Multigrade engine oil, viscosity SAE 20W/50	Duckhams Hypergrade
2 Cooling system	Antifreeze to BS 3151, 3152 or 6580	Duckhams Universal Antifreeze and Summer Coolant
3 Brake fluid reservoir	Hydraulic fluid to SAE J1703F, DOT 3 or DOT 4	Duckhams Universal Brake and Clutch Fluid
4 Wheel hub bearings	Multi-purpose lithium-based grease	Duckhams LB 10
General greasing	Multi-purpose lithium-based grease	Duckhams LB 10

Routine maintenance

Maintenance is essential for ensuring safety and desirable for the purpose of getting the best in terms of performance and economy from the car. Over the years the need for periodic lubrication – oiling and greasing – has been drastically reduced if not totally eliminated. This has unfortunately tended to lead some owners to think that because no such action is required the components either no longer exist or will last for ever. This is a serious delusion. If anything, there are now more places, particularly in the steering and suspension, where joints and pivots are fitted. Although you do not grease them any more you still have to look at them – and look at them just as often as you may previously have had to grease them. It follows therefore that the largest initial element of maintenance is visual examination. This may lead to repairs or renewals.

Every 250 miles/400 km (or weekly, whichever comes first)

Battery
Check the electrolyte level and top-up if necessary

Engine/transmission oil
Check the sump oil level (photo) and top-up if required. The oil required to top-up from MIN to MAX level on the dipstick is $1\frac{3}{4}$ Imp pints (1 litre)

Check for visible coolant leaks (check coolant level in expansion bottle)

Check for any fluid leak puddles left under the car overnight

Steering and suspension
Check tyre pressures (photo) – the ride and handling of this car can be severely altered by incorrect tyre pressures

Examine tyres for wear or damage

Is the steering smooth and accurate?

Take note of any unusual noises when travelling and visually check the suspension and shock absorbers

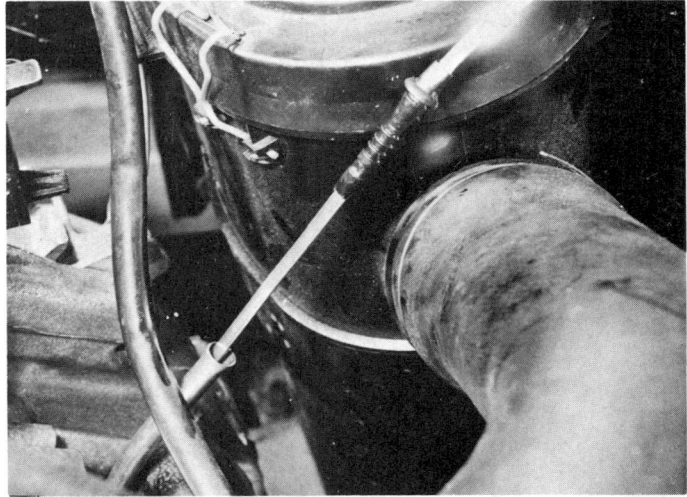
Check the oil level

Top-up oil level

Check tyre pressures

Routine maintenance

Top-up brake fluid reservoir

Change the oil filter

Fit new air filter

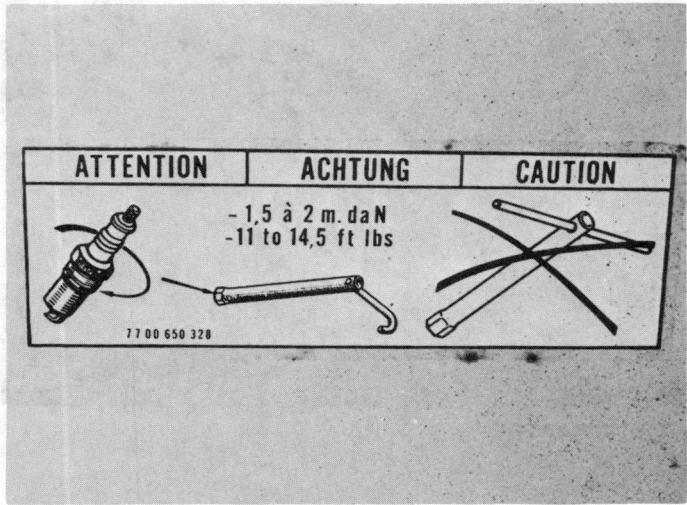
Tighten plugs to specified torque

Brakes
   Is there a fall-off of braking efficiency?
   Try an emergency stop.

Lights, wipers and horns
   Do all the bulbs work at the front, rear, interior: and work well?
   Are the headlight beams aligned properly?
   Do both wipers and horn work fully?

Every 3000 miles/5000 km (or four months, whichever comes first)

Engine/transmission oil
   Drain the sump; you will need a container of at least 7 Imp pints (4 litres) capacity and a suitable spanner with which to undo the sump drain plug. Allow to drain for at least five minutes, Re-install the sump plug and refill with 7 Imp pints (4 litres) of the recommended engine oil

Cooling system
   Check the cooling system. The level must be correct in the expansion bottle, and the respective hoses and their connections must be in good order

Brakes
   Check the fluid level (photo) in the brake master cylinder reservoir, and if necessary top up using the recommended fluid

Bodywork
   Lubricate all hinges including the bonnet, catches and locks with engine oil. Do not forget to oil the car's jack

Electrics
   Check that the electric cooling fan and the rear window demister are working correctly

Every 9000 miles/15 000 km (or annually, whichever comes first)

Engine
   Change the oil filter. Pierce the top of the cartridge prior to removing it, to allow the old oil to drain into the engine and prevent spillage. Lubricate the seal when installing the new filter and ensure that it seats correctly (photo).
   Check the engine generally for signs of oil leakage, especially

around the sump, timing cover and cam cover joints. Tighten fastenings to specified torque settings or renew gasket if leak persists
Check and if necessary adjust the valve clearances

Fuel system

Inspect the carburettor, fuel lines and their connections for condition and security. If any signs of leakage are present, these must be rectified. Refer to Chapter 3 and check the carburettor idle speed when the engine is at its normal operating temperature. Lubricate the throttle and choke linkages with a light oil, and ensure that they are working correctly. Fit a new carburettor air cleaner element (photo), and clean carburettor float chamber
Clean the fuel pump and filter gauze of sediment
Check the tightness of the exhaust manifold

Electrical system

Check the condition and tension of the alternator drivebelt and renew or adjust as necessary

Ignition system

Fit new contact points, and adjust the gap
Check static ignition timing
Check the condition of the distributor cap and plug leads
Fit new spark plugs set to the correct gap, and tighten to the specified torque (photo)

Clutch

Refer to Chapter 5 and check the clutch operating clearance. Adjust as described if necessary

Steering

Examine all steering linkage, joints and bushes for signs of wear or damage
Check the front and rear wheel hub bearings
Check tightness of steering rack mounting bolts
Check driveshafts for wear at the outer universal joint, and the rubber bellows for leaks
Have the front wheel track checked and adjusted at your Renault garage

Suspension

Examine all bolts and bushes securing the suspension and shock absorbers. Tighten as necessary
If the shock absorbers show any signs of leakage, renew them

Brakes

Examine the front disc brake pads for wear and renew if necessary (see Chapter 8)
Examine rear brake shoes and renew if necessary
Check the condition of the wheel cylinder rubbers for leaks
Carefully check the condition of the hydraulic rigid and flexible hoses, particularly the fronts, for chafing, dents and any other form of deterioration; rectify if necessary
Lubricate the handbrake linkage and adjust if necessary

Exhaust system

Carefully examine the exhaust system for signs of deterioration and leakage. Check for security and renew any defective section

General check underneath

Check the respective component connections on the underside of the car for security. Look for any signs of serious corrosion or damage. Clean drain holes in bodywork. If serious corrosion is in evidence, consult your Renault dealer or local body repair shop for expert advice

Electric circuits

Check the functioning of items operated electrically including all lights, the windscreen wipers and washers. Check condition of windscreen wiper blades and renew if the old ones are worn or damaged
Check the safety belt harness connections for security, and the belts for fraying or damage

---

Every 18 000 miles/30 000 km (or every 2 years, whichever comes first)

Brakes

Renew the brake servo unit filter (where fitted)

Every 27 000 miles/45 000 km (or every 3 years, whichever comes first)

Cooling system

Drain the cooling system and flush through. Refill and use new antifreeze

Brakes

Renew the hydraulic fluid by bleeding

Chapter 1 Engine

For modifications, and information applicable to later models, see Supplement at end of manual

Contents

Big end and main bearings – examination and renovation 19	Engine removal – with subframe 6
Camshaft and rocker arms – inspection and renovation 25	Engine/transmission unit – installation (less subframe) 42
Clutch housing and transfer gears – removal 10	Engine/transmission unit – installation (with subframe) 43
Connecting rods, pistons and piston rings – examination and renovation 21	Fault diagnosis – engine 45
Crankcase dismantling 13	Flywheel – examination and renovation 26
Crankshaft and main bearings – examination and renovation ... 18	Flywheel – installation 39
Crankshaft, main bearings and connecting rods – reassembly 33	General description 1
Cylinder head – dismantling, inspection and renovation 17	Gudgeon pins – removal 22
Cylinder head – installation 34	Inlet and exhaust manifold – inspection 28
Cylinder head – removal 9	Lubrication system – general description 15
Cylinder liner bores – examination and renovation 20	Major operations possible with engine installed 2
Engine and gearbox – reassembly 36	Major operations requiring engine removal 3
Engine and transmission – separation 12	Methods of engine removal 4
Engine components – examination and renovation (general) ... 14	Oil filter – removal and refitting 16
Engine dismantling – ancillary items 8	Oil pump – examination and renovation 27
Engine dismantling – general 7	Pistons, connecting rods and liners – installation 32
Engine – initial start-up after overhaul 44	Timing chain and sprockets – examination and renovation 23
Engine mountings – inspection 29	Timing cover (engine installed) – removal and installation 11
Engine – preparation for reassembly 31	Timing cover – installation 37
Engine reassembly – final stages 41	Timing sprockets and chain – installation 35
Engine reassembly – general 30	Transfer gear unit and clutch housing – reassembly 40
Engine removal – less subframe 5	Transfer gear unit – inspection and renovation 24
	Valve rocker clearances – check and adjustment 38

Specifications

General

Engine type	Four-cylinder, in-line, OHC water cooled, transverse mounting
Engine type reference	129 A7 and 145 A7
Capacity	1218cc
Bore	75mm
Stroke	69mm
Compression ratio	9.3 : 1
Oil type/specification	Multigrade engine oil, viscosity SAE 20W/50 (Duckhams Hypergrade)
Oil capacity without filter	7 pints (4 litres) (common system with transmission)
Oil filter capacity	0.5 pint (0.25 litre)

Valves

Clearance (cold):
Inlet	0.004 in (0.10 mm)
Exhaust	0.010 in (0.25 mm)

Seat angle:
Inlet	120°
Exhaust	90°

Seat width:
Inlet	0.047 to 0.074 in (1.2 to 1.9 mm)
Exhaust	0.059 to 0.087 in (1.5 to 2.2 mm)

Stem diameter	0.315 in (8 mm)

Head diameter:
Inlet	1.448 in (36.8 mm)
Exhaust	1.153 in (29.3 mm)

Timing:
	Type 129-7-00	Type 129-7-10	Type 145-7-00
Inlet opens	15° BTDC	8° BTDC	19° BTDC
Inlet closes	45° ABDC	40° ABDC	49° ABDC

| Exhaust valve opens | 46° BBDC | 40° BBDC | 49° BBDC |
| Exhaust valve closes | 15° ATDC | 8° ATDC | 19° ATDC |

Camshaft
Type	Overhead, running in 5 bearings
Camshaft endfloat	0.003 to 0.0055 in (0.07 to 0.14 mm)

Pistons
Type	Aluminium with two compression and one oil control ring. Gudgeon pin free to rotate in piston – press-fit in connecting rod.
Standard ring thickness:	
Top compression	0.069 in (1.75 mm)
2nd compression	0.079 in (2.0 mm)
Oil control ring	0.158 in (4.0 mm)
Ring gap	Supplied pre-set
Gudgeon pin length	2.562 in (65 mm)
Gudgeon pin outside diameter	0.767 in (19.5 mm)

Liners
Length (overall)	4.823 in (122.5 mm)
Liner protrusion – less O-ring	0.004 to 0.006 in (0.10 to 0.17 mm)
Liner O-ring diameter	0.145 to 0.053 in (1.15 to 1.35 mm)
Bore location diameter	3.129 in (79.5 mm)
Liner standard internal bore diameter	2.953 in (75 mm)

Crankshaft
Number of bearings	Five
Endfloat	0.003 to 0.010 in (0.07 to 0.27 mm)
Endfloat thrust washer thicknesses available	0.090 in (2.30 mm)
	0.094 in (2.40 mm)
	0.096 in (2.45 mm)
	0.098 in (2.50 mm)

Main bearing journals
Journal diameter	1.9671 to 1.9677 in (49.965 to 49.981 mm)
Maximum ovality	0.001 in (0.02 mm)
Regrind diameter	1.9553 to 1.9559 in (49.665 to 49.681 mm)

Crankpin
Big-end diameter	1.7706 to 1.7713 in (44.975 to 44.991 mm)
Regrind diameter	1.7589 to 1.7595 in (44.675 to 44.691 mm)
Shell bearing material	Aluminium/tin

Oil pump
Idle pressure (minimum)	14.5 lbf/in^2 (1 bar)
Pressure @ 4000 rpm (minimum)	43.5 lbf/in^2 (3 bars)

Torque wrench settings
	lbf ft	Nm
Cylinder head bolts (cold engine):		
Stage 1	30	41
Stage 2	56 to 60	75 to 80
Crankshaft pulley nut	105 to 112	130 to 150
Camshaft sprocket bolt	52 to 60	70 to 80
Connecting rod nuts	26 to 30	35 to 40
Chain tensioner bolts	3.75 to 5.5	5.0 to 7.5
Coolant drain plug (cylinder block)	22.5 to 33.75	30 to 45
Flywheel bolts (Loctite)	48.75 to 52.5	65 to 70
Oil pump Allen screws	3.75 to 5.5	5.0 to 7.5
Oil pressure switch	30.0 to 37.5	40 to 50
Main bearing cap housing bolts		
Inner:		
Stage 1	26	35
Stage 2	37.5 to 41.25	50 to 55
Outer	11.25	15
Sump plug	17.75 to 22.5	25 to 30
Sump cover	7.5	10
Transmission housing-to-engine fastenings	7.5	10
Transfer gear housing-to-engine bolts	7.5	10
Clutch housing-to-engine bolts:		
7 mm	11.25	15
8 mm	15	20
Rocker cover	3.75 to 5.5	5.0 to 7.5
Timing chain cover	4.5	6.0

Chapter 1 Engine

Fig. 1.1 General cutaway view of the engine and transmission assemblies

1 General description

The 1218 cc engine is of a four-cylinder, in-line, overhead camshaft type. The engine is mounted transversely and is inclined to the rear at an angle of 72°.

The manual gearbox is also mounted transversely in line with the engine, and the final drive to the roadwheels is via the differential unit which is integral with the gearbox. Drive from the engine to the transmission is by means of transfer gears which are separately encased in the clutch housing.

The crankcase, cylinder head, gearcase and clutch housing are all manufactured from aluminium alloy. Removable wet cylinder liners are fitted; the aluminium pistons each have two compression rings and one oil control ring. The valves are operated by the single overhead camshaft and rocker arms. The camshaft drives the distributor at the flywheel end, and the timing sprocket, located at the other end of the camshaft, incorporates a separate eccentric lobe which actuates the fuel pump. The timing chain is driven from the crankshaft sprocket. Next to the timing chain sprocket is the gear wheel which drives the oil pump. This is mounted low down against the crankcase face and is enclosed in the timing chain cover.

The crankshaft runs in five shell main bearings and the endfloat is adjustable via a pair of semi-circular thrust washers. Somewhat inconveniently, the lower half interconnects the engine with the transmission unit.

The engine and transmission units share the same mounting, one each side at the front and similarly at the rear.

Chapter 1 Engine

Special notes

Because of the unconventional layout of the engine and transmission systems, extra care and attention must be taken during the maintenance and overhaul procedures which, in many instances, differ considerably from the more conventional systems.

Read through the various Sections concerned to analyse the instructions so that any snags or possible difficulties can be noted in advance. Because the sub-assembly casings are manufactured from aluminium alloy, it is of utmost importance that all fastenings are tightened to the specified torque settings, which are contained in Specifications.

Very few jobs can be undertaken without removal of the engine and transmission unit; but in addition when they are extracted the design in many instances demands that other sub-assemblies, not directly associated with the offending item, will have to be removed to gain access to the part concerned.

In most instances, special tools will not be required and the majority of procedures are not particularly difficult, but extra care and attention must be exercised to ensure a successful conclusion to the particular task at hand.

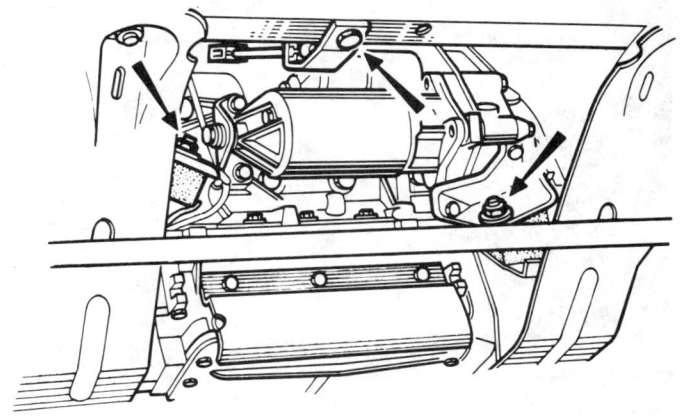

Fig. 1.2 View showing the engine stay and front engine mountings (arrowed)

2 Major operations possible with engine installed

(a) Renewal and installation of the transfer gears
(b) Removal and installation of the clutch unit
(c) Removal and installation of the timing cover
(d) Removal and installation of the timing chain, sprockets and oil pump unit

3 Major operations requiring engine removal

(a) Removal and installation of camshaft assembly
(b) Removal and installation of cylinder head
(c) Removal and installation of pistons, connecting rods and bearings
(d) Removal and installation of crankshaft bearings
(e) Removal and installation of the transmission and differential unit

4 Methods of engine removal

1 There are two basic methods of removing the engine from the Renault 14. The engine can be removed as a unit with the gearbox and differential through the top of the engine compartment after disconnecting the driveshaft and other attachments.

2 Alternatively the engine and transmission assembly can be left attached to the subframe and driveshafts, and removed as a unit by raising and wheeling the body away to the rear.

3 The method employed will probably depend on the work to be undertaken when it has been removed, workshop facilities, and of course personal preference. If you are removing the engine for the first time, read through Sections 5 and 6 to get an idea of what is involved. This should help you decide on the best course of action.

4 It is not possible to remove the engine without the gearbox and differential unit, or vice versa.

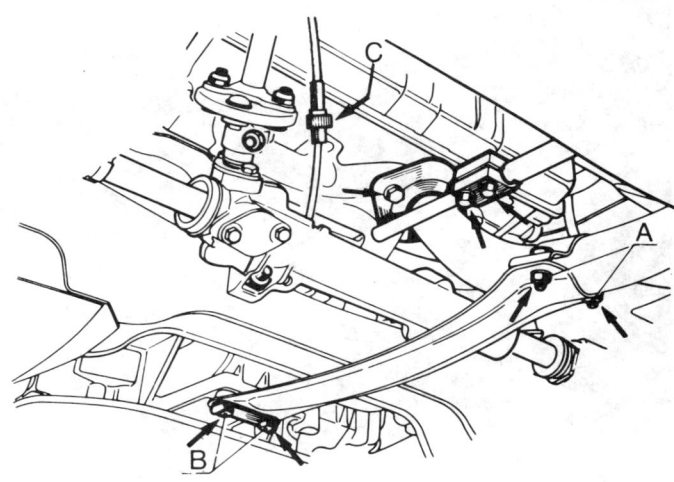

Fig. 1.3 Disconnect the speedometer cable at C and the exhaust pipe stay bracket at A and B (note spacers at A). The exhaust pipe flange and gear selector rod clamp are also arrowed

5 Engine removal – less subframe

1 The engine and transmission units must be removed as a complete assembly and cannot be separated until removed. Although the combined weight of the two components is not great, certain operations are awkward and care must be taken not to damage surrounding components in the engine compartment, particularly during removal. It therefore pays to have an assistant on hand whenever possible. On average, removal may be expected to take 1½ to 2 hours.

2 Position the car with the engine under the lifting tackle location and check that there is sufficient room around the car to work, particularly at the rear where space should be allowed for the body to be wheeled back, after raising the engine.

3 Chock the rear wheels.

4 Raise and support the bonnet. Mark the position of the bonnet hinge on the panel. Support the bonnet and unscrew the retaining

Fig. 1.4 Unscrew the selector quadrant retaining bolt (1) from the steering box

5.10 Extract the petrol pump operating plunger

5.22 Disconnect the exhaust manifold nuts

5.23 The exhaust pipe stay bracket and spacers – attachment to transmission housing

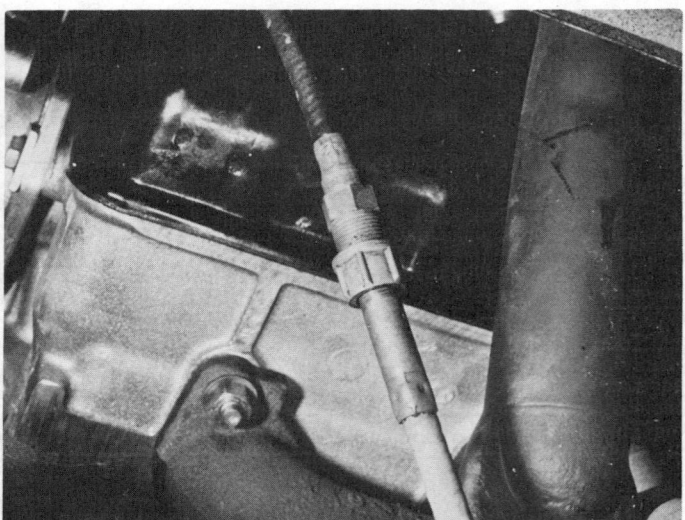
5.25 Disconnect the speedometer cable

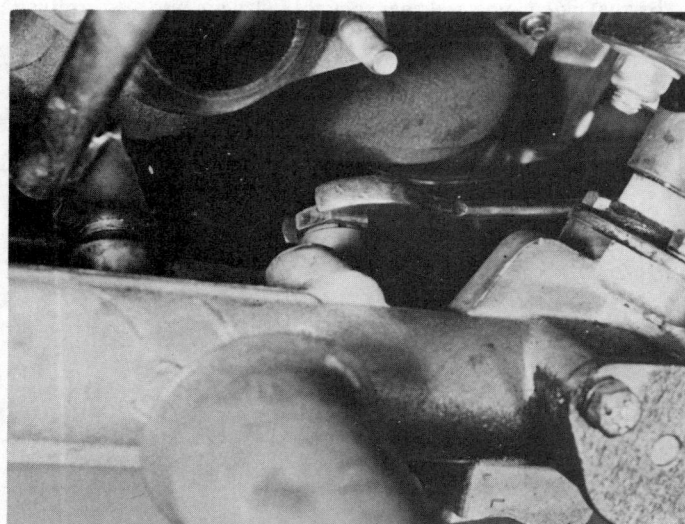
5.26 Unscrew the selector quadrant retaining bolt

5.27 Separate the gear selector control rods

bolts. Lift it clear and remove to a safe place.

5 Disconnect the battery leads; lift the battery clear and place safely aside.

6 Remove the spare wheel and its support bracket.

7 Remove the battery support tray which is retained by four bolts; note the earth strap positions.

8 Disconnect the air cleaner unit from the carburettor and detach the connecting hoses from the engine. Remove the air cleaner unit and hoses. The nylon hose retaining clips can be used again if in good condition, otherwise renew them on reassembly.

9 Detach the HT leads, remove the distributor cap, and withdraw the plastic cover and rotor arm.

10 Disconnect the fuel pump inlet and outlet hoses, and remove the pump from the cylinder head. Extract the operating plunger from the housing in the head (photo).

11 Drain the cooling system (see Chapter 2).

12 Drain the engine/transmission oil.

13 Disconnect the brake servo unit vacuum pipe from the cylinder head (where applicable).

14 Disconnect the radiator coolant hoses from the cylinder head at the water pump and at the thermostat housing.

15 Disconnect the heater hoses from the water pump and the thermostat housing; also the preheat hose from the cowling.

16 Take careful note of their respective positions and disconnect the following wire connectors:

    (a) *Alternator*
    (b) *Coolant temperature sender unit*
    (c) *Oil pressure switch*
    (d) *Coil*
    (e) *Radiator temperature sender unit (to electric fan)*
    (f) *Starter solenoid wires*
    (g) *Battery earth wire from clutch housing stud*
    (h) *Diagnostic socket assembly*

17 Disconnect the hose to the radiator from the coolant expansion bottle. Remove the expansion bottle.

18 Remove the radiator and cooling fan assembly which is retained by a single bolt at the top of the radiator.

19 Disconnect the accelerator and choke cables from the carburettor.

20 Disconnect the clutch cable.

21 Disconnect the engine stay retaining bolt.

22 Working underneath the car, unscrew the retaining nuts from the flange and detach the exhaust downpipe from the manifold (photo).

23 Disconnect the exhaust pipe stay bracket and remove it (photo).

24 Remove the preheat cowling from the exhaust manifold.

25 Unscrew the speedometer outer cable connector just to the rear and above the steering rack, and separate the cable connection (Refer to Chapter 9, Section 27).

26 Unscrew the selector quadrant retaining bolt from the steering box (photo).

27 Mark the relative positions then unscrew the rod retaining clip and separate the forward and rear gear selector rod (photo). Hinge the forward end of the rod around and in line with the engine to prevent it protruding during removal.

28 Unscrew and remove the engine mounting retaining nuts and washers. The large flat washer on the forward stud can be removed when the engine is lifted.

29 Check that all engine/transmission unit attachments are disconnected.

30 Connect the lifting hoist to the lifting brackets, and raise the engine/transmission unit so that the sump is just level with the engine front mounting. When lifting, take great care not to foul the master cylinder unit with the cam cover.

31 With the engine pulled as far to the right-hand side as possible extract the left-hand driveshaft, taking care not to damage the final drive oil seal. When the shaft is free rest it on the subframe.

32 Now move the engine over to the left-hand side of the compartment and disconnect the right-hand driveshaft in a similar manner.

33 Raise the engine/transmission unit sufficiently to clear the car, which should be wheeled back to allow the unit to be lowered for removal to workbench/work area.

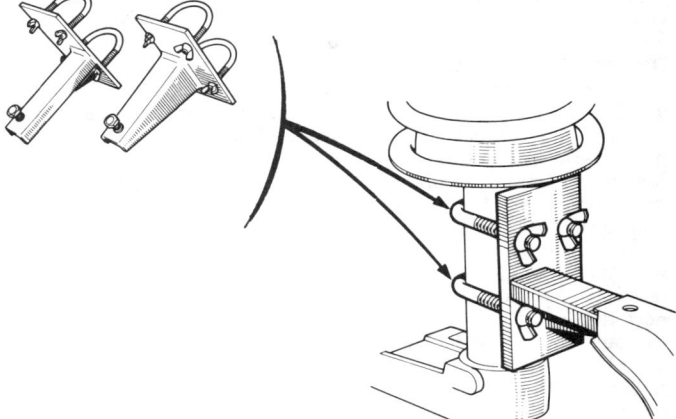

Fig. 1.5 Special tool MS 755 shown when in position

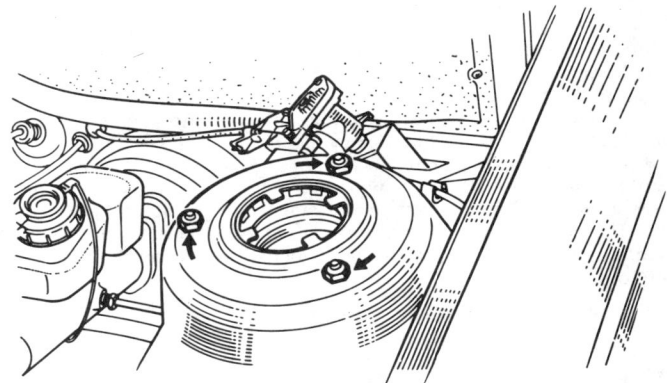

Fig. 1.6 Unscrew the suspension strut nuts

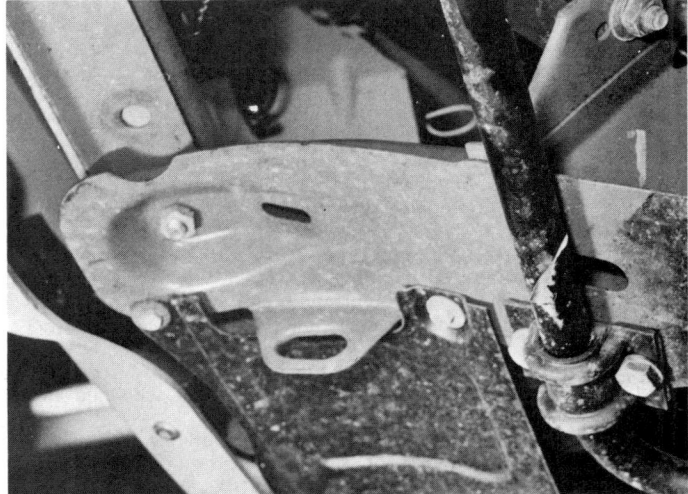

6.6 Side tray and sub-frame retaining bolts at front

6 Engine removal – with subframe

1 If removing the engine and subframe as a unit, equipment must be

available to lift the front of the bodywork sufficiently to allow the engine and subframe assembly to be withdrawn.
2   Refer to Section 5 and follow instruction 1 to 27 inclusive. Depending on the nature of the work being undertaken it may not be necessary to detach or remove all the items mentioned. If the engine and transmission are not being dismantled it is not necessary to remove the fuel pump, just disconnect the hoses. The battery tray does not have to be removed but the battery must be disconnected, and preferably removed to prevent spillage when lifting the body clear of the engine. When those items in Section 5 have been completed continue as follows:
3   Raise the car at the front and remove the roadwheels.
4   Refer to Chapter 8 and remove the front brake caliper units but do not disconnect the hydraulic fluid lines. When the calipers are disconnected support them, by tying to a body member, to prevent the hydraulic lines being distorted.
5   Mark the steering flexible coupling for reference and remove the retaining nuts and bolts.
6   Remove the side trays (photo).
7   Position a jack under the engine subframe unit and support it.
8   Disconnect the two rear engine mountings.
9   To locate the front suspension struts during the engine/transmission removal, they must be supported. This can be done by fabricating a special tool similar to that used by Renault dealers (Tool MS 755) which is shown in Fig. 1.5. This is located between the strut and subframe on each side.
10   Refit the front wheels.
11   Unscrew and remove the top bolts retaining the MacPherson struts to the inner wing panels on each side.
12   Lower the jack under the subframe and detach the subframe mountings at the front.
13   Check the engine/transmission unit attachments to ensure that all fittings are released and out of the way prior to removal.
14   Carefully raise the body at the front and, when high enough, wheel out the subframe complete with the engine and transmission unit. Lower the body and position on axle-stands or suitable supports.
15   The engine and transmission unit can now be removed from the subframe following the sequence contained in Section 5, paragraphs 28 to 32.

7   Engine dismantling – general

1   A good size clean work area will be required preferably on a bench. Before moving the engine/transmission unit to the area reserved for dismantling clean the outside of the various components as detailed below:
2   During the dismantling process, the greatest care should be taken to keep the exposed parts free from dirt. As an aid to achieving this thoroughly clean down the outside of the engine/transmission unit, first removing all traces of oil and congealed dirt.
3   A good grease solvent will make the job much easier, for, after the solvent has been applied and allowed to stand for a time, a vigorous jet of water will wash off the solvent and grease with it. If the dirt is thick and deeply embedded, work the solvent into it with a strong stiff brush.
4   Finally, wipe down the exterior of the engine/transmission unit with a rag and only then, when it is quite clean, should the dismantling process begin. As the engine is stripped, clean each part in a bath of paraffin or petrol.
5   Never immerse parts with oilways in paraffin (eg crankshaft and camshaft). To clean these parts, wipe down carefully with a petrol dampened rag. Oilways can be cleaned out with wire. If an air-line is available, all parts can be blown dry and the oilways blown through as an added precaution.
6   Re-use of old gaskets or oil seals is false economy. To avoid the possibility of trouble after the engine has been reassembled always use new items throughout.
7   Do not throw away the old gaskets, for sometimes it happens that an immediate replacement cannot be found and the old gasket is then very useful as a template. Hang up the gaskets as they are removed.
8   If this is the first time that you have dismantled your Renault 14 engine/transmission unit then special attention should be given to the location of the various components and sub-assemblies. This is especially necessary due to the slightly unconventional layout of the model.
9   Many of the component casings are manufactured in aluminium alloy and special care must therefore be taken not to knock, drop or put any unnecessary pressure on these components.
10   Wherever possible, refit nuts, bolts and washers from where they were removed in order not to mix them up. If they cannot be reinstalled lay them out in such a way that it is clear where they came from.
11   Do not remove or disturb the TDC plate on the clutch housing.

8   Engine dismantling – ancillary items

1   Irrespective of whether you are going to dismantle the engine completely and rebuild it, or are simply going to exchange it for a reconditioned unit, the ancillary components will have to be removed.
2   The only possible method of discovering the exact condition of the engine, and determining the extent of reconditioning required is to dismantle it completely.
3   Refer to the relevant Chapter and remove the following units:

(a) Generator – Chapter 9
(b) Distributor – Chapter 4
(c) Carburettor – Chapter 3
(d) Thermostat – Chapter 2
(e) Water pump – Chapter 2
(f) Starter motor – Chapter 9

4   Remove the exhaust manifold rear engine crossmember, and front mounting brackets.
5   Remove the diagnostic socket (mounted on the clutch housing).
6   If the block is to be stripped or exchanged, remove the oil pressure switch.
7   If the engine is to be exchanged check what ancillary items are included in an exchange unit. Make sure that the engine is cleaned before being exchanged.

9   Cylinder head – removal

1   It is necessary to remove the engine/transmission unit in order to remove the cylinder head. To do this refer to Sections 4, 5 or 6.
2   If not already removed, detach the alternator drivebelt.
3   Unscrew the crankshaft pulley nut. To prevent the crankshaft from turning, wedge a suitable bar through the timing aperture in the clutch housing into the teeth of the starter ring on the flywheel. Align the flywheel timing mark with the O-mark on the fixed plate (photo).
4   Remove the rocker cover which is retained by two bolts and a nut, each with flat and nylon washers.
5   Unscrew the timing cover bolts and withdraw the cover (complete with fuel pump and plunger, if still fitted).
6   Before removing the timing chain assembly, the chain tensioner must be locked in position to prevent it from springing open when released. To lock it use a suitable screwdriver to rotate the ratchet anti-clockwise as shown (Fig. 1.7).
7   Unscrew the bolt from the camshaft sprocket, and remove the bolt and fuel pump drive eccentric.
8   Withdraw the camshaft sprocket and release the chain.
9   Progressively unscrew the cylinder head bolts in the sequence shown (Fig. 1.8).
10   Withdraw the through-bolts and recover the nuts from their channels in the crankcase. Remove the rocker shaft assembly.
11   Before removing the cylinder head the following must be noted. The cylinder head is positioned during assembly by means of two dowels, located as shown in Fig. 1.9. When removing the cylinder head it is most important not to lift it directly from the cylinder block; it must be twisted slightly. This action prevents the cylinder liners from sticking to the cylinder head face and being lifted with it, thus breaking their bottom seals.
12   Before the cylinder head can be twisted, the dowel at the flywheel end must be tapped down flush with the top of the cylinder block, using a drift as shown in Fig. 1.10.
13   When the dowel is flush with the top of the cylinder block, twist the cylinder head by tapping at the point indicated in Fig. 1.11 using a block of wood or a soft hammer.

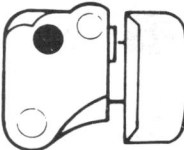

Fig. 1.7 Rotate the tensioner ratchet anti-clockwise to lock before removal

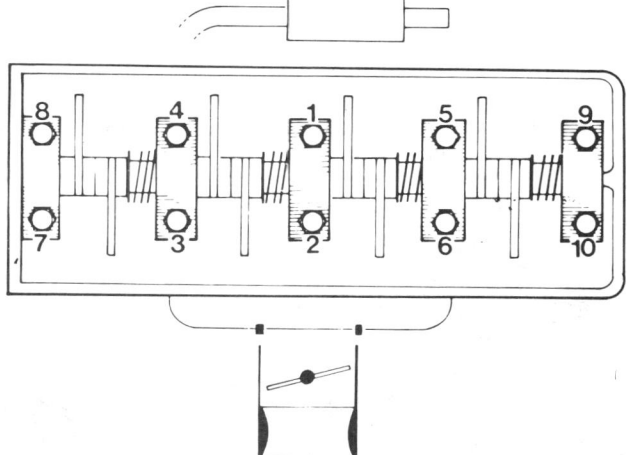

Fig. 1.8 The cylinder head bolts removal/tightening sequence

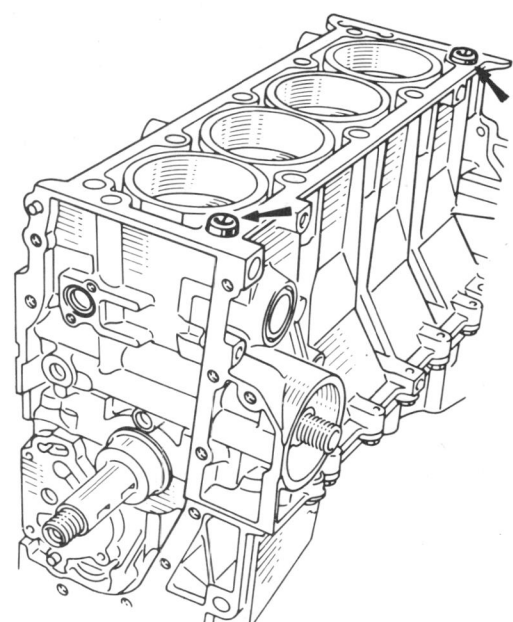

Fig. 1.9 The cylinder head locating dowel positions (arrowed)

8.1 The engine and transmission units (viewed from the front) removed and cleaned ready for dismantling. The main assemblies indicated are:
1 Timing case 3 Transmission 5 Clutch housing
2 Main bearing housing 4 Transfer gear housing

9.3 Align the flywheel timing mark with the O-mark on the fixed plate

9.16 Remove the camshaft retaining plate and bolt

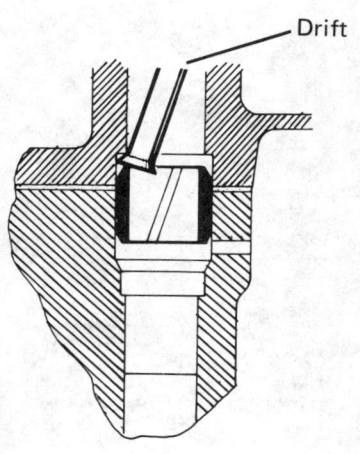

Fig. 1.10 Use a suitable drift to knock the dowel flush with the cylinder head surface

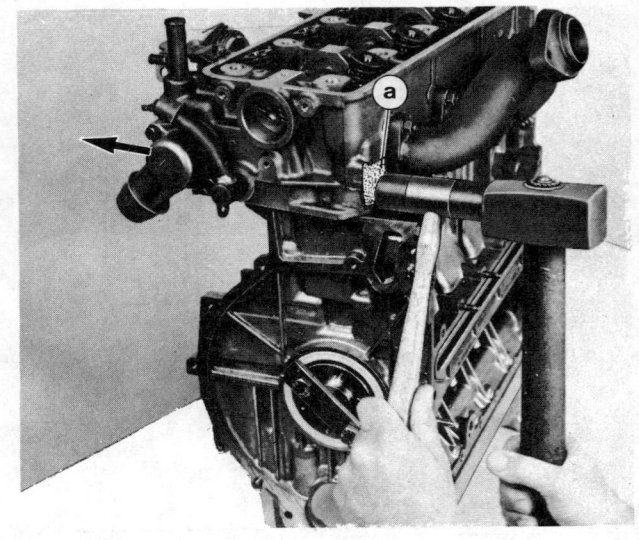

Fig. 1.11 Tap the cylinder head sideways in the direction of the arrow

Do not strike head outside area (a)

10.4 Note the engine lift bracket position

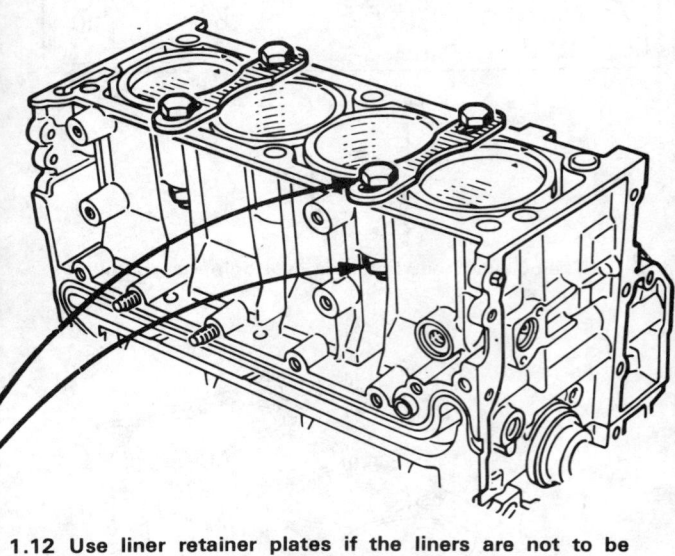

Fig. 1.12 Use liner retainer plates if the liners are not to be removed

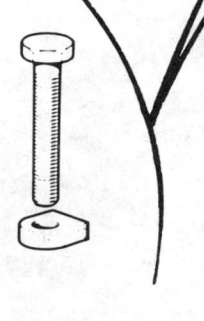

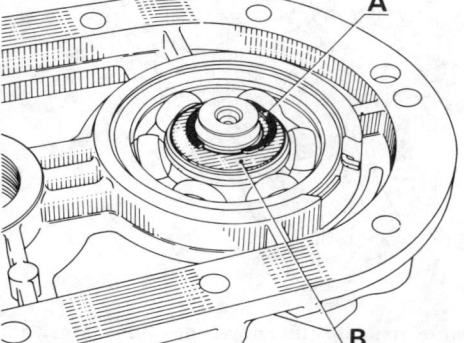

Fig. 1.13 Remove snap-ring A and washer B

14 When the seal between the top of the liners and the cylinder head face has been broken lift the head clear and remove the gasket.
15 If the cylinder head is to be removed for a prolonged period fit a liner retaining plate as shown (Fig. 1.12) to prevent any possibility of the liner's bottom seal being disturbed. On no account rotate the crankshaft whilst the cylinder head is off, if clamps have not been fitted.
16 To remove the camshaft, unscrew the location plate retaining bolt (photo). Remove the plate and extract the camshaft taking care not to score or damage the bearing surfaces in the cylinder head with the edges of the cam lobes.

10 Clutch housing and transfer gears – removal

1 The clutch and transfer gear housings are combined and can be removed with the engine installed in the car. It therefore follows that to repair or remove these items or the associated components (including the flywheel and main bearing oil seal), it is not necessary to remove the engine from the car.
2 If removing the clutch assembly with the engine installed, refer to Chapter 5.
3 To remove and dismantle the clutch housing and associated components with the engine removed, proceed as follows:
4 Unscrew and remove the twelve housing retaining bolts, noting their respective positions. Also note the engine lifting bracket location (photo). Detach the clutch lever spring and remove the actuating rod.
5 Carefully tap and prise the housing free and remove with gasket.
6 Mark the position of the clutch unit in relation to the flywheel, unscrew the clutch unit retaining bolts and remove from the flywheel.
7 Unscrew the retaining bolts and remove the flywheel (if required).
8 Unscrew the retaining bolts and separate the transfer pinion assembly from the clutch housing. Remove the intermediate pinion to prevent it falling out.
9 To dismantle the transfer gear assembly proceed as follows:
10 Support the intermediate plate and press or drive out the needle roller bearing.
11 Prise the circlips from their grooves in the engine output and gearbox input shafts. Remove the dished spring washers.
12 Support the intermediate plate to prevent any distortion, and press or drive the shafts out of the bearings using a piece of suitable diameter tube.
13 The ball bearings can be removed in the same manner having spread the locating circlips so they no longer engage the outer race grooves. You probably have to apply a little pressure to the bearing initially to free the circlip, spreading it as much as possible with a suitable pair of pliers while pressing the bearing through. Do not apply too much pressure initially as the circlip may not be free of the bearing groove, and damage or distortion may occur.

11 Timing cover (engine installed) – removal and installation

1 The timing cover is removed for access to the timing chain, tensioner and sprockets, also to the oil pump and its drive pinions.
2 Although the crankshaft oil seal is retained in the timing cover, directly behind the crankshaft pulley, special tools are necessary to extract the old seal and accurately insert the new replacement with the timing cover in-situ. Therefore, unless the cover is to be removed, it is generally recommended that this operation be left to your Renault dealer.
3 To remove the timing cover, raise the bonnet and disconnect the battery. Remove the spare wheel and carrier.
4 Drain the cooling system (see Chapter 2). Remove the fuel pump and plunger.
5 Loosen the alternator bolts, disconnect the drivebelt and remove the alternator.
6 For ease of access, remove the hose from the radiator to the water pump and also the expansion bottle.
7 Remove the plastic cover from the timing aperture in the clutch housing. If the timing chain or sprockets are to be removed, position the flywheel at TDC. Jam the starter ring gear on the flywheel with a large screwdriver or similar implement, and unscrew the crankshaft pulley nut. Remove the nut with the flat washer.
8 Withdraw the pulley. If levering is necessary, do not apply excessive pressure against the timing cover; being aluminium it is easy to damage it.

9 Remove the rocker cover.
10 Unscrew and remove the eighteen timing cover bolts. When all are removed, carefully tap the timing cover away from the crankcase assembly and remove.
11 Details of dismantling and inspection of the timing chain and associated components are given in Sections 9 and 23.
12 When reassembling ensure that all the old timing cover gasket is removed and of course use a new gasket when replacing the cover. When the cover is refitted and bolted in position, carefully trim the protruding section of gasket away flush with the rocker cover face.
13 Lubricate the crankshaft oilseal lips to ease assembly, and tighten all bolts to the specified torque.
14 Readjust the alternator drivebelt tension on assembly. Refit the fuel pump and plunger.
15 Refill the cooling system as described in Chapter 2.
16 Run the engine and check for oil/water leaks on completion.

12 Engine and transmission – separation

1 Unscrew and remove the crankcase-to-gearbox securing bolts. As the bolt lengths vary, take careful note of their locations on removal. Do not disturb or remove the TDC plate on the clutch housing.
2 In addition to the nuts and bolts on the top and bottom flanges of the casings there are two bolts and a nut to be removed from the flywheel end flange (Fig. 1.15).
3 Support the engine and gearbox as the last bolts are removed, and carefully separate the two units. Should they be stuck together prise them apart using a piece of wood as a lever between the differential housing and the exhaust manifold (if still fitted). If the manifold has been removed use a wooden block in its place.

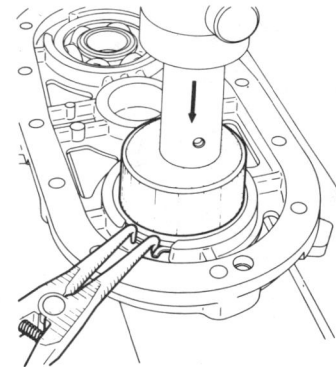

Fig. 1.14 Spread the snap-ring and press the bearing out

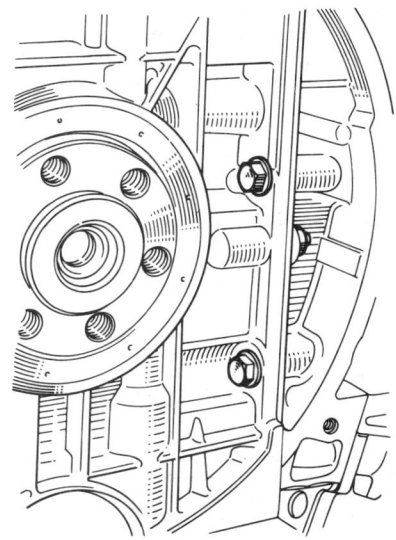

Fig. 1.15 Retaining bolts and nuts position at the flywheel end

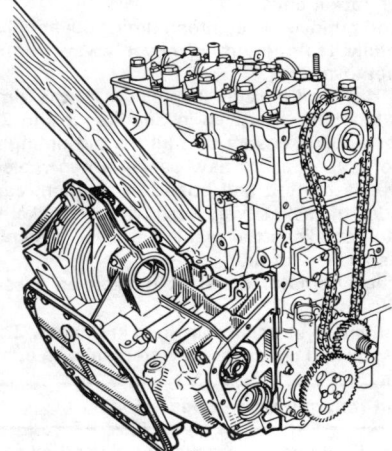

Fig. 1.16 Prise the engine and gearbox apart as shown

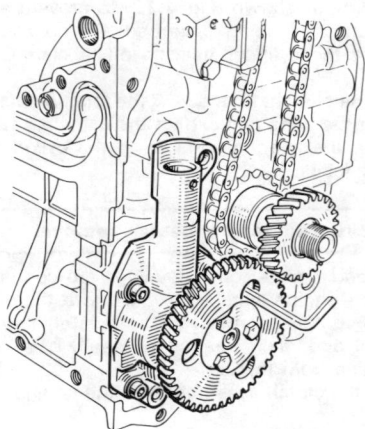

Fig. 1.17 Unscrew the oil pump retaining screws using key inserted through the holes in the gear

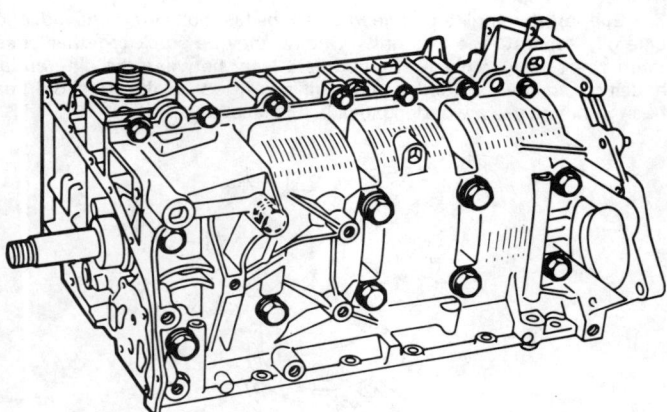

Fig. 1.18 The main bearing housing securing bolts

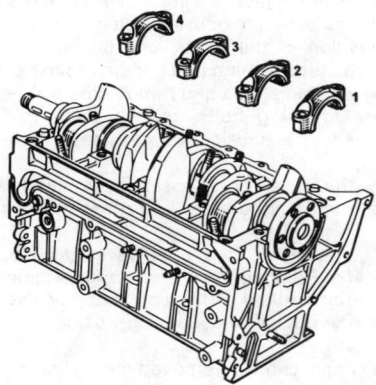

Fig. 1.19 Mark the rods and caps numerically

13 Crankcase – dismantling

Timing components and oil pump

1 With the timing case removed proceed as follows:
2 Use the correct size Allen key and remove the oil pump unit complete with drivegear. Access to the retaining screws can be gained through the holes in the gear.
3 Use a suitable puller and remove the oil pump drive pinion from the crankshaft. Also remove the spacer. Note that the gear has a recessed flange and this faces inwards. (On some models the spacer is integral with the gear).
4 Lock the chain tensioner (see Section 9, paragraph 6) and unscrew the retaining bolts. Remove the gauze filter.
5 Remove the timing chain and drive sprocket from the crankshaft.

Main bearing housing

6 Unscrew the combined main bearing cap and housing bolts and remove the housing (Fig. 1.18).
7 Note its location and remove the crankshaft oil seal.

Pistons and connecting rods

8 Inspect the big-end assemblies and ensure that the connecting rods and caps are marked with a file or dot punch to identify their location and orientation. They should be marked on the oil filter side with No. 1 at the flywheel end (Fig. 1.19).
9 Unscrew the big-end nuts and remove the caps. The caps and rods may have to be prised apart but take care not to damage the bearing shells in any way in case they are to be used again. Keep the shell bearings with their respective rods and caps.
10 Assuming that liner retaining plates have been fitted (see Section 9, paragraph 15), the pistons can be withdrawn from the top of each liner, although, to prevent damage to the liner, it is better to remove them with the liners and separate them on the bench.
11 If the liners are to be removed matching marks should be made on the piston crowns and on the liners (on the oil pump end of the engine) before they are withdrawn. In this way the respective liners, pistons, and rods will be kept together and may be set aside for inspection or reinstallation. Under no circumstances allow the liners to interchange positions if they are to be reinstalled. Mark them numerically on the outer surface to avoid confusion.

Crankshaft

12 Lift the crankshaft clear of the crankcase and remove the main bearing shells and thrust half washers, noting their locations and keeping them in order. Remove the crankshaft oil seal.
13 The crankcase is now dismantled and ready for cleaning and inspection.

14 Engine components – examination and renovation (general)

1 With the engine dismantled, all components must be thoroughly cleaned and examined for wear as described in the following Sections.
2 If a high mileage has been covered since new or the last engine rebuild, and general wear is evident, consideration should be given to renewing the engine assembly.
3 If a single component has malfunctioned and the rest of the

Chapter 1 Engine

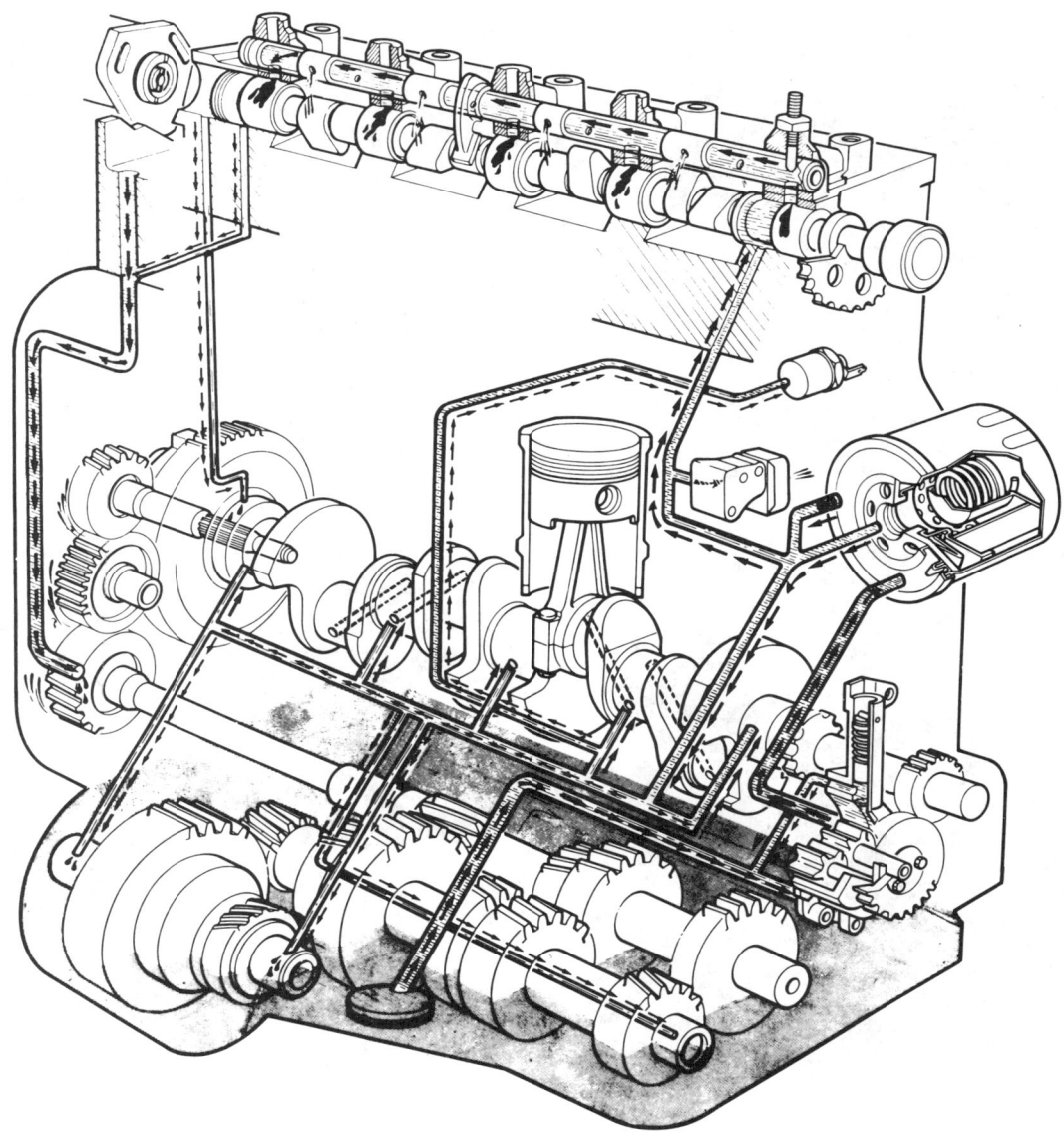

Fig. 1.20 The lubrication circuit

engine is in good condition endeavour to find out the cause of its failure if not readily apparent. For example, if a bearing has failed, check that the adjoining oilways are clear; the new bearing will not last long if it is not being lubricated!

4 If uncertain about the condition of any components, seek a second opinion, preferably from a Renault dealer/mechanic who will obviously have an expert knowledge of your model and be able to advise on the best course of action.

5 Check on the availability of replacement parts before discarding the old ones. Check the new part against the old to ensure that you have the correct replacement.

6 Many of the measurements required will need the use of feeler blades or a micrometer, but in many instances wear will be visually evident or the old component can be compared with a new one.

15 Lubrication system – general description

1 A forced feed lubrication system is employed and is shown in Fig. 1.20. The oil pump is attached to the main bearing housing in the lower section of the timing chest and it incorporates the pressure relief valve. The pump is driven by gears from the crankshaft.

2 Oil from the pump passes via an oilway to the oil filter, and thence to the crankshaft main bearings, connecting rod bearings and transmission components. Another oilway from the filter delivers oil to the overhead camshaft and rocker components. Oil from the cylinder head passes to the transfer gear housing and then back to the sump, contained within the transmission housing.

3 Apart from the standard replaceable canister filter located on the outside of the crankcase there is a gauze filter incorporated in the oil pump suction intake.

4 The oil level must be correctly maintained by reference to the dipstick. The oil filler cap is in the rocker cover. An oil pressure warning switch is fitted which illuminates a warning light in the instrument panel should a drop in pressure occur.

16 Oil filter – removal and refitting

1 To change the oil filter first pierce the canister to allow the oil it contains to drain back into the engine. Use a strap wrench to loosen the filter, or drive a drift through the canister to provide the leverage

17.2 Using a valve spring compressor to release the tension on the retaining collets for removal/insertion

17.5 If cylinder head is to be completely dismantled, note position of brackets. The core plugs should also be renewed at this stage if necessary

17.10a Insert the valve into its guide to check for wear

17.10b Check the exhaust valve guides and ...

17.10c ... the special inlet valve guides for wear or distortion

17.14 Replace the spring seating washers

needed to unscrew it.

2 Smear a little engine oil on the seal of the new filter and screw it into the cylinder block until the seal touches, then screw down an additional one half turn.

3 Top-up the oil level if required, start the engine and check that there are no leaks.

4 Always use an oil filter with a built-in pressure relief valve.

17 Cylinder head – dismantling, inspection and renovation

1 Having removed the cylinder head, place it onto a clean workbench where it can be dismantled and examined. Note that there is no separate inlet manifold; the inlet tracts are cast into the cylinder head.

2 Remove each valve and spring assembly using a valve spring compressor. Extract the split collets from between the spring retaining cup washer and valve stem (photo).

3 Progressively release the tension of the compressor until it can be removed, the spring and retainer withdrawn, and the valve extracted from the guide.

4 As the valves are removed, keep them in order by inserting them in a card having suitable holes punched in it, numbered from 1 to 8.

5 Wash the cylinder head clean and carefully scrape away the carbon build-up in the combustion chambers and exhaust ports, using a scraper which will not damage the surfaces to be cleaned. If a rotary wire brush and drill is available this may be used for removing the carbon. Take care to prevent foreign matter entering the inlet manifold; since it is cast into the cylinder head, cleaning is difficult.

6 The valves may also be scraped and wire-brushed clean in a similar manner.

7 With the cylinder head cleaned and dry, examine it for cracks or damage. In particular inspect the valve seat areas for signs of hairline cracks, pitting or burning. Check the head mating surfaces for distortion, the maximum permissible amount being 0.002 in (0.05 mm). Note that the cylinder head must not be resurfaced and if distorted beyond the specified amount it must be renewed.

8 Minor surface wear and pitting of the valve seats can probably be removed when the valves are reground. More serious wear or damage should be shown to your Renault dealer or a competent automotive engineer who will advise you on the action necessary.

9 Carefully inspect the valves, in particular the exhaust valves. Check the stems for distortion and signs of wear. The valve seat faces must be in reasonable condition and if they have covered a high mileage they will probably need to be refaced on a valve grinding machine; again, this is a job for your Renault dealer or local garage/automotive machine shop.

10 Insert each valve into its respective guide and check for excessive wear ((photo). Worn valve guides allow oil to be drawn past the inlet valve stem causing a smoky exhaust, while exhaust leakage through the exhaust valve guide can overheat the valve guide and cause sticking valves.

11 If the valve guides are to be renewed this is a job best left to your Renault agent who will have the required specialist equipment.

12 Assuming the valves and seats are in reasonable condition they should be reseated by grinding them using valve grinding carborundum paste. The grinding process must also be carried out when new valves are fitted.

13 The carborundum paste used for this job is normally supplied in a double-ended tin with coarse paste at one end and fine at the other. In addition, a suction tool for holding the valve head so that it may be rotated is also required. To grind in a valve, first smear a trace of the coarse paste onto the seat face and fit the suction grinder to the valve head. Then with a semi-rotary motion grind the valve head into its seat, lifting the valve occasionally to redistribute the grinding paste. When a dull matt continuous line is produced on both the valve seat and the valve then the paste can be wiped off. Apply a little fine paste and finish off the grinding process. If a light spring is placed over the valve stem behind the head this can often be of assistance in raising the valve from time to time against the pressure of the grinding tool so as to redistribute the paste evenly round the job. The width of the line which is produced after grinding indicates the seat width, and this width must be within the tolerance in Specifications. If, after a moderate amount of grinding, it is apparent that the seating line is too wide, it probably means that the seat has already been cut back one or more times previously, or else the valve has been ground several

17.15 Relocating the valve spring and cup

17.16a Refit the camshaft and ...

17.16b ... locate with slotted plate and bolt with shakeproof washer

Fig. 1.21 Checking the crankshaft centre main bearing for ovality using V-blocks and a dial gauge

times. Here again, specialist advice is best sought.

14 Examine all the valve springs to make sure that they are in good condition and not distorted. It will have been noticed when they were being removed whether any were broken, and if they are then they should be renewed. It is a good ideal to renew all the valve springs anyway. If you have reached this stage it is false economy not to do so. They are relatively cheap.

15 It is a good idea to renew the valve spring seating washers which sit directly on the cylinder head. These wear reasonably quickly (photo).

16 Before reassembling the valve and springs to the cylinder head make a final check that everything is thoroughly clean and free from grit, then lightly smear all the valve stems with engine oil prior to reassembly. The camshaft can now be re-installed in the cylinder head and located with the retaining plate. This is then secured with its bolt and a new shakeproof washer (photo).

18 Crankshaft and main bearings – examination and renovation

1 Carefully examine the crankpin and main journal surfaces for signs of scoring or scratches, and check the ovality and taper of each journal in turn. Use a dial gauge and V-blocks (Fig. 1.21) and check the main bearing for ovality. If any journals are found to be more than 0.001 in (0.02 mm) out of round then they will have to be reground. If the crankpins are scored or scratched, don't bother measuring them as they will have to be reground.

2 If a bearing has failed after a short period of operation look for the cause and rectify before reassembly.

3 If the crankshaft is to be reground this will have to be done by your Renault dealer or a competent automotive engineer. They will also be able to supply the new shell bearings to suit the undersize requirement.

4 If the crankshaft is found to be in good condition and new bearing shells of the same size are to be fitted, check that the size and markings on the replacements correspond to those of the old bearings.

19 Big-end and main bearings – examination and renovation

1 The main bearing shells themselves are normally a matt grey in colour all over and should have no signs of pitting or ridging or discolouration as this usually indicates that the surface bearing metal has worn away and the backing material is showing through. It is worthwhile renewing the main bearing shells anyway if you have gone to the trouble of removing the crankshaft, but they must, of course, be renewed if there is any sign of damage to them or if the crankshaft has been reground. When the crankshaft is reground the diameter is reduced and consequently one must obtain the proper undersize bearing shells to fit. Only one regrind size is specified for the 129 or A7 engine, 0.0118 in (0.30 mm) undersize, and care has to be taken that the roll-hardened zones of the journals remain intact. The advice of your Renault dealer should be sought.

2 If the crankshaft is not being reground, yet bearing shells are being renewed, make sure that you check whether or not the crankshaft has been reground before. This will be indicated by looking at the back of the bearing shell and this will indicate whether it is undersize or not. The same type of shell bearing must be used when they are renewed.

3 The big-end bearings are subject to wear at a greater rate than the crankshaft journals. Signs that one or more big-end bearings are getting badly worn is a pronounced knocking noise from the engine, accompanied by a significant drop in oil pressure due to the increased clearance between the bearing and the journal permitting oil to flow more freely through the resultantly larger space. If this should happen quite suddenly and action is taken immediately, and by immediately is meant within a few miles, then it is possible that the bearing shell may be renewed without any further work needing to be done.

4 If this happens in an engine which has been neglected, and oil changes and oil filter changes have not been carried out as they should have been, it is most likely that the rest of the engine is in a pretty terrible state anyway. If it occurs in an engine which has been recently overhauled, then it is almost certainly due to a piece of grit or swarf which has got into the oil circulation system and finally come to rest in the bearing shell and scored it. In these instances renewal of the shell alone accompanied by a thorough flush-through of the lubrication system may be all that is required.

20 Cylinder liner bores – examination and renovation

1 The liner bores may be examined for wear either in or out of the engine block; the cylinder head must, of course, be removed in each case. If the liners are still in the block (and it is hoped that they will not need renovation) the liner retainers must be left in a place so that relocation does not have to take place. However, if you have got to the stage where the pistons are out it is better to remove the liners for inspection even if they do not require renovation. Relocation itself does not take much time, skill or money. Each bore may be examined in turn with the piston at the bottom of its stroke.

2 First of all examine the top of the cylinder about a quarter of an inch below the top of the liner and with the finger feel if there is any ridge running round the circumference of the bore. In a worn cylinder bore a ridge will develop at the point where the top ring on the piston comes to the uppermost limit of its stroke. An excessive ridge indicates that the bore below the ridge is worn. If there is no ridge, it is reasonable to assume that the cylinder is not badly worn. Measurement of the diameter of the cylinder bore both in line and with the piston gudgeon pin and at right angles to it, at the top and bottom of the cylinder, is also another check to be made. A cylinder is expected to wear at the sides where the thrust of the piston presses against it. In time this causes the cylinder to assume an oval shape. Furthermore, the top of the cylinder is likely to wear more than the bottom of the cylinder. It will be necessary to use a proper bore measuring instrument in order to measure the differences in bore diameter across the cylinder, and variations between the top and bottom ends of the cylinder. As a general guide it may be assumed that any variations more than 0.010 inch (0.25 mm) indicates that the liners should be renewed. Provided all variations are less than 0.010 inch (0.25 mm) it is probable that the fitting of new piston rings will cure the problem of piston-to-cylinder bore clearances. Once again it is difficult to give a firm ruling on this as so much depends on the amount of time and effort which the individual owner is prepared, or wishes to spend, on the task. Certainly, if the cylinder bores are obviously deeply grooved or scored, the liner must be renewed regardless of any measurement differences in the cylinder diameter. If the engine has been removed from the car for overhaul anyway, any cylinder bore wear in excess of 0.005 inch (0.127 mm) indicates that renewal of the liners should be considered, to do otherwise would be a waste of time and effort. However, one must bear in mind again the fact that renewal will require the fitment of new pistons and the expense of this once again could affect the owner's decision.

21 Connecting rods, pistons and piston rings – examination and renovation

1 With the pistons removed from the liners, carefully clean them and remove the old rings, keeping them in order and the correct way up. The ring grooves will have to be cleaned out, especially the top, which will contain a burnt carbon coating that may prevent the ring from seating correctly. A blunt hacksaw blade or a broken piston ring will assist in groove cleaning. Take care not to scratch the ring bands or piston surface in any way.

2 The top ring groove is likely to have worn the most. After the groove has been cleaned out, refit the top ring and any excessive wear will by obvious by a sloppy fit. The degree of wear may be checked by using a feeler gauge.

3 Examine the piston surface and look for signs of any hairline cracks especially round the gudgeon pin area. Check that the oil relief holes below the oil control ring groove are clear, and if not, carefully clean them out using a suitable size drill.

4 If any of the pistons are obviously badly worn or defective they must be renewed. A badly worn top ring groove may be machined to accept a wider, stepped ring, the step on the outer face of this type of ring being necessary to avoid fouling the unworn ridge at the top of the cylinder bore.

5 Providing the engine has not seized up or suffered any other severe damage, the connecting rods should not require any attention other than cleaning. If damage has occurred or the piston/s show signs of irregular wear it is advisable to have the connecting rod alignment checked. This requires the use of specialised tools and should therefore be entrusted to a Renault agent or a competent automotive engineer, who will be able to check and realign any defective rods.

6 The ring gaps are pre-set during production and do not require checking.

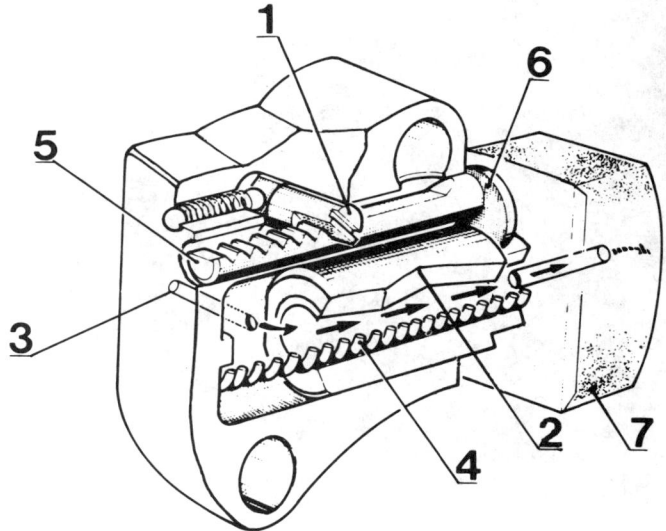

Fig. 1.22 The chain tensioner unit

1 Ratchet
2 Piston
3 Orifice for oil
4 Spring
5 Rack
6 Washer
7 Shoe

22 Gudgeon pins – removal

1 The gudgeon pins float in the piston and are an interference-fit in the connecting rods. This interference-fit between gudgeon pin and connecting rod means that heat is required (230–260°C/450–500°F) before a pin can be satisfactorily fitted in the connecting rod. If it is necessary to renew either the piston or connecting rod, we strongly recommend that the separation and assembly of the two be entrusted to someone with experience. Misapplied heat can ruin one, or all, of the components very easily.

23 Timing chain and sprockets – examination and renovation

1 Examine the teeth of both sprockets for wear. Each tooth on a sprocket is an inverted V-shape and wear is apparent when one side of the tooth becomes more concave in shape than the other. When badly worn, the teeth become hook-shaped and the sprockets must be renewed.

2 If the sprockets need to be renewed then the chain will have worn also and should also be renewed. If the sprockets are satisfactory, examine the chain and look for play between the links. When the chain is held out horizontally, it should not bend appreciably. Remember, a chain is only as strong as its weakest link, and being a relatively cheap item it is worthwhile fitting a replacement anyway.

3 Check the condition of the tensioner shoe for wear. It is not recommended that the tensioner be dismantled as it is a relatively cheap item and normally replaced as a unit.

4 Inspect the oil pump drive gears for wear or damage and renew if necessary.

24 Transfer gear unit – inspection and renovation

1 The condition of the transfer gears, their bearings and the input and output shafts, is obviously critical as they transmit the power of the engine to the transmission unit, and are liable to be a source of noise on this model.

2 Clean the input and output shaft ball bearings and check them for excessive play and/or signs of damage. Inspect the intermediate shaft needle roller bearings. Renew any suspect or worn bearings. If a bearing has collapsed due to general wear and fatigue, then the chances are that the other bearings are close to failure and it is therefore advisable to renew all of the bearings.

3 Carefully inspect the transfer gears. If excessive transmission noise has been experienced it may be reduced by changing the transfer gears. If the teeth are worn or damaged, then the gears should be renewed. Renew the gear set rather than a single gear; it is not good practice to mesh new gears with old as the wear rate of both is increased and they will be noisy in operation.

4 Check the input and output shafts, and inspect their splines for wear or damage. Renew them if necessary.

5 Note that if the intermediate gear or clutch housings are damaged, they must be renewed as a unit and not individually.

25 Camshaft and rocker arms – examination and renovation

1 The camshaft lobes should be examined for signs of flats or scoring or any other form of wear and damage. At the same time the rocker arms should also be examined, particularly on the faces where they bear against the camshaft, for signs of wear. If the case-hardened surfaces of the cam lobes or rocker arm faces have been penetrated it will be quite obvious as there will be a darker, rougher pitted appearance to the surface in question. In such cases, the rocker arm or the camshaft will need renewal. Where the camshaft or rocker arm surface is still bright and clean, and showing slight signs of wear, it is best left alone. Any attempt to reface either will only result in the case-

25.4 Unscrew and remove the stud (with shakeproof washer) ...

25.5 ... and dismantle the rocker assembly

hardened surface being reduced in thickness with the possibility of extreme and rapid wear later on.
2   The camshaft bearing journals should be in good condition and show no signs of pitting or scoring as they are relatively free from stress.
3   If the bearing surfaces are scored or discoloured it is possible that the shaft is not running true, and in this case it will have to be renewed. For an accurate check get your Renault agent to inspect both the camshaft and cylinder head.
4   The rocker arm assembly can be dismantled on removing the fixing bolt/stud from the No 1 support.
5   When removing the various rocker components from the shaft take careful note of the sequence in which they are removed. In particular note that the No 2 and 4 rocker bearings are identical. Keep the components in order as they are removed from the shaft for inspection.
6   Check the rocker shaft for signs of wear. Check it for straightness by rolling it on a flat surface. It is unlikely to be misaligned but if this is the case it must either be straightened or renewed. The shaft surface should be free of wear ridges caused by the rocker arms. Check the oil feed holes and clear them out if blocked or sludged-up.
7   Check each rocker arm for wear on an unworn part of the shaft. Check the end of the adjuster screw and the face of the rocker arm where it bears on the camshaft. Any signs of cracks or serious wear that may have penetrated the case-hardening will necessitate renewal of the rocker arm.

26 Flywheel – examination and renovation

1   There are two areas in which the flywheel may have been worn or damaged. Firstly, is on the driving face where the clutch friction plate bears against it. Should the clutch plate have been permitted to wear down beyond the level of the rivets, it is possible that the flywheel has been scored. If this scoring is severe it may be necessary to have it refaced or even renewed.
2   The other part to examine is the teeth of the starter ring gear around the periphery of the flywheel. If several of the teeth are broken or missing, or the front edges of all teeth are obviously very badly chewed up, then it would be advisable to fit a new ring gear.
3   The old ring gear can be removed by cutting a slot with a hacksaw down between two of the teeth as far as possible, without cutting into the flywheel itself. Once the cut is made a chisel will split the ring gear which can then be drawn off. To fit a new ring gear requires it to be heated first to a temperature of 220°C (435°F), no more. This is best done in a bath of oil or an oven, but not, preferably, with a naked flame. It is much more difficult to heat evenly and control it to the required temperature with a naked flame. Once the ring gear has attained the correct temperature it can be placed onto the flywheel making sure that it beds down properly onto the register. It should then be allowed to cool down naturally. If by mischance, the ring gear is overheated, it should not be used. The temper will have been lost, therefore softening it, and it will wear out in a very short space of time.
4   Although not actually fitted into the flywheel itself, there is a bush in the centre of the crankshaft flange onto which the flywheel fits. Whilst more associated with gearbox and clutch it should always be inspected when the clutch is removed. The main bearing oilseal is revealed when the flywheel is removed. This can be prised out with a screwdriver but must always be renewed once removed. The spigot bush is best removed using a suitable extractor. If a suitable extractor is not available to get it out another method is to fill the recess with grease and then drive in a piece of close-fitting steel bar. This should force the bush out. A new bush may be pressed in.

27 Oil pump – examination and renovation

1   The oil pump will have to be removed from the crankcase before any inspection can be made, and therefore the timing case will have to be removed.
2   The pump unit is retained by four Allen screws and these can be unscrewed using a suitable Allen key inserted through the holes in the pump drivegear where necessary.
3   The drivegear can be removed after bending flat the ears of the washer and unscrewing the retaining bolts.
4   Remove the split-pin from the relief valve housing and withdraw the cup, spring, guide and piston.
5   Clean and examine all components. Any damaged, scored or badly worn parts must be renewed. If the oil pump gears are obviously worn and the car has covered a high mileage, consideration should be given to renewing the complete pump unit, as the body will have worn with the other components.

28 Inlet and exhaust manifolds – inspection

1   The inlet manifold is integral with the cylinder head casting. Check that the mating face to which the carburettor is attached is smooth and undamaged. Check the exhaust manifold for cracks or other damage.
2   Use a straight-edge to check the faces of the exhaust manifold for distortion. If there should be any sign of pitting or warping of the faces, they must be refaced by a competent machinist, or renewed.
3   Any accumulations of carbon within the exhaust manifold ports

29.1a Check the condition of the mountings at the front ...

29.1b ... and the rear of the engine

can be removed using a flexible wire brush and/or scraper.

29 Engine mountings – inspection

1 The engine mounting rubbers are often ignored simply because they do not normally present any problems. However their work rate is probably equal to any other engine component and therefore if the engine is removed at any time, it is worthwhile checking the condition of the mounting rubbers. If they show signs of deterioration due to oil impregnation, heat or simply age, they should be renewed. Mountings that have lost their resistance to shocks will cause engine/transmission vibrations and increase the fatigue rate of other associated engine/transmission connections, as well as for the driver and occupants of the car (photos).

30 Engine reassembly – general

It is during the process of engine reassembly that the job is either made a success or a failure. From the word go there are certain basic rules which it is folly to ignore namely:

(a) *Absolute cleanliness. The working area, the components of the engine and the hands of those working on the engine must be completely free of grime and grit. One small piece of carborundum dust or swarf can ruin a big-end in no time, and nullify all the time and effort you have spent. No matter what the pundits say this engine and its other components can be reconditioned and rebuilt very successfully and continue working efficiently*

(b) *Always, no matter what the circumstances may be, use new gaskets, locking tabs, seals, nyloc nuts and any other parts mentioned in the Sections in this Chapter. It is pointless to dismantle an engine, spend considerable money and time on it and then to waste all this for the sake of something as small as a failed oil seal. Delay the rebuilding if necessary*

(c) *Don't rush it. The most skilled and experienced mechanic can easily make a mistake if he is rushed*

(d) *Check that all nuts and bolts are clean and in good condition and ideally renew all spring washers, lockwasher and tab washers as a matter of course. A supply of clean engine oil and clean cloths (to wipe excess oil off your hands only!) and a torque spanner are the only things which should be required in addition to all the tools used in dismantling the engine*

(e) *The torque wrench is an essential requirement when reassembling the engine (and transmission) components, especially with the Renault 14. This is because the various housings are manufactured from aluminium alloy and, whilst this gives the advantage of less weight, it also means that the various fastenings must be accurately tightened as specified to avoid distortion and/or damage to the components. Cracked or distorted housings are not cheap to renew so beware*

31 Engine – preparation for reassembly

1 Assuming that the engine has been completely stripped for reconditioning and that the block is now bare, before any reassembly takes place it must be thoroughly cleaned both inside and out.

2 The ideal situation is to dip the block in a garage's cleaning tank usually filled with a mixture of paraffin and cleaning fluid, and then to leave it submerged for an hour or so. Then get to work on it with a wire brush and screwdriver. Clean out all the crevices, do not scratch any machined surfaces, and scrub both inside and out. A great deal of sediment often collects around the liner seatings. Chip this away if necessary.

3 Hose down the block with a garden hose and if possible dry it off with an air jet. Dry and thoroughly clean out the block with a lint-free cloth until it is spotless.

4 Clean out the oilways using a bottle brush, pipe cleaner or other suitable implement, and blow through with compressed air. Squirt some clean engine oil through to check that the oilways are clear.

5 If the core plugs are defective and show signs of weeping, they must be renewed at this stage. To remove, carefully drive a punch through the centre of the plug and use the punch to lever the plug out. Clean the aperture thoroughly and prior to fitting the new plug, smear the orifice with sealant. Use a small-headed hammer and carefully drive the new core plug into position with the convex side outwards. Check that it is correctly seated on completion.

6 As the components are assembled, lubricate them with clean engine oil and use a suitable sealant as and where applicable.

32 Pistons, connecting rods and liners – installation

1 As explained previously, the pistons and connecting rods are best separated and reassembled by your Renault agent, due to the difficulty of removing and refitting the gudgeon pin. Check that the pistons have been correctly fitted to the connecting rod so that the arrow mark on the piston crown will face the timing gear when the connecting rods and caps are assembled to the crankshaft with their identifying

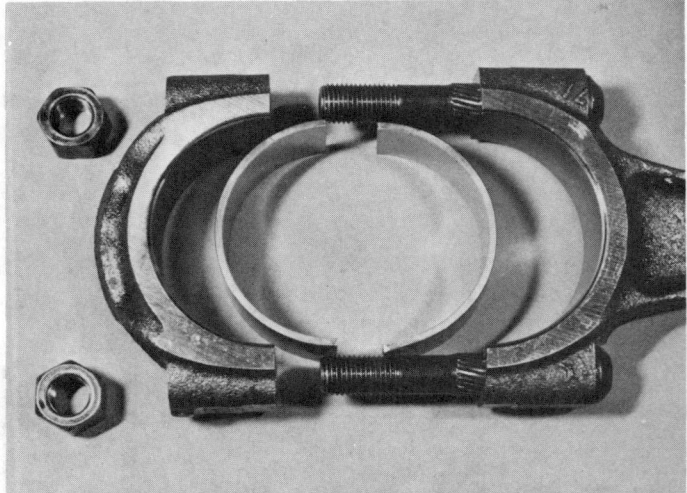

31.6 An example of good assembly preparation — connecting rod bearings and cap laid out ready for lubrication and fitting

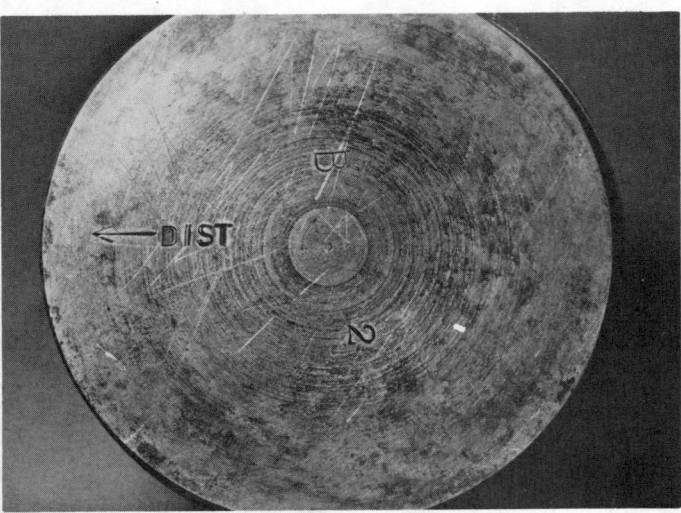

32.1 Arrow mark to point to timing case end

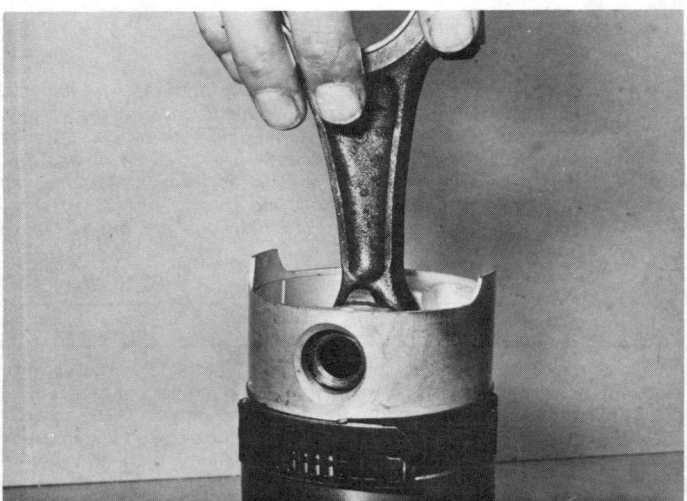

32.4 Installing the piston assembly into the cylinder liner using a ring compressor

32.9 Fit a new O-ring seal to each liner

32.10 Install the piston and liner assemblies into the crankcase

numbers on the oil filter side.

2   Fit the piston rings to each piston using a ring expander if available, or by carefully spreading the ends and lowering over the piston ring lands using a feeler gauge as a guide. Do not fit them from the bottom upwards as you may score the piston skirt. Keep the ring square to the piston as it is lowered. Do not distort or over expand it. Fit the scraper ring first, then the 2nd compression ring which is fitted with the TOP or O-mark on the ring facing upwards, and finally the top compression ring. This may also be marked TOP. If oversize rings are fitted, follow the manufacturers instructions carefully.

3   Lubricate the pistons and rings, and position the ring gaps so they are staggered relative to each other (Fig. 1.23).

4   Lubricate each liner bore in turn, and insert the piston and connecting rod into the liner, using a suitable ring clamp to compress the rings (photo).

5   When the liners have their respective piston/connecting rod assemblies refitted, check that the machined face of each big-end is parallel to the liners machined edge as in Fig. 1.24.

6   Before fitting the new O-ring seals to the liners, place the liner assemblies into the crankcase with the piston-to-liner matching marks in line, and the piston arrows facing the timing gear end.

7   When in position, the liner protrusions above the top face of the

crankcase must be measured. This can be achieved by placing a straight-edge over the top of each liner in turn and inserting a feeler gauge between the cylinder block top face and rule. The liners must of course be fully seated and the liner protrusion should be between 0.004 and 0.006 in (0.10 and 0.17 mm). Check the clearance on each side of the liner and ensure that the difference between each side does not exceed 0.002 in (0.05 mm). The protrusion difference between adjacent liners must never exceed 0.0016 in (0.04 mm).

8 If the tolerances given cannot be achieved, check that the liners are seated correctly and that no dirt particles are preventing this. If difficulty is found with new liners it may be possible to interchange their positions to suit but in this case, the piston and connecting rods must be kept in their respective positions in the crankcase and the liners changed accordingly. If the liner protrusions still do not meet the specified tolerances they may possibly have been incorrectly machined and a check should be made by your Renault agent.

9 When the liner protrusions have been checked and are in order, remove the liner and piston assemblies and carefully fit a new O-ring seal to each liner (photo). Smear the seal seating flange with sealant to ensure a good joint when the liner is installed.

10 Insert each liner into its respective position in the crankcase (photo). Check that previously used liners have their matching marks in line on the oil pump side of the cylinder block. If new liners are being fitted mark them to identify their location.

11 To retain the liners in position during further assembly procedures, fit liner clamps (Fig. 1.12). These will prevent the liners from moving when the connecting rods and caps are fitted to the crankshaft.

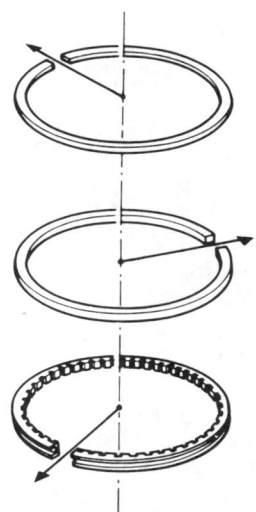

Fig. 1.23 Stagger the ring gaps as shown

33 Crankcase, main bearings and connecting rods – reassembly

1 Invert the crankcase, and make sure that the bearing shell housings are perfectly clean in the crankcase and connecting rods.

2 Insert the main bearing upper shells and ensure that the location tongue is fully engaged. Note that all the upper shells are grooved (photo). Lubricate them with clean engine oil.

3 Insert the connecting rod big-end bearing shells and lubricate with clean engine oil (photo).

4 Lubricate the big-end and main bearing journals with clean engine oil and carefully lower the crankshaft into position in the crankcase (photo).

5 Insert the crankshaft thrust washers with the oil grooves towards the shaft (Fig. 1.25).

6 Check the crankshaft endfloat using a dial gauge or feeler gauges as shown (photo). If the endfloat is not within the tolerance specified, alternative half thrust washers are available to suit (see Specifications).

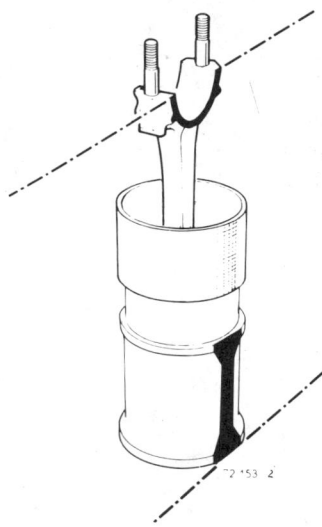

Fig. 1.24 Piston assembly and liner correctly fitted showing machined faces parallel

33.2 Install the grooved main bearing shells into the crankcase

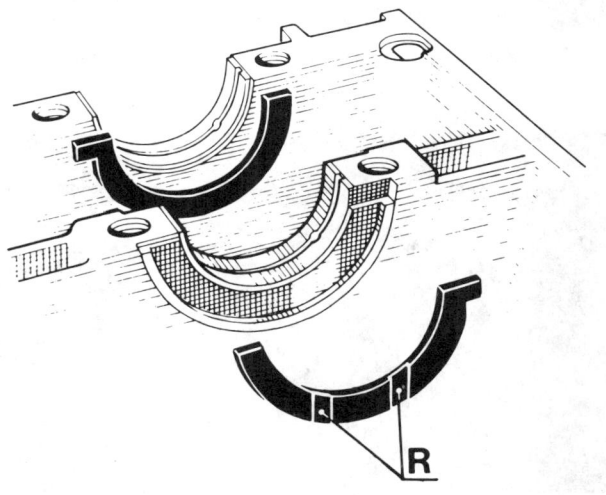

Fig. 1.25 Select correct thrust washers – fit with oil groove (R) towards crankshaft

33.3 Install the connecting rod bearings (note location tab slots into groove in rod)

33.4 Carefully install the crankshaft

33.5 The crankshaft thrust washer location. It can be fitted before or after fitting crankshaft

33.6 Check the crankshaft endfloat

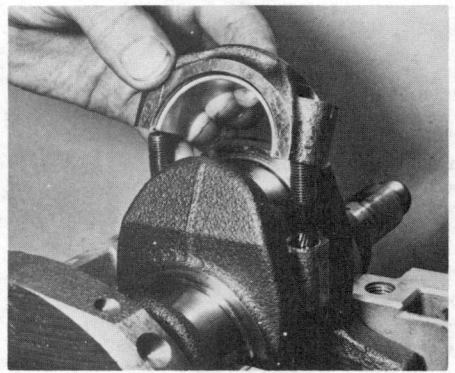

33.7a Refit the connecting rod caps and bearings and ...

33.7b ... tighten to the specified torque

33.8 Fit a new O-ring to the oilway sleeve

33.10 Install the main bearings into the housing – grooved bearings to Number 2 and 4 positions

33.11 Carefully install the housing

33.13a Carefully install the new oil seal

33.13b Prise out the old output shaft oil seal for renewal

33.14 A method of turning the crankshaft using two flywheel bolts and a suitable lever

Chapter 1 Engine 35

7 The connecting rods and caps can now be fitted to the crankshaft (photo). Check that the connecting rod identification marks are towards the oil filter side and that the bearing shell is properly located. Fit the big-end to the crankshaft journal and place the bearing cap and shell in position. The cap and rod identification numbers must align. Fit the nuts and tighten to the specified torque setting (photo).

8 Check that the main bearing cap housing location dowels are in position in the mating face of the crankcase and fit a new O-ring to the oilway sleeve (photo).

9 Smear the joint faces of the crankcase and main bearing housing with sealant and locate the gasket over the dowels on the crankcase.

10 Insert the bearing shells into the main bearing housing, the grooved bearings being fitted to No 2 and 4 bearing positions (photo).

11 Lubricate the bearings with clean engine oil and carefully fit the main bearing housing into position.

12 Reinsert the retaining bolts and hand tighten. Now referring to Fig. 1.26, initially tighten bolts 1 to 10 in the sequence shown to a torque setting of 26 lbf ft (35 Nm). Then retighten in the same sequence to 41 lbf ft (55 Nm). The seven outer bolts can be tightened to 11.25 lbf ft (15 Nm).

13 Fit the new crankshaft oil seal into position. Lubricate the sealing lips to ease assembly and ensure correct seating (photo). Drift the seal into position using a suitable tube drift. Renew the output shaft oil seal in the centre of the crankshaft and if necessary the inner bush.

14 Now that the crankshaft is initially assembled check that the crankshaft is free to rotate (photo) without excessive binding but take care that the liners are not pushed upwards with the pistons if retaining plates are not fitted.

34 Cylinder head – installation

1 Having checked that the mating surfaces of the cylinder head and crankcase are perfectly clean, ensure also that the location dowels are correctly positioned as shown in Fig. 1.27 and that they protrude by 0.276 in (7 mm) above the surface of the crankcase.

2 Rotate the crankshaft and position it so that the pulley keyway is in alignment with the crankcase-to-main bearing housing joint as shown in Fig. 1.28. This is necessary to position the pistons at the halfway position in the bores to prevent the possibility of the pistons and valves touching, and to enable the correct valve timing to be set.

3 Place a new cylinder head gasket into position on the cylinder block (photo). Do not apply any form of sealant.

4 Lower the cylinder head into position and locate over the protruding dowels. The dowels must not be pushed down into the crankcase when fitting the cylinder head (photo).

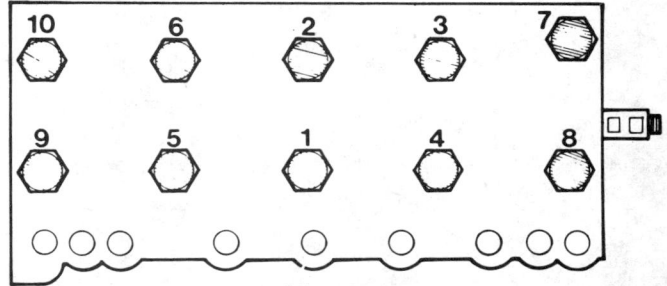

Fig. 1.26 Main bearing housing retaining bolts – tightening sequence

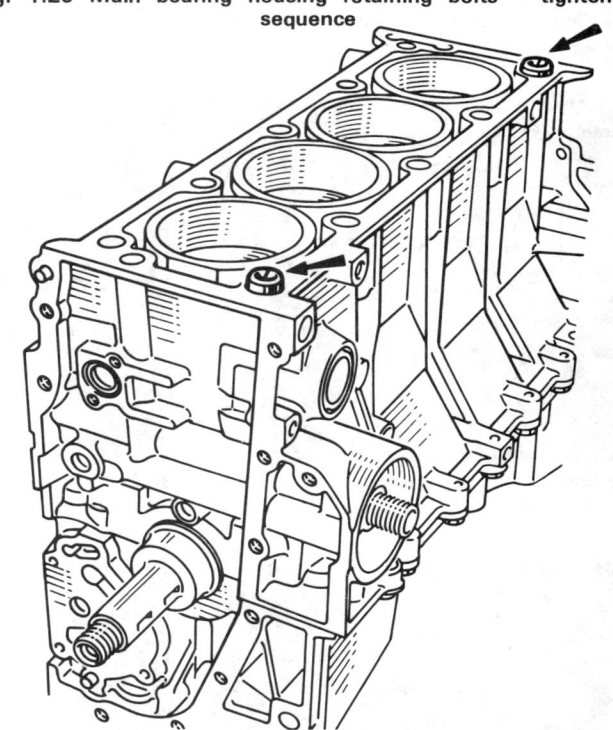

34.3 Position cylinder head gasket over locating dowels

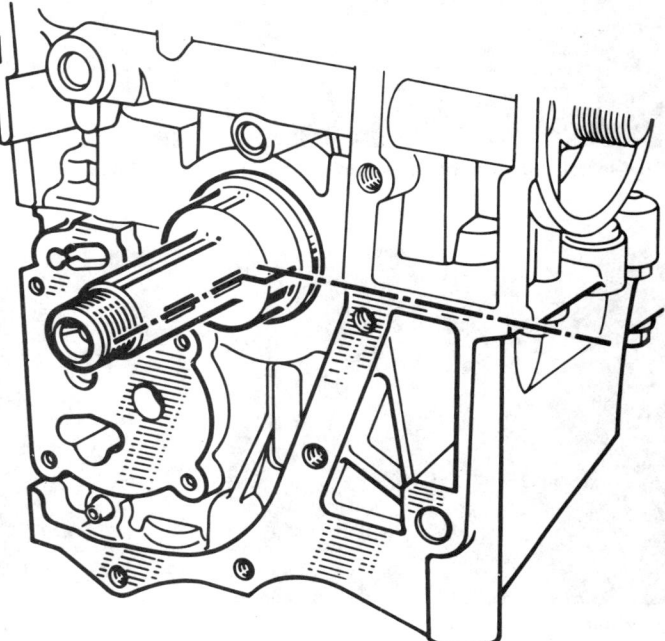

Fig. 1.27 Check the cylinder head locating dowels are in position

Fig. 1.28 Align the crankshaft keyway as shown

34.4 Carefully lower the cylinder head into position

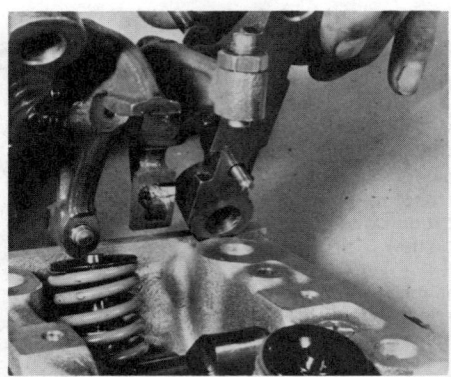
34.6a Refit the rocker shaft assembly with dowel as shown

34.6b Rocker shaft assembly refitted

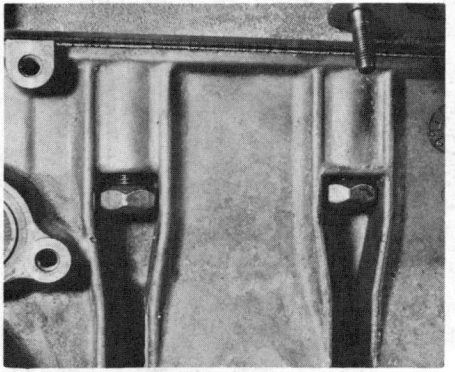
34.8 Relocate the cylinder head nuts

35.2a The camshaft sprocket and chain timing marks

35.2b The crankshaft sprocket and chain timing marks

35.4 Install the tensioner filter into its housing

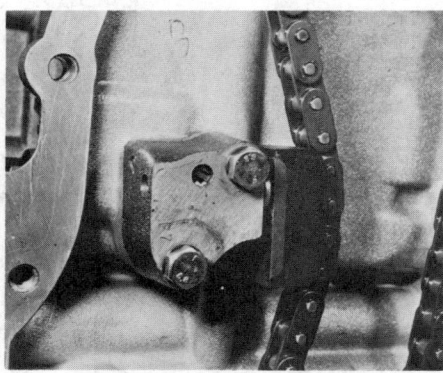
35.5 Chain tensioner reinstalled

35.7 Refit the fuel pump operating cam and retaining bolt

35.8 Refit the oil pump unit

35.10 Refit the oil pump drivegear to the crankshaft ...

35.11 ... and the driven gear to the pump spindle flange

Chapter 1 Engine

5 Rotate the camshaft so that the sprocket keyway is aligned as shown in Fig. 1.29.
6 Refit the rocker shaft assembly to the cylinder head and check that the alignment dowels are properly located (photo). The rocker finger wear faces should be lubricated with multi-purpose grease.
7 Reinstall the cylinder head bolts together with their flat washers. Smear both faces of the flat washers with sealant to prevent oil leaking past the bolt stems.
8 Relocate the nuts to the cylinder head bolts (photo) and then tighten the bolts to the specified torque settings. Initially tighten all bolts in the sequence shown to 30 lbf ft (41 Nm). Then retighten in sequence to a setting of 56 to 60 lbf ft (75 to 80 Nm).
9 Run the engine as described in Section 44. Allow it to cool for at least two hours. Unscrew No. 1 bolt a quarter of a turn and then re-tighten to 56 to 60 lbf/ft (75 to 80 Nm). Repeat on all other bolts following the bolt tightening sequence diagram.

35 Timing sprockets and chain – installation

1 Check that the keyways of the crankshaft and the camshaft are positioned as shown in Figs. 1.28 and 1.29, and insert the keys.
2 Fit the respective camshaft and crankshaft sprockets into the timing chain so that the timing marks on the chain and sprockets correspond, see Fig. 1.30 (photos).
3 Supporting the sprockets in position in the chain, assemble them onto their respective shafts and tap evenly into position over the keys.
4 Install the chain tensioner filter into its recess in the end of the crankcase (photo).
5 Refit the chain tensioner unit, retaining with two bolts which must be tightened to the torque figure specified (photo).
6 Using a screwdriver, release the tensioner lock ratchet by screwing in a clockwise direction and allow the tensioner spring to automatically take up the chain tension.
7 Check that the sprockets are in alignment then refit the fuel pump operating cam and bolt (photo). Tighten the bolt to the specified torque.
8 The oil pump unit is fitted next and should be lubricated with clean engine oil prior to installation (photo). Fit the pump together with its spacer plate and secure with the four Allen screws, which must be progressively tightened to the specified torque.
9 When tightened, check that the pump spindle is free to rotate. If it is stiff or binds, loosen the retaining bolts and relocate the pump unit. Retighten the bolts and recheck the pump spindle action. It is essential that the pump rotates freely.
10 Fit the key into the crankshaft and slide the oil pump drive gear into position (or spacer and drive gear – dependent on the type fitted) (photo).
11 If not already fitted to the pump unit, refit the driven gear to the spindle flange, and retain with the special tab washer and three bolts. Tighten the bolts and then bend over the tab washer to lock the bolt heads (photo).

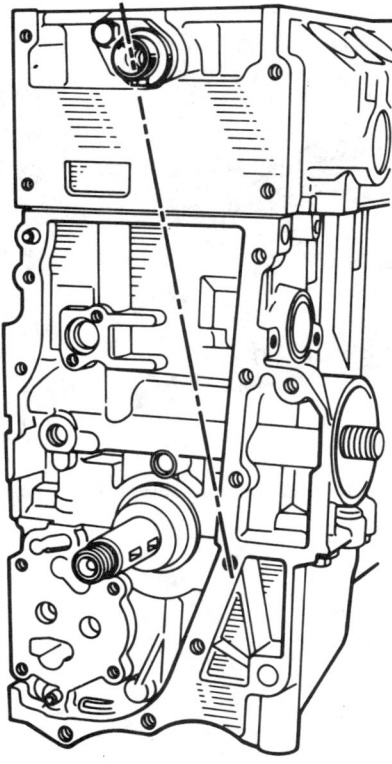

Fig. 1.29 Align the camshaft keyway as shown

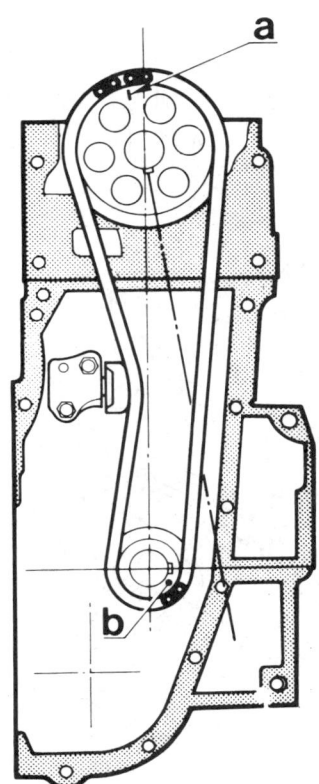

Fig. 1.30 Align the chain and sprocket timing marks, a and b as shown

36.2 Clean and refit the pump filter

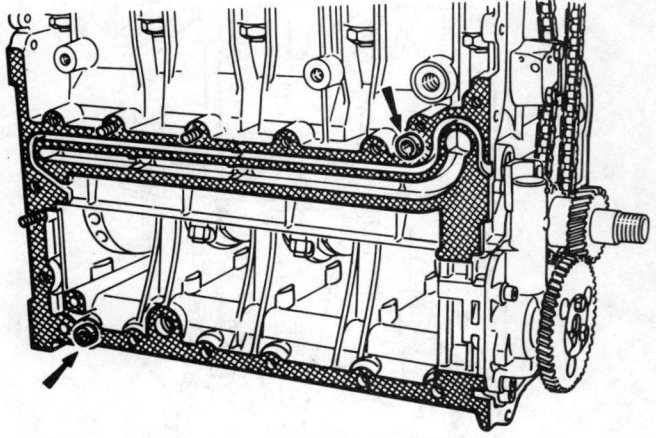

Fig. 1.31 Check that the location dowels are in position (arrowed)

36 Engine and gearbox – reassembly

1 Before assembling the two units, check that the respective mating faces are perfectly clean.
2 If the transmission unit has not been dismantled it is advisable to remove the bottom cover plate and extract the gauze oil pump suction filter, retained by four bolts (photo). Clean the filter thoroughly using petrol, and dry with compressed air. Refit the suction filter unit and sump cover in the reverse sequence using a new gasket. Tighten the bolts to the specified torque. This additional chore is well worthwhile, ensuring that any sludge or particles trapped in the sump tray or suction filter will not be circulated around the rebuilt engine.
3 Install a new O-ring into position where the oil pump suction pipe passes through the engine/transmission joint face.
4 Check that the two locating dowels are in position on the joint face of the engine (Fig. 1.31) and then smear a layer of sealer over the joint face. Be careful not to get carried away applying the sealer as it must not be allowed to enter the oil supply channel.
5 Fit the gearbox to the engine taking care not to press the location dowels inwards (photo). Insert four convenient bolts to hold the casings together and do them up finger tight. Do not fully tighten until all the fastenings are in position. These consist of four top bolts and one nut at the timing end, one top bolt and one nut at the flywheel end, six bottom bolts, and two bolts and one nut on the flywheel face (Fig. 1.32). When all the fastening are installed and finger tight use a torque wrench to tighten to the specified torque. **Do not overtighten.**

37 Timing cover – installation

1 Check that the mating surfaces of the timing cover and crankcase are clean.
2 If it has been removed, fit a new oilseal into the timing cover with the cavity side inwards. Use a tube drift to tap it squarely into position. Ensure that the seal is fitted correctly and do not distort or damage it during assembly. Lubricate the seal lips.
3 Position the timing cover gasket on the cylinder block but do not use any sealant. Fit the timing cover and install the retaining bolts, but do not tighten (photo).
4 Fit the pulley to the crankshaft and centralise the cover before tightening the timing cover bolts fully to the specified torque.
5 Carefully trim off the portion of gasket protruding above the

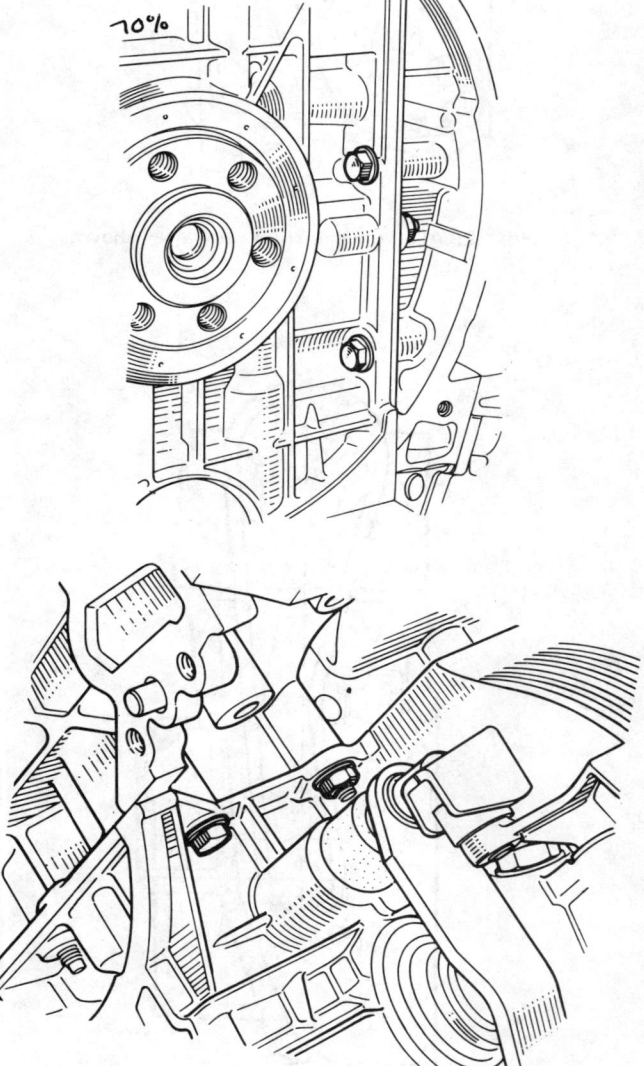

Fig. 1.32 Flywheel end fixing nuts and bolts

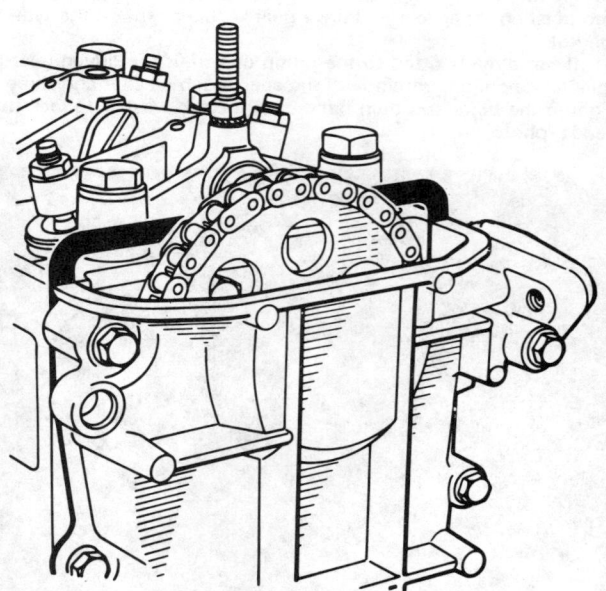

Fig. 1.33 Trim off the protruding gasket

cylinder head flush with the rocker cover face, using a suitably sharp knife.

6 Fit the pulley retaining nut and flat washer. Smear the nut thread with Loctite and tighten the nut to the specified torque. Prevent the crankshaft from turning during the tightening operation, using the bolts in the flange and a suitable lever.

38 Valve rocker clearances – checking and adjustment

1 This operation is the same whether carried out with the engine installed or removed. When performed as a maintenance task with the engine installed, the work must only be undertaken when the engine is cold. The exhaust valves can be adjusted from the top of the engine using a stub screwdriver.

2 The engine can be turned by using a spanner on the crankshaft pulley nut – rotation is made easier if the spark plugs are removed.

3 Remove the rocker cover and then turn the engine until the valves on No 1 cylinder are rocking (ie inlet valve opening and exhaust valve closing). The rocker arm clearances of both valves of No 4 cylinder can now be checked and if necessary adjusted. Remember that No 1 cylinder is at the flywheel end of the engine.

4 The feeler gauge of the correct thickness is inserted between the valve stem and rocker arm. When the clearance is correctly set the feeler gauge should be a smooth sliding fit between the valve stem and rocker arm.

5 If the feeler gauge is a tight or loose fit then the clearance must be adjusted. To do this, loosen the locknut of the adjustment stud and screw the adjuster stud in or out until the feeler gauge blade can be felt to drag slightly when drawn from the gap (photo).

6 Hold the adjuster firmly in this position and tighten the locknut. Recheck the gap on completion to ensure that it has not altered when locking the nut and stud.

7 Check each valve clearance in turn in the following sequence:

Valves rocking on cylinder	Adjust clearances on cylinder
1	4
3	2
4	1
2	3

8 Refit the rocker cover using a new gasket.

39 Flywheel – installation

1 Prior to refitting the flywheel to the crankshaft flange the mating faces must be examined for any signs of dents or burrs and if necessary these must be carefully removed using a fine file. Any oil or dirt must also be wiped away before the flywheel is fitted.

2 Fit the flywheel to the flange and align the bolt holes. These are

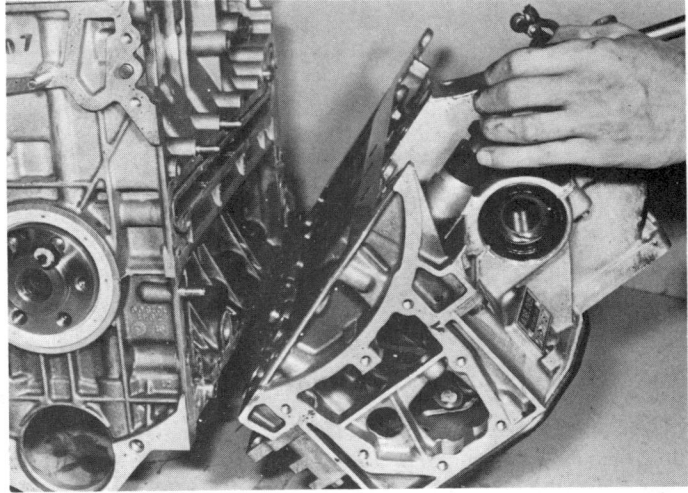

36.5 Carefully refit the transmission to the engine

37.3 Refit the timing cover

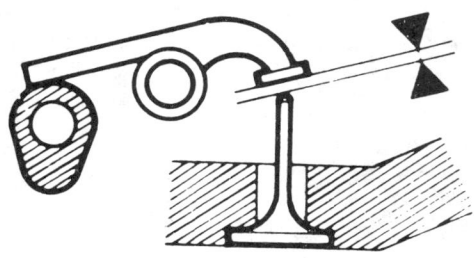

Fig. 1.34 The rocker to valve clearance

38.5 Checking the valve-to-rocker clearance

39.2 Refit the flywheel

40.8 Shaft installed and retained with washer and snap-ring

40.11 Refit the transfer gear assembly to the clutch housing

41.3 Refit the exhaust manifold using a new gasket

41.6a Refit the distributor. Use a new O-ring seal and note offset drive pegs

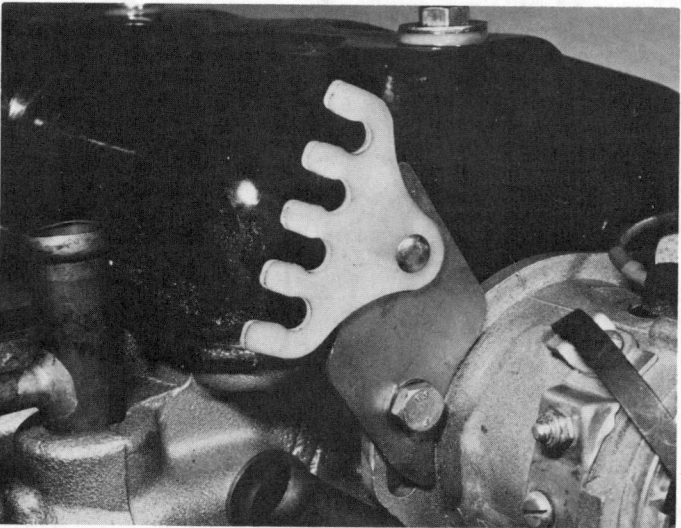

41.6b Note position of HT lead location plate

asymmetrically spaced to ensure correct alignment of the two components (photo).
3 Insert the bolts, which must be smeared with Loctite to prevent them working loose, and tighten them to the specified torque.
4 Refit the clutch unit as described in Chapter 5.

40 Transfer gear unit and clutch housing – reassembly

1 Fit the snap-rings to the grooves in the bearing housing apertures.
2 Ensure that the bearing housings are perfectly clean before inserting the bearings.
3 Lubricate the bearings and press or drift them into their apertures in the housing. As the bearings are inserted the snap-rings will have to be expanded to allow the bearing entry. Check that the bearings are inserted with the snap-ring groove in the outer race offset to the top.
4 Use a suitable diameter tube to press or drive the bearing home. If a press is used take care not to use too much pressure as the bearing aperture may be damaged or crack if the bearing is not in correct alignment with the housing. Press the bearing in until the snap-ring locates in both the bearing and housing grooves.
5 The input and output shafts are fitted to their respective bearings so that the gears are located on the opposite side of the bearing to the snap-ring.
6 The gearbox input shaft is located at the widest end of the housing. Each shaft should be pressed or drifted in until it butts against its respective bearing.
7 Each shaft is retained by a dished spring washer and a snap-ring. The washer is fitted with its concave side towards the bearing.
8 Press or drift a snap-ring into position on each shaft so that it locates in the shaft groove against the pressure of the spring washer (photo).
9 Lubricate the needle roller bearings and insert one into the clutch housing and the other into the transfer gear housing using the intermediate gear as an aid to installation. Align the intermediate gear to mesh with the input and output gears and fit it into the transfer gear housing.
10 When the intermediate gear is in position the teeth of the gear must be in exact alignment with those of the output and input gears. Place a straight edge along the side of the three gears to check that their heights are the same. Also check that the gears rotate freely.
11 Check that the dowels are in position in the clutch housing, apply sealant to the joint faces and locate the transfer gear housing over the dowels (photo).
12 Before refitting the cover plate apply sealant to both joint faces. Refit the cover using a new gasket. Tighten the retaining bolts to the specified torque.
13 Make sure the locating dowels are positioned in the crankcase and fit a new gasket over them. Do not use sealant. Offer up the clutch housing and transfer gear housing assembly to the engine – turn the flywheel and input shaft alternately to enable the splines to mesh. When correctly positioned retain with the bolts. Do not forget that the engine bracket, the diagnostic socket and the clutch cable sleeve stop are attached to the engine by these bolts. Tighten the bolts to the specified torque.

41 Engine reassembly – final stages

1 Having reassembled the major engine components, the various ancillary components can now be refitted. We chose to reassemble most of those components prior to engine installation. However you may for some reason prefer to delay fitting some of the items mentioned below until the engine is reinstalled. In this case ensure that it will be possible to refit the component once the engine is in position.
2 Refit the rear engine mounting crossmember.
3 Refit the exhaust manifold using a new gasket and tighten the retaining nuts to the specified torque (photo).
4 Fit the thermostat into its housing. See Chapter 2.
5 Check that the seating surfaces are perfectly clean and fit the new oil filter – hand tight only.
6 Refit the distributor making sure its offset drive pegs engage into the corresponding slots in the end of the camshaft. Use a new O-ring seal (photo) and secure with three bolts and two special washers. Note the HT lead clamp plate which is attached by a distributor bolt as shown in the photo.
7 Bolt the carburettor directly onto the cylinder head, using a new gasket.
8 Place the starter motor into position and retain with the three bolts at the clutch housing and two bolts in the bracket at the tail end.
9 Refit the alternator and adjusting bracket (photo). Fit the drivebelt and adjust the tension.
10 Refit the forward engine mounting brackets and tighten the retaining bolts to the specified torque.
11 Reconnect the gearbox end of the speedometer cable and tighten retaining bolt.
12 The engine is now ready for installation.

41.9 Refit the alternator and drivebelt. Adjust tension

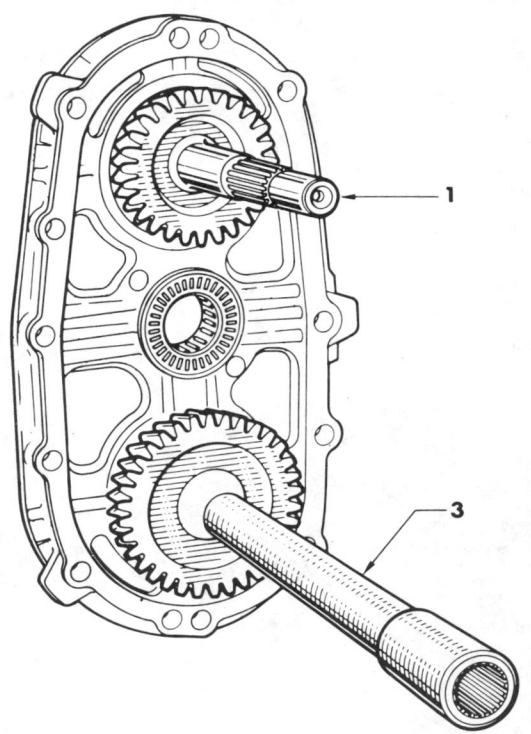

Fig. 1.35 The output shaft (1) and the input shaft (3) positioned in the housing

42.4 Lower engine/transmission unit carefully into position

42.11 Refit the preheat duct

42.20 Refit the return spring bracket and throttle rod

42 Engine/transmission unit – installation (less subframe)

1   Generally speaking the installation of the engine is a reversal of the removal procedure and is as follows:
2   Check that the lifting sling is securely located and is adequate for the job.
3   Before lowering the engine into the car check that the surrounding components in the engine compartment are out of the way.
4   Lower the engine slowly and guide it carefully into a position, offset to the left-hand side of the car, at a suitable height to allow the right-hand driveshaft to enter the differential housing (photo). Guide the driveshaft carefully into position and take care not to damage the oil seal in the differential housing. This may prove fiddly and patience will be necessary. Do not force the shaft home.
5   When the right-hand side driveshaft has slid into position slowly swing the engine over to the right-hand side of the car taking care not to damage the hydraulic fluid reservoir.
6   Guide the left-hand driveshaft into place in a similar manner.
7   When both driveshafts are re-engaged, slowly centralize the engine over its mountings. Replace any washers that have been removed before lowering the engine. Refit the mounting nuts and tighten to the specified torque.
8   Refit the gear control linkage, inserting the retaining bolt into the steering rack housing (just to the right of centre).
9   Reconnect the exhaust pipe to the manifold.
10 Reconnect the gear change control rods and clamp, aligning the reference marks. Assuming that the gearbox is in neutral, restrain the gear lever in the corresponding position and tighten the clamp. Check the gear lever engagement positions, and if necessary adjust accordingly until all gears can be selected satisfactorily. (See Chapter 6 for further details). Alternatively, engage 1st gear, move the gear lever into the corresponding position and tighten the rod connecting clamp. Recheck for satisfactory gear engagement.
11 Refit the preheat duct which is attached to the exhaust manifold and the crossmember (photo).
12 Reconnect the speedometer cable.
13 Attach the exhaust pipe stay bracket to the differential housing and retain with spacers and bolts.
14 Refit the clutch actuating rod and connect the spring over the lever – see Chapter 5 for further information and adjustment procedure.
15 Lubricate the fuel pump actuating rod and insert it into the housing. Reconnect the fuel supply line to the fuel pump inlet pipe. Refit the fuel pump using a new gasket, then connect the carburettor fuel line to the pump outlet.
16 Refit the distributor rotor arm, plastic cover and cap. Reconnect the HT leads to the spark plugs and coil.
17 Reconnect the heater hoses to the thermostat housing and water pump. Reconnect the water pump top hose.
18 Refit the brake servo hose to the cylinder head (where fitted).
19 Connect the choke cable to the carburettor and attach the outer cable (in rubber protector sleeve) to the vertical bracket on the cylinder head, adjacent to the temperature sender unit.
20 Attach the accelerator cable and the throttle connector rod to the throttle return spring bracket. Check the operation of the choke and accelerator controls or assembly (photo).
21 Reconnect the hose from the cowling to the preheat ducting.
22 Connect the battery earth cable to the clutch housing stud and retain with nut, and reconnect the yellow earth cable to the inner wing panel.
23 Refit the radiator and cooling fan unit, inserting the pins into the bottom location holes in the body crossmember. Secure the retaining bolt and reconnect the hoses.
24 Refit the battery support and secure. Refit the battery and secure with the clamp plate (on its lower ledge).
25 Replace the engine stay retaining bolt.
26 Reconnect the following electrical components:

(a) Alternator
(b) Coolant temperature sender switch
(c) Oil pressure switch
(d) Coil
(e) Radiator temperature sender unit and fan
(f) Starter solenoid
(g) Diagnostic socket assembly

27 Refit the air filter unit.

Chapter 1 Engine

28 Replace the expansion bottle and secure with the rubber strap. Reconnect the hose to the radiator and refill the cooling system — see Chapter 2 for special filling instructions.
29 Before refitting the spare wheel and support bracket refill the engine/transmission with oil to the correct level.
30 Reconnect the battery terminals and check that all the operations listed have been completed. Check the operation of the electrical circuits. The engine is now ready for starting.

43 Engine/transmission unit – installation (with subframe)

1 The engine installation into the subframe and the installation of the subframe into the car is a direct reversal of the removal sequence, but note the following:
2 When reconnecting the driveshafts take care not to damage the seals in the differential housing.
3 Tighten the engine mounting bolts, the steering shaft flexible coupling bolts and nuts, the subframe mountings, the suspension and the brake component retaining nuts and bolts to the specified torque values.
4 Reconnect the various engine and associated components as described in Section 42, paragraph 8 onwards.
5 Before using the car on the road, check that the brakes are fully effective — if any fluid has been lost, top up the reservoir and if necessary bleed the system. This should not normally be required if the caliper hoses were not disconnected, but it is essential to check.

44 Engine – initial start-up after overhaul

1 Make sure that the battery is fully charged and that all lubricants, coolant and fuel are replenished.
2 It will require several revolutions of the engine on the starter motor to pump the petrol up to the carburettor.
3 As soon as the engine fires and runs, keep it going at a fast tickover only (no faster), and bring it up to the normal working temperature.
4 As the engine warms up there will be odd smells and some smoke from parts getting hot and burning off oil deposits. The signs to look for are leaks of water or oil which will be obvious if serious. Check also the exhaust pipe and manifold connections, as these do not always 'find' their exact gas tight position until the warmth and vibration have acted on them, and it is almost certain that they will need tightening further. This should be done of course, with the engine stopped.
5 When normal running temperature has been reached adjust the engine idling speed, as described in Chapter 3.
6 Stop the engine and wait a few minutes to see if any lubricant or coolant is dripping out when the engine is stationary. The cylinder head will also require retightening after allowing the engine to cool for at least two hours. Refer to Section 34.
7 Road test the car to check that the timing is correct and that the engine is giving the necessary smoothness and power. Do not race the engine — if new bearings and/or pistons have been fitted it should be treated as a new engine and run in at a reduced speed.

45 Fault diagnosis – engine

Symptom	Reason/s
Engine will not turn over when starter switch is operated	Flat battery Bad battery connections Bad connections at solenoid switch and/or starter motor Starter motor jammed Defective solenoid Starter motor defective
Engine turns over normally but fails to fire and run	No spark at plugs No fuel reaching engine Too much fuel reaching engine (flooding)
Engine starts but runs unevenly and misfires	Ignition and/or fuel system faults Incorrect valve clearances Burnt out valves Blown cylinder head gasket, dropped liners Worn out piston rings Worn cylinder bores
Lack of power	Ignition and/or fuel system faults Incorrect valve clearance Burnt out valves Blown cylinder head gasket Worn out piston rings Worn cylinder bores
Excessive oil consumption	Oil leaks from crankshaft oil seal, timing cover gasket and oil seal, rocker cover gasket, sump gasket Worn piston rings or cylinder bores resulting in oil being burnt by engine (smoky exhaust is an indication) Worn valve guides and/or defective valve stem oil seals
Excessive mechanical noise from engine	Wrong valve to rocker clearance Worn crankshaft bearings Worn cylinders (piston slap) Slack or worn timing chain and sprockets Worn transfer gears and/or bearings

Chapter 1 Engine

Symptom	Reason/s
Unusual vibration	Broken engine/gearbox mounting Misfiring on one or more cylinders

Note: *When investigating starting and uneven running faults do not be tempted into a snap diagnosis. Start from the beginning of the check procedure and follow it through. It will take less time in the long run. Poor performance from an engine in terms of power and economy is not normally diagnosed quickly. In any event the ignition and fuel systems must be checked first before assuming any further investigation needs to be made*

Chapter 2 Cooling system

Contents

Coolant antifreeze mixture ... 6	Radiator and electric cooling fan – removal and refitting ... 7
Cooling system – draining ... 3	Radiator cooling fan – removal and refitting ... 8
Cooling system – filling ... 5	Thermal switch (electric cooling fan) – removal and refitting ... 13
Cooling system – flushing ... 4	
Expansion bottle – general ... 11	Thermostat – removal, testing and refitting ... 9
Fault diagnosis – cooling system ... 15	Water pump/alternator drivebelt – renewal and adjustment ... 2
General description ... 1	Water pump – removal and refitting ... 10
Heating system – general ... 14	Water temperature sender unit – removal and refitting ... 12

Specifications

Type of system	Pressurised and sealed with centrifugal circulation pump, thermostat and electric cooling fan
Coolant capacity	$10\frac{1}{2}$ Imp pints/6 litres/$12\frac{1}{2}$ US pints
Coolant type/specification	Antifreeze to BS 3151, 3152 or 6580 (Duckhams Universal Antifreeze and Summer Coolant)
Expansion bottle	Attached to inner wing panel. Incorporates the cooling system pressurising valve
Radiator	Corrugated fin and tube type
Fan	Electric, thermostatically controlled
Water pump	Belt-driven, centrifugal type

Drivebelt tension
On assembly	$\frac{5}{32}$ to $\frac{3}{16}$ in (4 to 4.5 mm)
In service	$\frac{1}{4}$ in (6 mm)

Thermostat
Type	Wax
Opening temperature:	
Initial	82°C (180°F)
Fully open	94°C (201°F)

Torque wrench settings
	lbf ft	Nm
Cylinder block drain plug	22 to 33	30 to 45
Thermal switch on cylinder head	30 to 37	40 to 50

1 General description

The cooling system is pressurised and sealed, and has an expansion bottle to accept displaced coolant from the system when hot and return it when cool.

The coolant is circulated by thermosyphon action, assisted by the impeller in the belt-driven water pump.

A thermostat is fitted and is located in the cylinder head at the flywheel end of the engine. When the engine is cold the thermostat valve remains in the closed position to restrict the coolant flow and thus accelerate the warming up process.

As the engine warms up, the thermostat valve is progressively opened allowing the coolant to flow to the radiator.

The coolant circulates around the engine's coolant passages absorbing heat as it travels in an upward direction to exit via the top hose into the radiator header tank.

The coolant then passes through the radiator and is cooled by the airflow caused by the forward motion of the car, or the fan when it is in operation. The coolant then exits from the radiator and returns to the cylinder block via the water pump, which assists the coolants circulation.

When the airflow through the radiator provides insufficient cooling, the cooling fan cuts in by means of a thermostatic switch located in the radiator. When the temperature of the coolant has dropped sufficiently the switch will cause the fan to cut out.

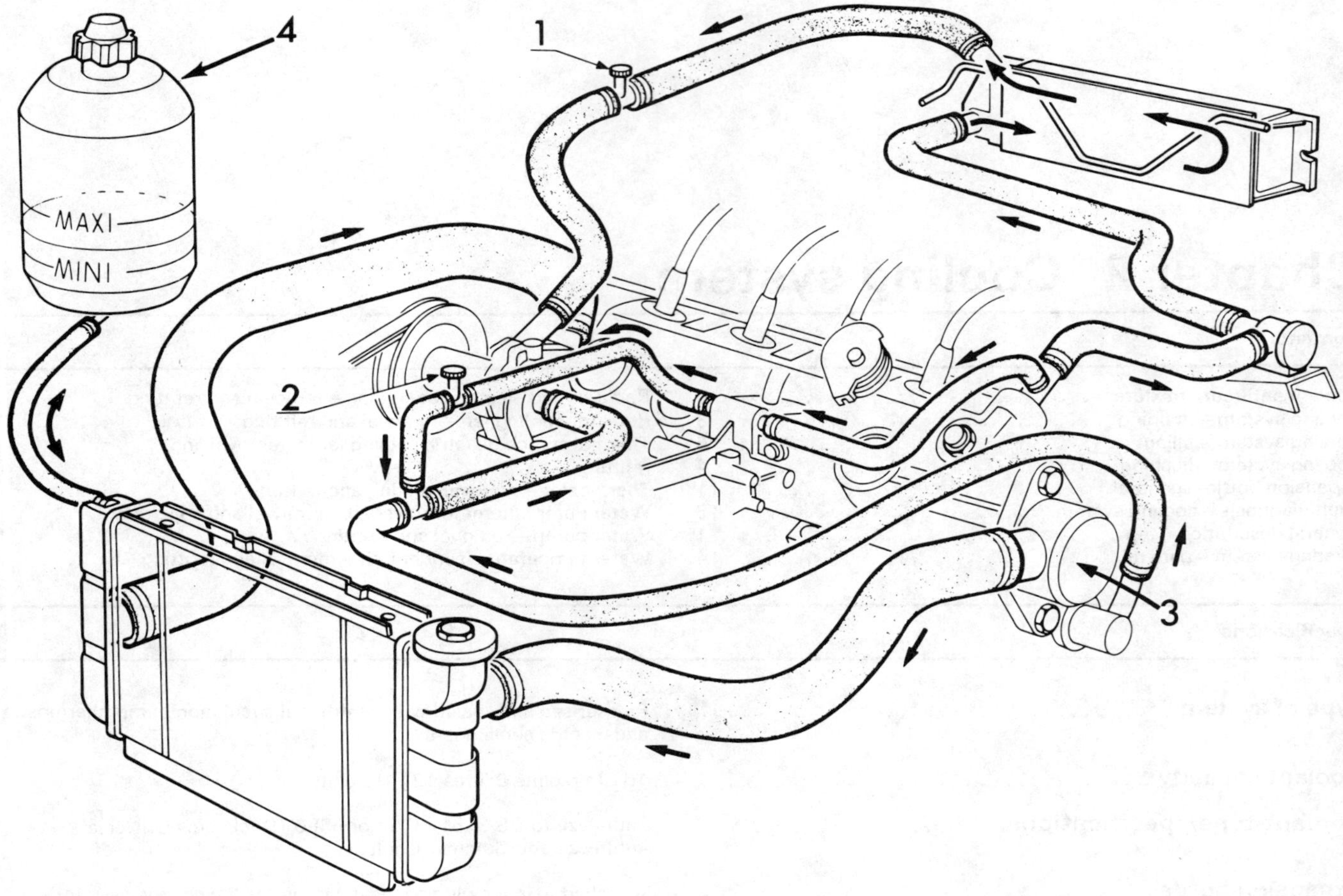

46 Chapter 2 Cooling system

Fig. 2.1 The cooling circulation system between the radiator, engine and heater unit

1 Bleed screw   2 Bleed screw   3 Thermostat housing   4 Expansion bottle

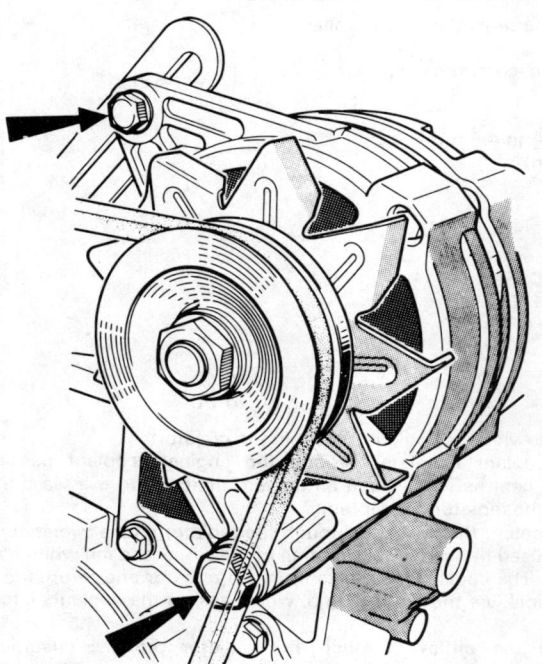

Fig. 2.2 The alternator mounting bolts (arrowed). The top bolt tightens onto the adjustment strap

The car interior heater is operated from the cooling system. A cable-operated valve may be opened or shut as required to allow coolant circulation through the heater matrix.

The engine coolant is also used to circulate through the carburettor hot-spot.

Air bleed screws are fitted to enable any air trapped in the system to be released. The air bleeds are normally only opened during the coolant refilling process and this is explained later in this Chapter.

Normal maintenance of the cooling system requires regular checking of the coolant level, and the various hoses and connections for leaks or signs of deterioration. Always use the correct grade and quantity of antifreeze, and this should of course be a non-corrosive type suitable for use in aluminium engines.

2 Water pump/alternator drivebelt – renewal and adjustment

1 The water pump and alternator are belt-driven from the crankshaft pulley. It is essential that the drivebelt is kept in good condition and is correctly adjusted.

2 If the drivebelt is allowed to operate with insufficient tension then it will probably slip on the pulleys. This will cause excessive belt wear and prevent the alternator and water pump from operating to their full capacity. However, should the drivebelt be overtightened, the water pump and alternator drive bearings will be overloaded.

3 To remove the drivebelt, loosen but do not remove the alternator

Chapter 2 Cooling system

mounting bolts. Pivot the alternator towards the engine to release the tension on the drivebelt, enabling it to be removed from the pulleys.

4 Place the new drivebelt over the pulleys and pivot the alternator away from the engine to tension the belt. Before fully tightening the alternator bolts, check the drivebelt tension by measuring the amount of deflection, midway between the alternator and water pump pulleys, under finger pressure. With a new belt the initial deflection should be $\frac{5}{32}$ to $\frac{3}{16}$ in (4 to 4.5 mm) (photo).

5 Run the engine for ten minutes and then recheck the deflection which should be $\frac{1}{4}$ in (6 mm).

6 If the existing belt is to be adjusted, loosen the alternator mounting bolts just sufficiently to enable it to be pivoted as required to set the tension as described above. Check that the alternator mounting bolts are retightened on completion.

3 Cooling system – draining

1 If the engine coolant is known to be in good clean condition and contains the correct ratio of antifreeze to distilled water, it can be used again. Therefore obtain a clean container, of a suitable capacity, into which the coolant can be drained.

2 If the coolant is being drained to allow the system to be flushed out, it is preferable to run the engine up to its normal operating temperature so that loose sludge and dirt particles are in suspension and more readily removed.

3 Set the heater control to *HOT*.

4 Remove the expansion bottle filler cap, and the pressurising valve.

5 If the radiator is fitted with a drain plug in its base remove it. If there is no drain plug fitted, unscrew the bottom hose connection at the radiator and disconnect the hose (photo).

6 Unscrew the drain plug in the cylinder block. This is not particularly convenient, being located on the exhaust manifold side of the block at the timing case end above the rear engine mounting (photo).

7 Open the bleed screws to ensure complete draining of the system.

4 Cooling system – flushing

1 Remove the thermostat from its location in the cylinder head (see Section 9) and refit the housing. Fabricate a temporary gasket for use between the housing and its mating flange in the cylinder head to prevent leakage during the flushing operation.

2 Check that the drain plugs are open and that the heater control is on *HOT*. Remove the radiator cap.

3 Insert a hose in the radiator filler neck and flush through until the water emerges clean. Use hot water if the engine was drained hot.

4 The accumulation of sludge or scale may necessitate removal of the radiator (Section 7) and reverse flushing. This is carried out by inverting the radiator matrix and placing a hose in the outlet pipe so that water flows in a reverse direction to normal.

5 The use of chemical descaler should only be used in a cooling system if scale and sludge formation are severe and then adhere strictly to the manufacturers instructions.

6 Leakage of the radiator or cooling system may be temporarily stopped by the use of a proprietary sealant but in the long term, a new cylinder head or other gasket, water pump, hoses or radiator matrix must be installed. Do not attempt to solder a radiator yourself. The amount of local heat required will almost certainly melt adjacent joints. Take the radiator to a specialist or exchange it for a reconditioned unit.

7 On completion, refit the thermostat and refill the system as described in Section 5.

5 Cooling system – filling

1 Prepare a suitable quantity of coolant. Mix the correct proportions of antifreeze and distilled water for the climatic zone (see Section 6).

2 Turn the heater control lever to *HOT*.

3 Remove the expansion bottle lid, and the pressurising valve.

4 Check that the radiator drain plug or bottom hose connection have been fully retightened. Also ensure that the engine drain plug is refitted and secure.

5 Referring to Fig. 2.1, unscrew bleed screws 1 and 2 (photo).

6 Remove the rubber strap securing the expansion bottle. Raise the

2.4 Check the drivebelt tension

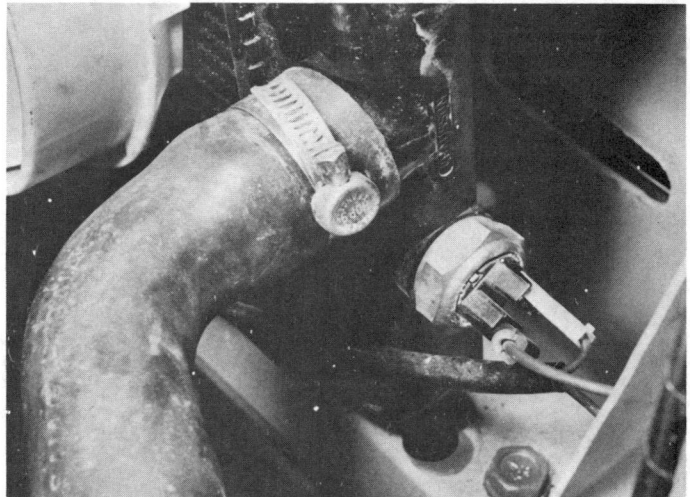
3.5 Disconnect the radiator bottom hose to drain when no plug is provided. Note position of the electric fan thermal switch

3.6 The cylinder block drain plug position (A)

5.5 Bleeder screw

7.8a Radiator retaining bolt

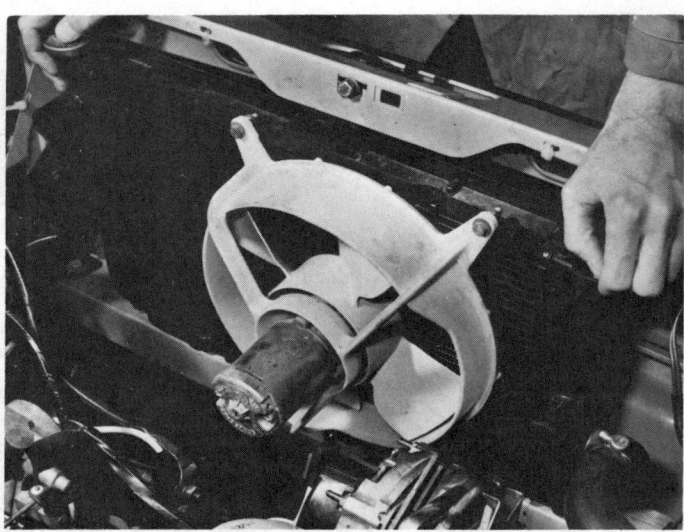

7.8b Lift out the radiator complete with fan assembly

7.9 Radiator bottom location peg

9.3 Unscrew the thermostat housing bolts (A). Note also the coolant temperature sender unit (B)

9.6 Installing the thermostat

Chapter 2 Cooling system

bottle as high as possible.

7 Unscrew the radiator cap and commence filling the radiator. When it is full refit the cap.

8 The remainder of the cooling system is filled via the expansion bottle. Continue filling until the mixture flows from each air bleed screw in turn, which should then be retightened.

9 Initially the expansion bottle must be filled to a level of about $2\frac{3}{4}$ in (70 mm) above the MAXI mark.

10 Refit the expansion bottle cap and the pressurising valve. Then run the engine for a few minutes up to its normal operating temperature.

11 Unscrew the bleed screws to allow any air in the system to escape, and as soon as the coolant is seen to flow out, screw the bleed screws down tight to seal the system.

12 Refit the expansion bottle in its mounting and secure with the rubber strap.

13 When the engine has cooled down, recheck the level of coolant in the expansion bottle, and top-up if required.

6 Coolant antifreeze mixture

1 The original coolant in the sealed system is of long-life type with antifreeze characteristics. It is preferable to mix a new coolant solution every year however, as apart from the antifreeze properties, rust and corrosion inhibitors (essential to a light alloy engine) will after this period of usage, tend to lose their effectiveness.

2 Mix the antifreeze with either distilled or rain water in a clean container. Drain and flush the system as previously described and pour in the new coolant. Do not allow an antifreeze mixture to come in contact with vehicle paintwork, and check the security of all cooling system hoses and joints as antifreeze mixture has a searching action.

3 The following amounts of antifreeze (as a percentage of the total coolant capacity – see Specification) will give positive protection down to the specified temperatures.

Percentage of antifreeze	Protection level (approx)
35%	–23°C (–9°F)
50%	–40°C (–40°F)

A mixture of less than 20% strength is not recommended as anti-rust and anti-corrosion action is not effective.

7 Radiator and electric cooling fan – removal and refitting

1 Drain the cooling system as described in Section 3.
2 Disconnect the battery earth terminal.
3 Detach the bottom radiator hose.
4 Detach the radiator top hose.
5 Detach the hose between the radiator and the expansion bottle.
6 Disconnect the radiator temperature switch.
7 Disconnect the fan motor leads.
8 Unscrew and remove the single radiator retaining bolt (photo), then carefully lift the radiator clear complete with fan assembly (photo).
9 Refitting is a direct reversal of the removal procedure. Ensure that the radiator location pegs (photo) slot into their respective holes. Renew defective hoses or securing clips on assembly. Refill the cooling system as described in Section 5 and check for leaks on completion.

8 Radiator cooling fan – removal and refitting

1 Disconnect the battery earth terminal.
2 Disconnect the fan motor leads.
3 Unscrew and remove the four securing screws retaining the fan assembly to the radiator. Remove the fan unit. Further dismantling of the fan unit is dependent on the problem. If the fan motor is defective, have it looked at by your Renault agent or local automotive electrician, who will advise you on the best course of action.
4 Refitting is a direct reversal of the removal procedure, but check that the electrical connections are correctly made. Run the engine up to its normal operating temperature and check the fan for correct operation.

9 Thermostat – removal, testing and refitting

1 The housing containing the thermostat is located in the cylinder head at the flywheel end of the engine (adjacent to the distributor).
2 Before removing the thermostat, drain off sufficient coolant to allow the level to drop below the radiator top hose. The draining proce-

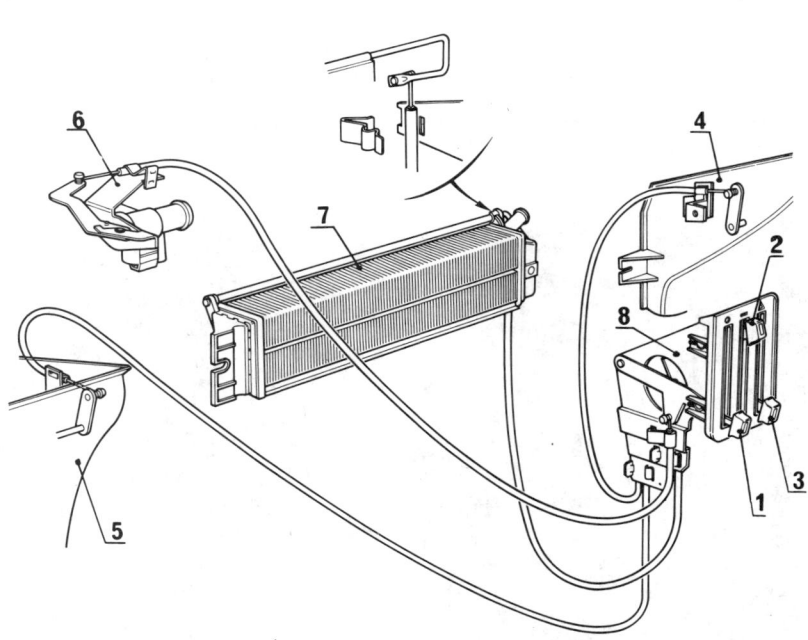

Fig. 2.3 The heater controls and matrix (left-hand drive version shown)

1 Heater valve control lever
2 Air inlet flap lever
3 Heated air distribution lever
4 The right-hand air distribution box
5 The left-hand air distribution box
6 Heater valve
7 Matrix
8 Lever control panel

dure is described in Section 3.

3 Unscrew the thermostat housing retaining bolts and remove the housing (photo). The thermostat can now be extracted.
4 To test the thermostat, suspend it in a container of water and heat. As the water temperature increases the thermostat should open. The opening temperatures are given in the Specifications. When fully open, transfer the thermostat to a container of cold water and immerse. The valve should close fully within 20 seconds.
5 If the thermostat is found to be faulty, it must be renewed.
6 Refitting is the reversal of the removal process but check that the sealing rings are in good condition and that the sealing surfaces are perfectly clean (photo).
7 Top-up the cooling system and bleed as described in Section 5. On completion run the engine and check for leaks.

10 Water pump – removal and refitting

1 Disconnect the battery earth terminal.
2 Drain the cooling system – see Section 3.
3 Loosen the alternator mounting bolts and pivot the alternator towards the engine to slacken the drivebelt.
4 Detach the drivebelt from the water pump pulley.
5 Loosen the securing clips and disconnect the hoses from the water pump.
6 Unscrew the retaining bolts and lift the pump unit clear of the cylinder block.
7 The water pump is not designed for repair or overhaul. Should leakage occur from the shaft gland or the bearings become worn or noisy, the pump should be renewed on an exchange basis.
8 The housing joint gasket can be renewed by removing the housing-to-pump mounting flange retaining bolts. Scrape away all traces of the old gasket and ensure that the respective mating surfaces are perfectly clean.
9 When replacing the pump, the O-ring seal between the pump mounting flange and the cylinder block must be renewed. Prise the old O-ring from its groove in the cylinder block face and insert the new one making sure that it seats correctly. Clean the mounting flange mating face before assembly (photo).
10 Bolt the mounting flange to the block face and then refit the impeller unit with a new gasket (photo).
11 The other reassembly procedures are a direct reversal of the removal instructions. Re-adjust the pump drivebelt as described in Section 2.

11 Expansion bottle – general

1 The expansion bottle incorporates the cooling system pressurising valve, situated beneath the cap. The pressure of the system depends on the vehicle specification but the value is stamped on the valve. **815** indicates that the cooling system is pressurised to 815 millibars.
2 Removal of the expansion bottle simply requires the bottle to be supported with one hand while the rubber retaining strap is released, allowing the bottle to be lifted from its mounting on the inner wing panel.
3 If it is to be removed completely the connecting hose must be detached at the radiator or bottle end.
4 Refit in the reverse order. If necessary it can be cleaned with hot water and soap, but should be well rinsed.

12 Water temperature sender unit – removal and refitting

1 The temperature gauge sender unit is located in the cylinder head just above the thermostat housing.
2 Prior to removal, drain the cooling system sufficiently to avoid loss of fluid when the sender unit is unscrewed (refer to Section 3).
3 Detach the battery earth cable.
4 Disconnect the wires from the sender unit terminals.
5 Unscrew the sender unit and remove.
6 Refitting is a direct reversal of removal; top-up and bleed the cooling system on completion.

13 Thermal switch (electric cooling fan) – removal and refitting

1 The thermal switch for the electric fan is located on the side of the radiator below the return hose.
2 Drain the radiator coolant (refer to Section 3).
3 Detach the battery earth cable.
4 Disconnect the electrical connections from the thermal switch and unscrew the unit to remove.
5 Refit in the reverse order. Top-up and bleed the cooling system on completion. Check the fan for correct operation by running the engine up to its normal operating temperature.

14 Heating system – general

1 The heating system for the interior of the car incorporates the windscreen demisting system. A system of hoses connect the engine cooling system to a special heater matrix mounted on the bulkhead. Flow of hot coolant to the heater matrix is controlled by a hot water valve. An electric fan is fitted to force air through the hot matrix and into the car, when the car is stationary or moving slowly. Normally the force of the air as the car runs is sufficient to heat the car.

10.9 Water pump mounting flange (A). Note O-ring in position (B)

10.10 Refit pump unit with new gasket

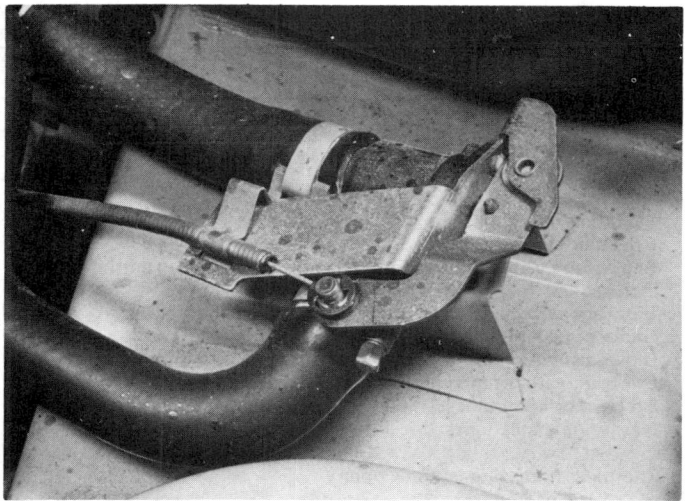

14.10 The heater valve unit. Unclip snap-ring to release cable

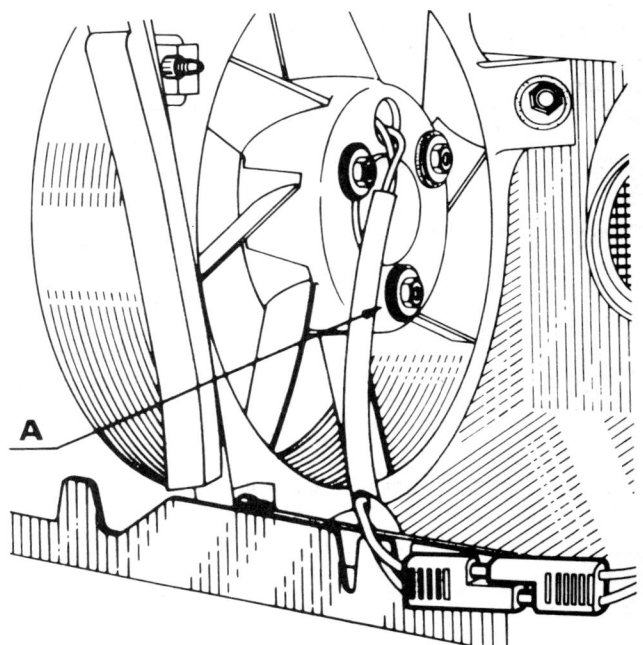

Fig. 2.4 The heater fan motor and shroud units installed. Check that vibration washers (A) are in good condition

2 Normally the heater components give very little trouble and require little maintenance, an occasional check of the hoses and their connections being all that is usually required. Should it be necessary to remove any part of the system, proceed as follows:

Heater fan motor

3 Disconnect the battery earth lead.
4 Unscrew and remove the heater support closure panel retaining screws. Withdraw the panel towards the engine as much as possible, sliding the panel along the hoses.
5 Detach the connecting wire at the socket.
6 Unscrew the fan unit retaining screws and the cap nut that retains the fan to the motor shaft.
7 Finally unscrew and remove the three retaining nuts from the motor shroud and remove the fan unit.
8 Reassembly is the reverse of removal but ensure that the vibration washers are correctly located under the fan-to-motor retaining nuts. Ensure that the seals are in good order before refitting the closure panel.

Heater valve

9 The cooling system will either have to be partly drained to facilitate disconnection of the hoses to the heater valve or, alternatively, use a suitable pair of clamps to compress the hoses to prevent leakage. If using clamps take care not to damage the hoses.
10 Unscrew the hose clips and detach the hoses from the valve unit. Disconnect the operating cable (photo).
11 To remove the valve unit carefully pull upwards to free from the clip.
12 Refit in the reverse order and adjust the cable. Top-up the coolant level and bleed the system if necessary.

Heater valve and air distribution control cables

13 Withdraw the control panel, as described in Chapter 9, to gain access to the cable connections at the control levers. Disconnect the appropriate cable, and remove from the retaining clamp. Disconnect the cable from the heater valve or air distribution box and withdraw.
14 When reassembling, check that the retaining clamp firmly secures the cable. On completion check that the sliding lever has not completed its full travel when the air distribution flaps are closed.
15 If the valve control cable has been renewed, ensure that the sliding lever has not completed its full travel when the valve is fully closed.

Air entry control cable

16 Withdraw the heater strut closure panel sufficiently to detach the cable from the air entry control flap. Withdraw the control panel, as described in Chapter 9, to gain access to the cable connections at the control levers. Disconnect the appropriate cable, remove from the retaining clamp and withdraw.
17 Refit in the reverse sequence. When the cable is reconnected to the control panel lever the sleeve must be adjusted so that $\frac{5}{64}$ in (2 mm) of upward lever travel remains when the flap is closed.

Heater matrix

18 Disconnect the battery earth cable and then drain the cooling system, with the heater valve open, sufficiently to remove the coolant from the heater unit. Alternatively, clamp the inlet and outlet hoses of the heater matrix to prevent coolant spillage.
19 Disconnect the outlet hose from the air bleed screw unit. Unclip the heater valve and then remove the heater strut closure panel. Detach the heater air inlet flap cable and then detach the matrix inlet hose. The matrix unit retaining bolt can now be removed and the matrix withdrawn complete with return hose.
20 Refit in the reverse sequence, re-adjusting the air flap inlet cable as previously described. Ensure that the seals are in good order before refitting the strut closure panel. Refill/top-up the cooling system and bleed as described in Section 3. Finally check the hoses and connections for leaks.

15 Fault diagnosis – cooling system

Symptom	Reason/s
Loss of coolant but no overheating	Expansion bottle empty Small leaks in system
Overheating and loss of coolant only when overheated	Faulty thermostat Drivebelt slipping/electric fan faulty Engine out-of-tune due to ignition and/or fuel system settings being incorrect Blockage or restriction in circulation of cooling water Radiator cooling fins clogged up Blown cylinder head gasket or cracked cylinder head Sheared water pump impeller shaft Cracked cylinder block New engine still tight
Engine runs too cool and heater inefficient	Thermostat missing or stuck open

Chapter 3 Fuel and exhaust systems

For modifications, and information applicable to later models, see Supplement at end of manual

Contents

Air filter element – removal and installation	2	Fault diagnosis – fuel system and carburettor	11
Carburettor – adjustments	4	Fuel pump – removal, servicing and refitting	8
Carburettor – dismantling, overhaul and reassembly	6	Fuel tank – removal and refitting	10
Carburettor – removal and installation	5	General description	1
Choke cable – removal and replacement	7	Solex 32 SHA carburettor – description	3
Exhaust system – general	9		

Specifications

Fuel pump type .... Mechanical, diaphragm, actuated by eccentric on end of camshaft

Carburettor
Type .... Solex, Type 32 SHA
Mark 621
Float level adjustment datum .... 0.693 to 0.732 in (17.6 to 18.6 mm)
Choke tube .... 26 mm
Accelerator pump stroke .... 0.197 in (5.00 mm)
Idle jet 39
Main jet 122.5
Constant CO jet 30
Accelerator pump jet 40
Air compensation jet 140
Econostat 90
Needle valve .... 1.5 mm
Defuming valve stroke .... 0.019 to 0.039 in (0.5 to 1.00 mm)
Initial throttle opening (gauge rod diameter) .... 0.039 in (1.00 mm)
Engine idle speed .... 875 rpm ± 25

Fuel tank capacity
Early R1210 only .... 38 litres (8.36 Imp gals)
All other models .... 48 litres (10.55 Imp gals)

Fuel octane rating .... 97-99 RON (4 star)

1 General description

The Renault 14 fuel system is conventional in layout and operation. The fuel tank is situated at the rear of the car beneath the luggage compartment. The fuel is drawn from the tank by a mechanically operated diaphragm pump actuated via a pushrod from an eccentric cam located at the timing gear end of the camshaft. The fuel pump incorporates a filter.

The carburettor is a horizontal single choke instrument with an accelerator pump and manual choke. The carburettor body is warmed by a connection to the engine cooling system.

Carburettor air is drawn through an air filter with a removable paper element. A thermostatic valve selects a warm air source in cold weather.

2 Air filter element – removal and installation

1 The air filter element must be periodically renewed at the mileage intervals prescribed in the Routine Maintenance Section at the beginning of the book. In severe environmental conditions such as hot dusty conditions the element should be changed more frequently.
2 If the filter element is not changed regularly it may become clogged and consequently restrict the air passage to the carburettor air intake. This will upset the air-to-fuel mixture ratio, normally governed by the carburettor settings, causing the engine to lose tune.
3 The element is easily changed. First, unclip the four filter canister lid clips.
4 Lift the lid and remove it sufficiently to allow the element to be extracted from the canister.
5 Before inserting the new filter element, wipe the inside of the container and lid clean using a non-fluffy cloth.
6 Insert the new element and check that it is seated correctly. Refit the cover and secure the clips.

3 Solex 32 SHA carburettor – description

1 The Solex 32 SHA carburettor is of relatively simple design. The fuel/air ratio supplied to the engine under the various operating conditions is governed by the air intake calibration orifice influencing the supply of fuel passing through the main jet and emulsion tube system.
2 An Econostat system allows a leaner basic fuel/air ratio setting to be used. A tube projects into the choke tube and supplies fuel from the float chamber when the airflow is high, thus enriching the mixture at high rpm without affecting it at wide throttle openings at low rpm.
3 A mechanical accelerator pump is incorporated into the float chamber. When the engine is idling the pump diaphragm is pushed down under the spring pressure and the pump chamber is filled. When

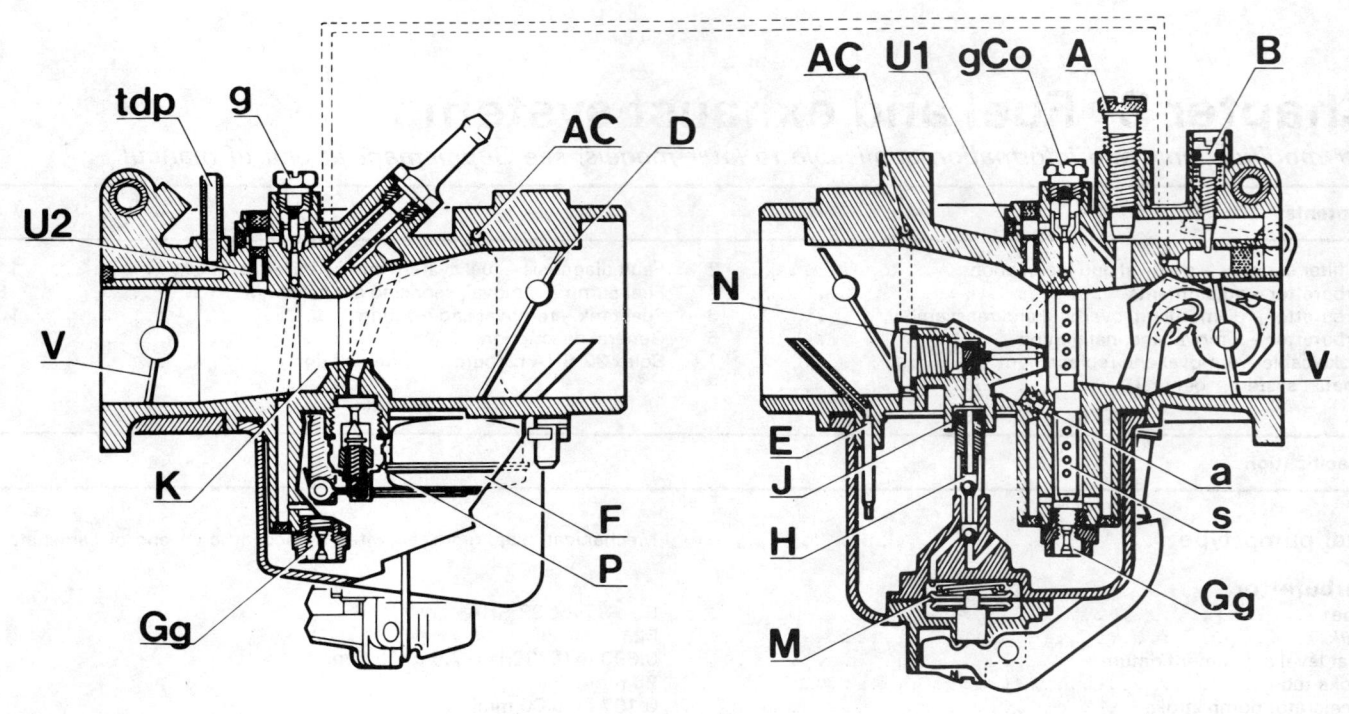

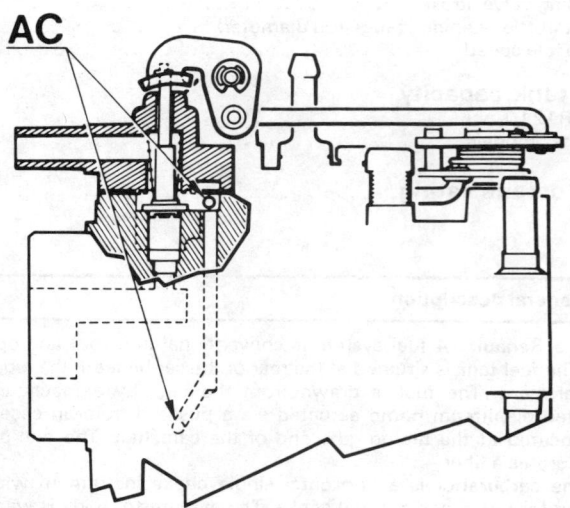

Fig. 3.1 Schematic diagram of the Solex 32 SHA carburettor

E	Econostat
H	Accelerator pump ball valve
M	Pump diaphragm
S	Jet assembly
tdp	Vacuum advance outlet
U2-U1	Air calibrating orifices (Idle + constant CO)
V	Butterfly
D	Choke flap
A	Idle volume screw
B	Idle fuel screw
P	Ball type needle valve
F	Float
Gg	Main jet
gCO	Constant CO idle jet
g	Idle jet
K	Choke tube
a	Calibrated jet
N	Accelerator pump calibrated jet
J	Accelerator pump O-ring
AC	Float chamber vent

Chapter 3 Fuel and exhaust systems

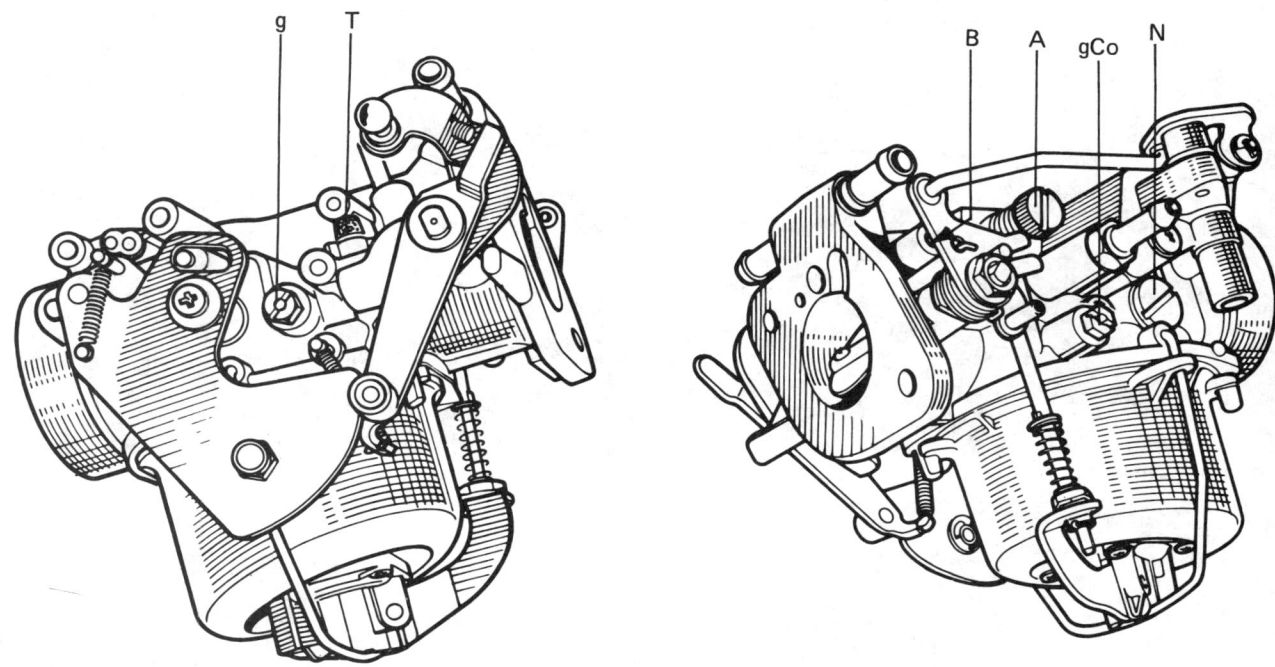

Fig. 3.2 Carburettor adjusting screws and jet positions

g	Idle jet	T	Throttle stop screw	B	Idle fuel screw (behind linkage)
		A	Idle volume screw	N	Accelerator pump calibrated jet
				gCO	Constant CO idle jet

the throttle is opened the throttle spindle actuates the diaphragm and fuel is injected into the choke tube via the ball valve and calibrated jet. The rate of injection is governed by the pump jet calibration.

4 For normal engine idling, a dual circuit constant richness system is employed. An idle speed control screw is fitted, located directly above the float chamber. The butterfly valve angle is preset at the factory, and must not be altered or the idle adjustment will be adversely affected. Similarly the throttle stop screw and the idle mixture richness screw must not be adjusted as they are set at the factory and the plastic cap is fitted to prevent interference.

5 A mechanical choke is fitted for use when the engine is cold. The choke cable operates the choke flap lever which incorporates a cam which partially opens the throttle butterfly when the choke control is extended. The actual choke flap is connected to the choke flap lever by a calibrated spring so that the vacuum created when the engine starts may open the choke flap slightly, against the pressure exerted by the spring, thus ensuring a correct mixture for the engine.

6 The carburettor is heated by the engine coolant which circulates within the carburettor throttle block close to the cylinder head.

4 Carburettor – adjustments

1 Generally speaking unless the carburettor is obviously out of tune or is malfunctioning it is not advisable to tamper with it. On any account *do not* touch the throttle stop screw or the idle fuel screw as they have been preset at the factory and are accurately adjusted to give optimum results.

2 The only adjustment normally required is to the idle adjustment screw.

3 Correct adjustment of the carburettor cannot be achieved unless the engine is in generally good condition. The valve clearances must be correct and the ignition system must also be in good condition and adjusted correctly.

4 Run the engine up to its normal operating temperature before any

2.5 Filter canister and element (out of the car)

adjustments are made. The air filter must be connected during adjustments.

5 If a tachometer is available connect this up so that the correct engine idle speed can be achieved.

6 Assuming the carburettor is in reasonable condition, the idle speed adjustment is set by turning the volume screw A as required. When unscrewed the idle speed is increased, whilst tightening will reduce the idle speed.

5.7 Removing the carburettor

6.7 Drive out the float hinge pin in direction of arrow

6.8 If required, remove the main jet (1) and ball valve (2). Check rubber seal (3) which must be in good condition and correctly seated

6.12 Accelerator pump rod showing the retaining circlip, nylon guide and spring

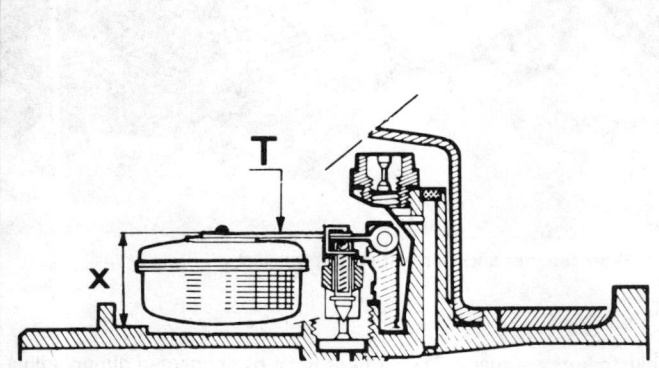
Fig. 3.3 Check distance X. Bend the lever (T) if necessary to adjust to the correct figure

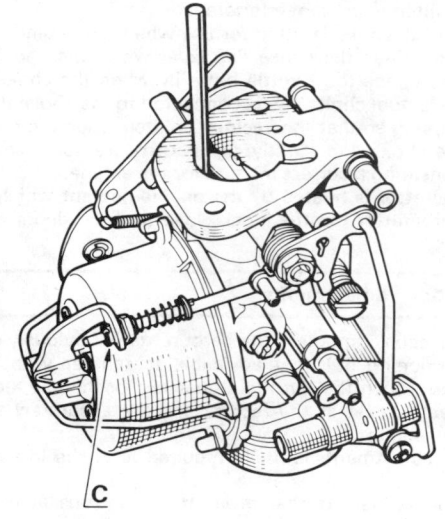
Fig. 3.4 Insert gauge rod as shown and adjust pump stroke at C

Chapter 3 Fuel and exhaust systems

7 Turn the screw slowly and progressively and when the correct idle speed is obtained, open the throttle a couple of times and check that when it closes again the idle speed is correct. If not, check that the throttle and choke cable connections are operating correctly and do not stick open.

8 If a new or overhauled carburettor has been installed, the procedure is slightly different as the fuel screw B may also need adjustment. First turn the idle screw A to give an engine speed of 900 rpm, then turn the fuel control screw progressively to increase the engine speed as much as possible. Re-set the idle screw to reduce the idle speed to 900 rpm. Repeat this procedure until the optimum setting of the fuel screw is reached. Then tighten the fuel screw to weaken the mixture until the engine speed has dropped by approximately 25 rpm and the specified idle speed (875 rpm) has been obtained with the engine running smoothly.

5 Carburettor – removal and installation

1 Carburettor removal requires either partial draining of the cooling system or clamping the inlet and outlet coolant hoses as near as possible to their connections to the carburettor, so that coolant is not spilled when they are disconnected.
2 Detach the air cleaner tube by unscrewing the retaining clip.
3 Unscrew the fuel supply pipe retaining clip and pull the pipe from its connection at the carburettor. Plug the pipe to prevent spillage of fuel and the ingress of dirt.
4 Pull free the vacuum advance pipe which leads to the distributor.
5 Unscrew the choke cable retainer and pull the cable free.
6 Disconnect the throttle linkage from the balljoint.
7 Unscrew the retaining nuts and carefully lift the carburettor from the manifold (photo). Plug the manifold inlet with some clean rag to prevent the ingress of dirt etc.
8 Refitting the carburettor is a direct reversal of the removal procedure. A new gasket must be used. When the carburettor is installed reconnect the choke cable and throttle linkage. Ensure that the choke cable is correctly adjusted giving a small amount of free play when the choke is off but allowing it to be fully applied.
9 Top up the coolant as applicable and bleed the system as described in Chapter 2.
10 Adjust the carburettor to give the correct idle speed as described previously.

6 Carburettor – dismantling, overhaul and reassembly

1 The carburettor should not normally need to be dismantled except for cleaning and checking the float level.
2 If the carburettor is to be dismantled, remember that it is a relatively delicate instrument and therefore requires careful handling. Use the correct tools for the job and do not interchange jets or clean them out with wire or any other such item which could score and damage them permanently.
3 Before dismantling any part of the carburettor, first clean it on the outside and prepare a clean work area where the parts can be laid out in order of dismantling for inspection.
4 *DO NOT* remove or disturb in any way the throttle stop screw (covered with plastic cap) or the idle fuel screw.
5 To remove the float chamber, prise the retaining clip free and lift the chamber away from the main body taking care not to snag the accelerator pump lever on the rod.
6 The accelerator pump unit is enclosed in the base of the float chamber. If it is thought to be defective have it checked out by your Renault dealer.
7 To remove the float, use a suitable fine drift and drive the float hinge pin out. It can only be removed toward the main jet as shown (photo).
8 The float needle valve can now be removed, as can the main jet (photo).
9 If a new replacement is available, remove the rubber O-ring seal from the rim of the chamber.
10 Unscrew the idle jet from the body of the carburettor, on the choke mechanism side.
11 The choke flap and throttle butterfly valves must not be disturbed. Their actuating mechanisms are external and do not normally require any attention.

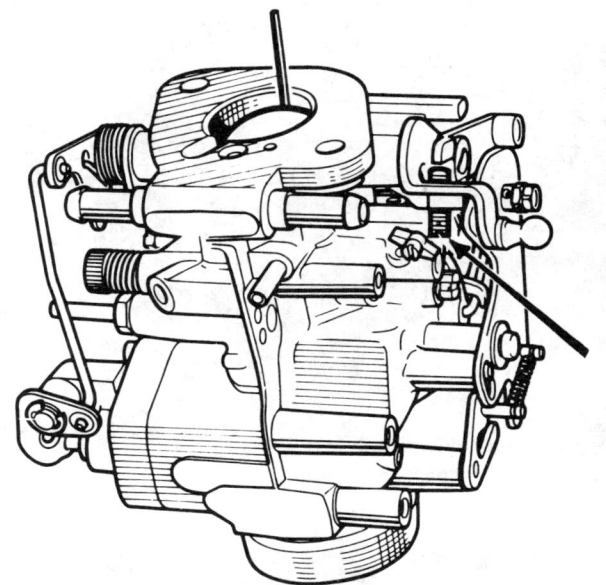

Fig. 3.5 Check initial throttle opening using gauge rod and adjust screw accordingly

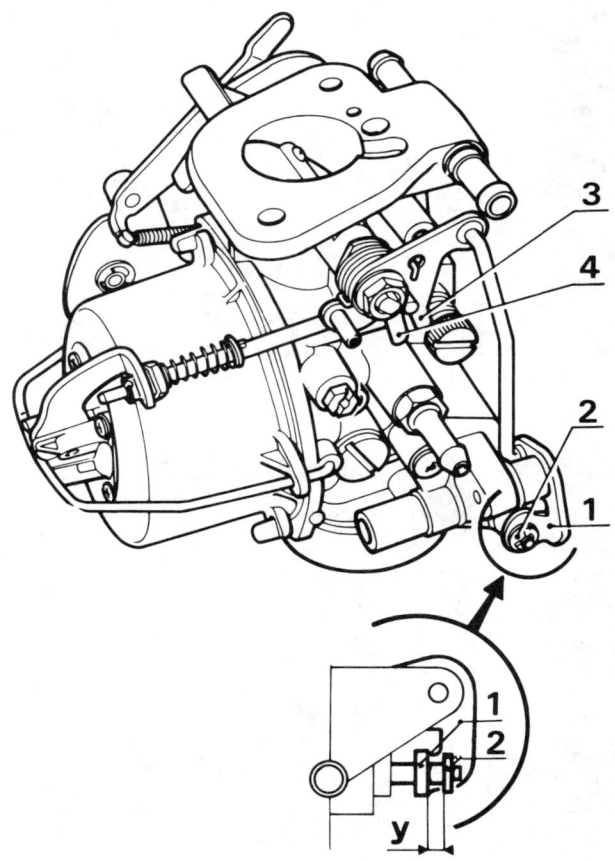

Fig. 3.6 Defuming valve check showing:

1 Lever 3 Lug
2 Flat washer 4 Lug

Measure clearance y

7.1 Choke cable connection at carburettor

8.2 Removing the fuel pump

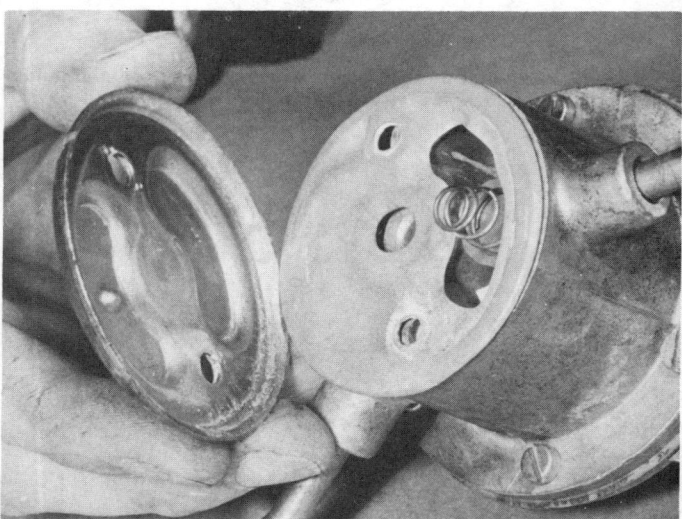
8.3 Removing the top cover

8.4 Remove filter and spring

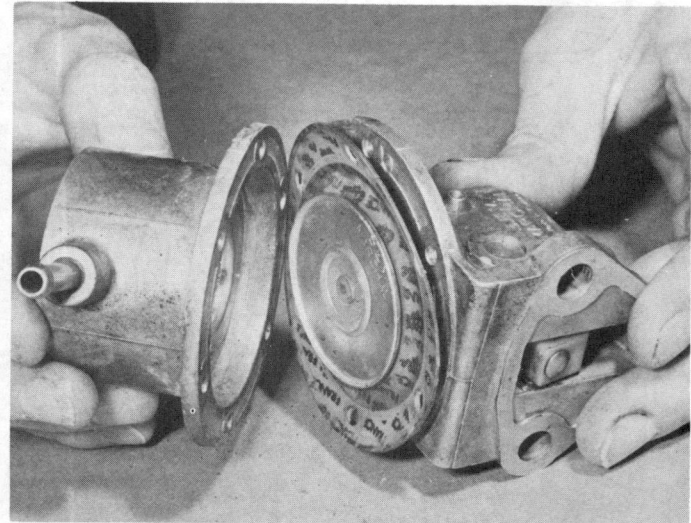
8.5 Separate the upper and lower body sections

8.6 Remove the diaphragm

Chapter 3 Fuel and exhaust systems

12 The accelerator pump rod spring can be removed after the circlip has been removed and the nylon lever guide withdrawn (photo). On refitting the accelerator pump stroke must be checked and adjusted if required. This is explained in paragraph 20.
13 The choke control calibrated spring is also external and for removal can be simply unhooked. Take care not to overstretch it during removal, or on refitment.
14 The respective chambers, passages and jet seatings can be brush cleaned using clean petrol and should then be blown dry with compressed air. Clean and blow through the jets in a similar manner. Do not under any circumstances use emery paper, wire wool or hard scrapers for cleaning!
15 Inspect for blockages and check the float is not punctured. Ensure that the needle valve operates freely. Renew any defective or suspect items but check that replacement parts are available before disposing of the old parts – they may be useable as a temporary repair.
16 If the throttle butterfly spindle or choke flap spindles are worn in the body then serious consideration should be given to renewing the complete carburettor. This wear is an indication that the carburettor is due for replacement, and it would be false economy to refit the original carburettor. Air leaks around worn spindles make it impossible to tune the carburettor correctly and poor performance and economy will result.
17 Reassembly is a reversal of the dismantling sequence, whenever possible use new gaskets and washers where fitted.
18 Check and adjust as necessary the following items during assembly:

Float level check

19 Invert the carburettor and support it horizontally. Measure the clearance between the bottom face of the float lever and the roof of the choke chamber (dimension X in Fig. 3.3). Check the measurement against that given in the Specifications and if necessary adjust to the correct figure by bending the float lever, taking care not to apply pressure on the needle.

Accelerator pump stroke

20 To carry out this check a special gauge rod will be required. The rod is inserted between the carburettor bore and the butterfly valve and in this position the accelerator pump must be at the limit of its stroke. If not the case, the retaining circlip on the rod must be moved up or down accordingly. A drill shank having a diameter of the specified accelerator pump stroke can be used if the special gauge rod is not available.

Initial throttle opening

21 To check the initial throttle opening you will need a gauge rod of 0·039 in (1·0 mm) diameter. To carry out the check fully shut the choke flap by operating the choke lever and then check with the gauge rod between the throttle butterfly and the carburettor body. If the clearance is incorrect, turn the screw indicated in Fig. 3.5 to adjust accordingly.

Defuming valve check

22 Open the choke flap fully and set the throttle butterfly to the idle position. Check that the specified clearance exists between the lever and flat washer of the valve operating rod. If not, adjust accordingly by bending the lugs (marked 3 and 4 in Fig. 3.6) towards or away from each other as necessary.

7 Choke cable – removal and replacement

1 Disconnect the cable from the carburettor by loosening the retaining nut (photo) and withdrawing cable.
2 Pull the cable through from the dashboard end.
3 Install in the reverse sequence. Lubricate the cable before fitting and ensure that the cable is correctly adjusted allowing the choke to open and close fully without binding.
4 To remove the outer cable, unscrew nut at dash panel, detach choke light cable (inside dash panel at cable sleeve), and pull cable through bulkhead and location clips. Replace in reverse order.

8 Fuel pump – removal, servicing and refitting

1 Disconnect the inlet and outlet pipes from the pump.
2 Unscrew the pump retaining bolts and carefully lift it clear of the flange and remove the old gasket (photo).
3 Unscrew the top retaining screws and remove the top cover (photo).
4 Extract the filter and coil spring (photo).
5 To remove and inspect the diaphragm assembly undo the five screws and separate the upper and lower body sections but remember to mark their alignment first (photo).
6 The diaphragm can now be removed for inspection (photo). Renew if it is torn or damaged in any way.
7 Use an air line and blow through the filter. Remove any sludge build up from the main chambers.
8 Reassemble in the reverse order ensuring that the screws are evenly tightened and the diaphragm is located correctly between the body section flanges. Refit the top cover using a new gasket.
9 Check for leaks on starting up.

9 Exhaust system – general

1 The exhaust system is conventional in its working and extremely simple to repair. It is wise only to use original type exhaust clamps and proprietary made systems (photos).

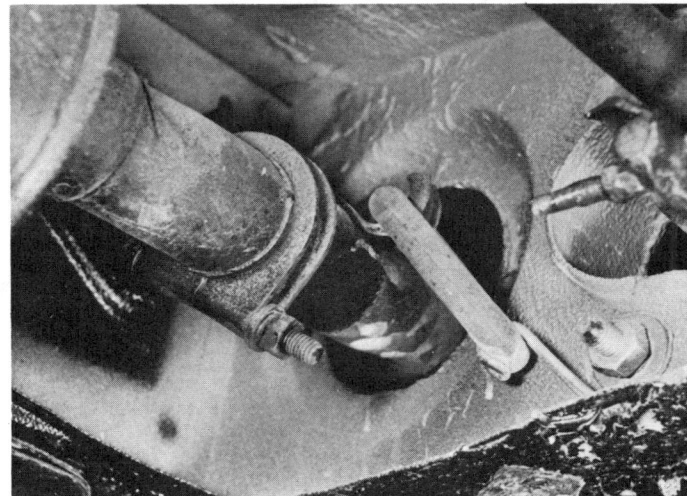

9.1a Typical pipe clamp and rubber location strap

9.1b View of downpipe and location bracket to transmission

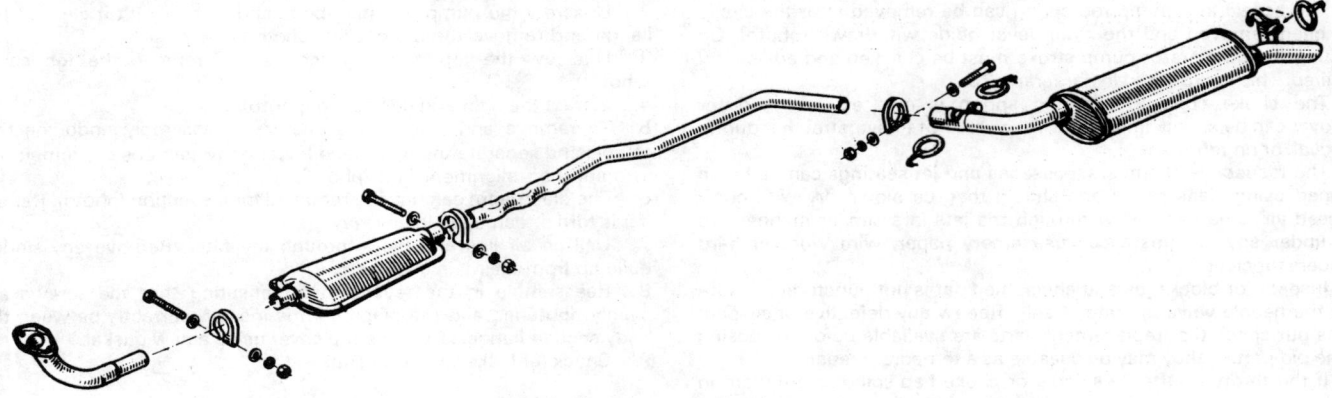

Fig. 3.7 The original exhaust system layout

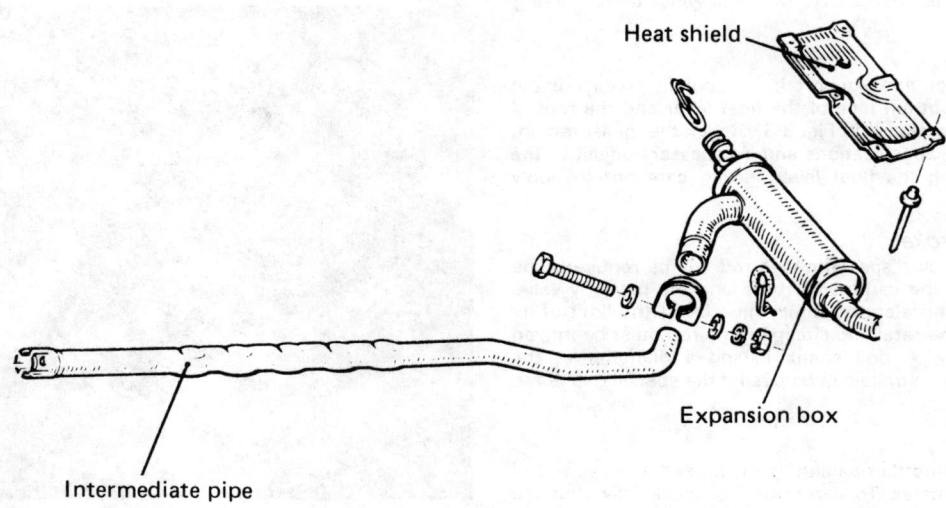

Fig. 3.8 The intermediate pipe and transverse expansion box system as fitted to 1979 models

2 When any one section of the exhaust system needs renewal it often follows that the whole lot is best replaced.
3 It is most important when fitting exhaust systems that the twists and contours are carefully followed and that each connecting joint overlaps the correct distance. Any stresses or strain imparted, in order to force the system to fit the hanger rubbers, will result in early fractures and failures.
4 When fitting a new part of a complete system it is well worth removing ALL the system from the car and cleaning up all the joints so that they fit together easily. The time spent struggling with obstinate joints whilst flat on your back under the car is eliminated and the likelihood of distorting or even breaking a section is greatly reduced. Do not waste a lot of time trying to undo rusted and corroded clamps and bolts. Cut them off. New ones will be required anyway if they are that bad.
5 Use an exhaust joint sealant when assembling pipe sections to ensure that the respective joints are free from leaks.
6 When fitting the new system, only semi-tighten the retainers initially until the complete system is fitted, then when you have checked it for satisfactory location, tighten the securing bolts/nuts. If the rubber retaining rings are stretched or perished they must be renewed otherwise the system will vibrate, leading to leaks and premature wear at the critical retaining points.

10 Fuel tank – removal and refitting

1 The fuel tank is located at the rear of the car directly below the boot floor. Its removal is simple if a little messy.
2 Disconnect the battery and detach the fuel gauge sender unit (see Chapter 9).
3 Place a suitable container below the drain plug in the bottom of the tank and drain the fuel into it.
4 Jack up the rear of the car and support with axle stands or blocks.
5 Undo the hose clips visible on the filler neck, remove and squeeze off the rubber centre hose. Remove the filler cap and take out the screws which hold its captive head to the bodyshell. These are visible once the cap is removed. Remove the overflow pipe from inside the rear wing.
6 Remove the four bolts that retain the tank in position underneath and carefully lower it from the car.
7 Refit in the reverse order and when refilled check for leaks.

Chapter 3 Fuel and exhaust systems

11 Fault diagnosis – Fuel system and carburation

Unsatisfactory engine performance and excessive fuel consumption are not necessarily the fault of the fuel system or carburettor. In fact they more commonly occur as a result of ignition and timing faults. Before acting on the following it is necessary to check the ignition system first. Even though a fault may lie in the fuel system it will be difficult to trace unless the ignition is correct. The faults below, therefore, assume that this has been attended to first (where appropriate).

Symptom	Reason/s
Smell of petrol when engine is stopped	Leaking fuel lines or unions Leaking fuel tank
Smell of petrol when engine is idling	Leaking fuel line unions between pump and carburettor Overflow of fuel from float chamber due to wrong level setting, ineffective needle valve or punctured float
Excessive fuel consumption for reasons not covered by leaks or float chamber faults	Worn jets Over-rich jet setting Sticking mechanism
Difficult starting, uneven running, lack of power, cutting out	One or more jets blocked or restricted Float chamber fuel level too low or needle valve sticking Fuel pump not delivering sufficient fuel

Chapter 4 Ignition system

For modifications, and information applicable to later models, see Supplement at end of manual

Contents

Condenser – testing, removal, and refitting	12
Contact breaker points – adjustment	3
Contact breaker points – removal and replacement	4
Distributor – dismantling, inspection and reassembly	6
Distributor – removal and installation	5
General description	1
Ignition coil – general	8

Ignition diagnostic socket and pick up	10
Ignition system – fault diagnosis	13
Ignition switch – removal and installation	11
Ignition timing – adjustment	7
Routine maintenance	2
Spark plugs and leads	9

Specifications

Spark plug
Type .... AC 42 LTS or Champion BN9Y
Electrode gap .... 0.022 to 0.026 in (0.55 to 0.65 mm)

Coil
.... SEV or Ducellier 12 volts

Distributor
Type .... Ducellier R299/D71
Rotation .... Anti-clockwise
Firing order .... 1 – 3 – 4 – 2
Contact points gap .... 0.016 in (0.4 mm)

Ignition timing
Initial advance setting .... $4° \pm 1$ BTDC (Type 129), $3° \pm 1$ BTDC (Type 145) or as indicated on equipment
Cam angle .... $57° \pm 3$
Dwell percentage .... $63\% \pm 3$

Torque wrench settings
	lbf ft	Nm
Spark plugs	11 to 14	15 to 20

1 General description

Ignition of the fuel/air mixture in the Renault engine is conventional in that one spark plug per cylinder is used and the high voltage required to produce the spark across the plug electrodes is supplied from a coil (transformer) which converts the volts from the supply battery to the several thousand necessary to produce a spark that will jump a gap under the conditions of heat and pressure that occur in the cylinder.

In order that the spark will occur at each plug in the correct order and at precisely the correct moment the low voltage current is built up (in the condenser) and abruptly discharged through the coil when the circuit is broken by the interrupter switch (contact points). This break in the low voltage circuit, and the simultaneous high voltage impulse generated from the coil, is directed through the selector switch (rotor arm) to one of four leads which connect to the spark plugs.

Due to different spark timing requirements under certain engine conditions (of varying speed or load) the distributor also has an automatic advance device (advancing the spark means that it comes earlier in relation to the piston position. Centrifugal weights moving against the tension of two springs advance the cam in relation to the main shaft according to the speed of the engine. A vacuum unit coupled to the carburettor also advances the ignition timing under part throttle conditions.

An additional feature of the Renault 14 ignition system is the diagnostic socket. This is bolted to the flywheel housing and enables a Renault dealer to use specialised diagnostic equipment to check the ignition timing and to quickly pin-point any defective part of the circuit. *Do not tamper with or attempt to use the socket as serious damage may result!*

2 Routine maintenance

Spark plugs
1 Remove the plugs and thoroughly clean away all traces of carbon. Examine the porcelain insulation round the central electrodes inside the plug and if damaged discard the plug. Reset the gap between the electrodes. Do not use a set of plugs for more than 9000 miles. It is false economy.
2 At the same time check the plug caps. Always use the straight tubular ones normally fitted. Good replacements can come from a Renault agency.

Distributor
3 Every 9000 miles remove the cap and rotor arm and smear the surfaces of the cam itself with petroleum jelly. Do not overlubricate as

Chapter 4 Ignition system

any excess could get onto the contact points surfaces and cause ignition difficulties.

4 Every 9000 miles examine the contact point surfaces. If there is a build up of deposits on one face and a pit on the other it will be impossible to set the gap correctly and they should be refaced or renewed. Set the gap when the contact surfaces are in order.

5 Check the proper functioning of the vacuum advance mechanism.

General

6 Examine all leads and terminals for signs of broken or cracked insulation. Also check all terminal connections for slackness or signs of fracturing of some strands of wire. Partly broken wire should be renewed.

7 The HT leads are particularly important as any insulation faults will cause the high voltage to jump to the nearest earth and this will prevent a spark at the plug. Check that no HT leads are loose or in a position where the insulation could wear due to rubbing against part of the engine.

3 Contact breaker points – adjustment

1 The distributor contact point clearance is set by means of an adjustment nut on the outside of the distributor body. If equipment is available for measuring the dwell percentage and cam-angle then the nut is simply turned in the direction required to achieve the correct dwell percentage and cam-angle as given in Specifications. If employing this method, the engine must be ticking over at its normal idle speed when making these adjustments.

2 To check and adjust the points using feeler gauges is not quite as simple due to the distributor location, but it can still be achieved as follows:

3 Unclip and remove the distributor cap.

4 Lift off the rotor arm and then the plastic cover.

5 Turn the engine over (by engaging a gear and pushing the car) so that the breaker points are open.

6 Now insert the feeler gauge down through spokes of the retainer flange and insert between the contact points (photo). Access is not particularly good and the use of a suitably positioned lead-light and a small mirror will help to guide the feeler gauge into position and make internal inspection of the contact point assembly possible.

7 If the contact gap is correct the feeler gauge will be a light touch between the contact faces. If it is too tight or loose turn the adjuster nut accordingly to achieve the correct clearance.

8 Reassemble the plastic cover and rotor arm. Wipe the distributor cap clean (inside and out) and inspect for hairline cracks. The four segments and the central carbon brush must be in good condition. Refit the cap and retain with clips.

4 Contact breaker points – removal and replacement

1 Changing the contact points will necessitate removal of the distributor. This is because of the location of the distributor and the position of the points within it. Removal of the distributor is covered in Section 5.

2 With the distributor removed, unscrew the retaining flange screws and withdraw the flange (photo).

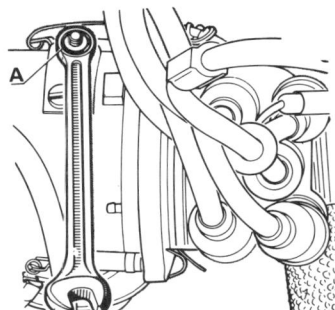

Fig. 4.1 Contact breaker adjustment nut (A) and spanner (cap still in position)

3.6 Insert feeler gauge to check contact points clearance

4.2 Unscrew and remove the flange

4.3 General view of distributor internal components showing:
1 Contact adjustment rod and spring
2 Contact points
3 Fixed contact retaining screw
4 Cam
5 Moving contact point pivot showing retaining washer and clip
6 Vacuum advance control rod and spring assembly

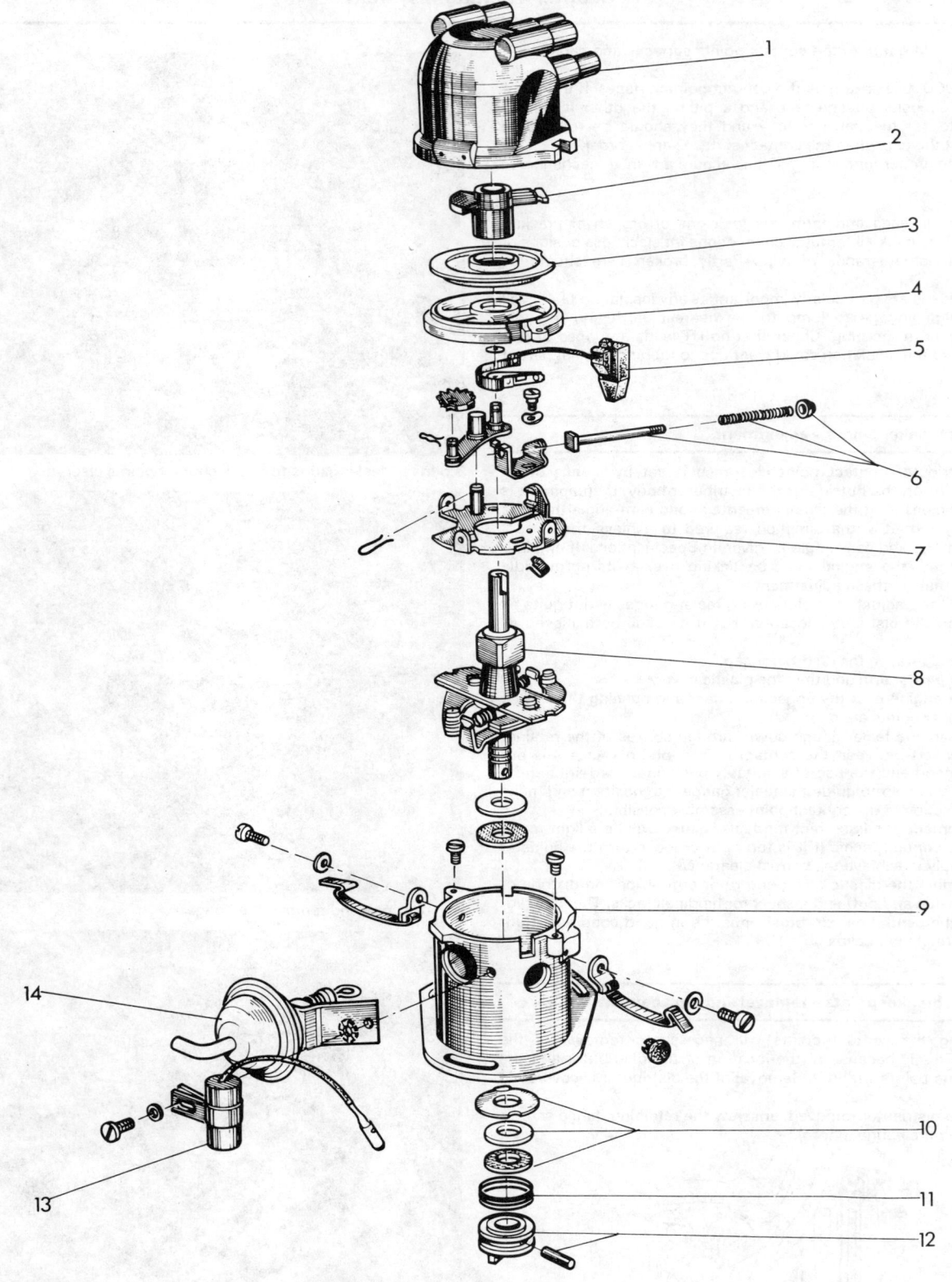

Fig. 4.2 The distributor components

1 Cap
2 Rotor
3 Cover
4 Flange
5 Moving points and LT lead
6 Adjustment nut/spring/rod
7 Base plate
8 Rotor shaft and centrifugal weight assembly
9 Distributor body
10 Thrust washers
11 Spring
12 Spigot and pin
13 Condenser
14 Vacuum advance control unit

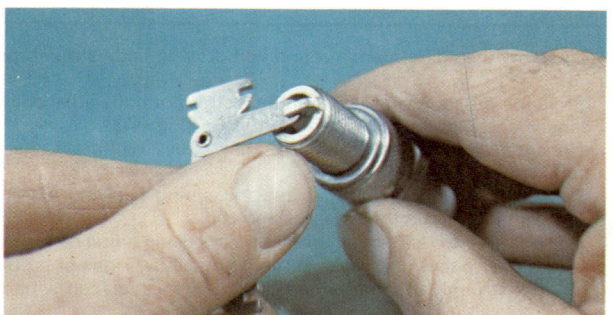

Measuring plug gap. A feeler gauge of the correct size (see ignition system specifications) should have a slight 'drag' when slid between the electrodes. Adjust gap if necessary

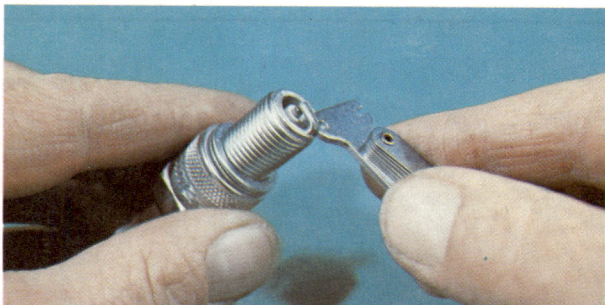

Adjusting plug gap. The plug gap is adjusted by bending the earth electrode inwards, or outwards, as necessary until the correct clearance is obtained. Note the use of the correct tool

Normal. Grey-brown deposits, lightly coated core nose. Gap increasing by around 0.001 in (0.025 mm) per 1000 miles (1600 km). Plugs ideally suited to engine, and engine in good condition

Carbon fouling. Dry, black, sooty deposits. Will cause weak spark and eventually misfire. Fault: over-rich fuel mixture. Check: carburettor mixture settings, float level and jet sizes; choke operation and cleanliness of air filter. Plugs can be re-used after cleaning

Oil fouling. Wet, oily deposits. Will cause weak spark and eventually misfire. Fault: worn bores/piston rings or valve guides; sometimes occurs (temporarily) during running-in period. Plugs can be re-used after thorough cleaning

Overheating. Electrodes have glazed appearance, core nose very white – few deposits. Fault: plug overheating. Check: plug value, ignition timing, fuel octane rating (too low) and fuel mixture (too weak). Discard plugs and cure fault immediately

Electrode damage. Electrodes burned away; core nose has burned, glazed appearance. Fault: pre-ignition. Check: as for 'Overheating' but may be more severe. Discard plugs and remedy fault before piston or valve damage occurs

Split core nose (may appear initially as a crack). Damage is self-evident, but cracks will only show after cleaning. Fault: pre-ignition or wrong gap-setting technique. Check: ignition timing, cooling system, fuel octane rating (too low) and fuel mixture (too weak). Discard plugs, rectify fault immediately

5.7 Note the offset spindle lugs and check that the O-ring is in good condition and correctly seated

7.4a The initial advance setting shown on clip attached to HT lead

7.4b Align the flywheel timing mark to the specified initial advance setting

3   Unscrew the adjustment nut and withdraw the adjustment rod and spring.
4   Prise the rubber grommet from the distributor body adjacent to the condensor and extract the retaining lug (C). (Fig. 4.3).
5   Unscrew the contact retaining screw and withdraw the fixed contact.
6   Detach the connector, prise free the spring clip and extract the moving contact. Retain the washers and clip.
7   It is possible to reface the contact points using a fine carborundum stone or emery paper. However, if the points show signs of burning or pitting, it is strongly recommended that they are replaced with a new set.
8   Reassembly of the contact points is a direct reversal of the removal procedure. Check that the fixed contact is free to slide under the retaining washer when being adjusted. Set the contact point clearance before refitting the distributor to the engine.

5 Distributor – removal and installation

1   Apart from replacing the contact points, the distributor will need removal if there are any indications that the rotor shaft bearings are worn, (giving contact gap adjustment difficulties) or if it is to be dismantled for general cleaning and checking.
2   Before removing the distributor it will help prevent future confusion if the engine is positioned with Nos 2 and 3 pistons at TDC and the rotor arm pointing towards the segment in the distributor cap leading to No 2 spark plug lead.
3   Detach the plug leads from the spark plugs and the coil HT lead from the distributor cap or coil. Remove the cap by unclipping the leaf spring clip at each side.
4   Undo the LT lead at the coil – this should be a screw on connector, and pull off the vacuum advance pipe from the distributor.
5   Before removing the distributor scribe an alignment mark between the distributor flange and the cylinder head. This will ensure correct timing on reassembly.
6   Unscrew the retaining screws and withdraw the distributor. As it is removed note the offset drive lugs on the base of the rotor shaft which enables it to be correctly positioned on reassembly.
7   Installation is a direct reversal of the removal procedure. Check that the rubber O-ring between the flange and cylinder head surface is in good condition and renew if necessary. Align the offset spindle lugs (photo) with the slots in the end of the camshaft and fit the distributor, turning the rotor shaft a little either way to enable the lugs and slot to engage. Do not forget to refit the HT lead location bracket under the retaining bolt head. Align the timing mark of the flange and cylinder head before tightening the retaining bolts. If a new distributor unit is being installed, set the timing as described in Section 7.

6 Distributor – dismantling, inspection and reassembly

1   If the distributor is known to be giving problems and you have checked the basic items (contact points, plug leads and their connections, distributor cap, carbon brush and condenser) then further dismantling for more detailed inspection will be necessary.
2   Before dismantling, check on the availability of replacement parts. Possible replacements include the vacuum advance unit, rotor shaft assembly and centrifugal weight springs. No repairs are possible to any of the distributor components and therefore unless spares are readily available it is probably necessary to replace the distributor as a unit.
3   Without proper test equipment it is difficult to diagnose whether or not the centrifugal advance mechanism is performing as it should. However, play in the shaft bushes can be detected by removing the rotor arm and gripping the end of the shaft and trying to move it sideways. If there is any movement then it means that the cam cannot accurately control the contact points gap. This must receive attention.
4   With the distributor removed take off the rotor and remove contact points as described in Section 4.
5   Unscrew and remove the condenser and detach the low tension lead.
6   Withdraw the vacuum advance control unit. The control rod of the vacuum unit is retained in position on the pivot by the semi-circular knurled plate secured by a wire clip. Note the position of the plate then

Chapter 4 Ignition system

prise the clip free, remove the semi-circular plate and lift the control rod over the pivot to disengage and remove the vacuum unit.

7   Extract the pivot pin plate.

8   Unscrew the baseplate retaining screws and carefully extract the baseplate noting its relative position in the distributor body.

9   Note the relationship of the offset lugs to the rotor arm locating groove, then, using a suitable drift, support the shaft assembly and drive out the bottom spigot retaining pin. Withdraw the spigot, spring and thrust washers.

10 Withdraw the rotor shaft and centrifugal weight assembly.

11 With the distributor dismantled, clean the respective components and inspect for wear and/or damage and renew as necessary. If in doubt regarding the condition of any parts have them checked by your Renault agent.

12 Reassembly of the distributor is a direct reversal of the dismantling procedure. Do not stretch the centrifugal springs. Smear the contact baseplate with a thin film of oil or grease between it and the moving pivot plate. Use a new retaining pin to locate the bottom spigot and ensure that the drive lugs are correctly positioned relative to the rotor arm locating groove. When fully assembled ensure that the rotor shaft rotates freely and the contact points are correctly adjusted and operating correctly.

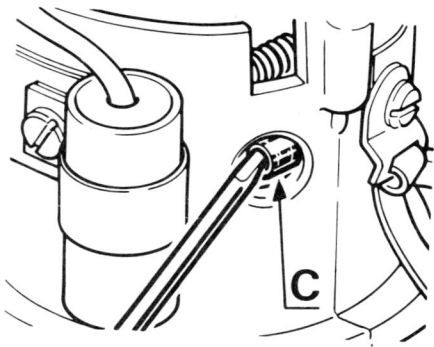

Fig. 4.3 Remove the grommet and extract lug (C)

7 Ignition timing – adjustment

1   It is necessary to time the ignition when it has been upset owing to overhauling. Dismantling may have altered the relationship between the position of the pistons and the moment at which the distributor delivers the spark. Also, if maladjustments have affected the engine performance it is very desirable, although not always essential, to reset the timing starting from scratch. In the following procedures it is assumed that the intention is to obtain standard performance from the standard engine which is in reasonable condition. It is also assumed that the recommended fuel octane rating is used.

2   To set the initial advance (static timing), first check that the contact breaker points are correctly adjusted (Section 3).

3   Rotate the engine so that the TDC mark on the periphery of the flywheel is aligned with the O-mark on the timing plate. Nos 2 and 3 pistons will then be at TDC and the distributor rotor arm will be pointing towards the distributor cap segments to either No 2 or 3 spark plug leads.

4   The initial ignition advance setting is usually marked on a clip attached to one of the HT leads (photo) or it may be stamped on the body of the distributor. If in doubt, refer to the Specifications. Turn the crankshaft so that the TDC mark on the flywheel is aligned with the appropriate initial advance setting (photo). Loosen the distributor retaining bolts and turn the distributor gently until the points are just breaking. Tighten the distributor retaining bolts.

5   The accuracy of this operation can be improved by employing a test light to indicate when the points break. Connect a 12 volt bulb in parallel with the contact breaker points (one lead to earth and the other from the distributor low tension terminal). Switch on the ignition and turn the distributor body clockwise until the bulb lights up, indicating the points have just opened.

6   Set in this manner the timing should be approximately correct but further minor adjustment may be necessary following a road test.

7   For more accurate setting of the distributor, including the relative amounts of vacuum and mechanical advance the car should either be taken to your Renault dealer who will be able to use the diagnostic socket to check the timing setting, or use a stroboscopic timing light.

8   Do not attempt to use the diagnostic socket yourself as this requires specialised equipment.

8.2 Check coil connections for security and keep clean and dry

9.1 Removing spark plug for service check

8 Ignition coil – general

1   The maintenance of the coil is minimal and is limited to periodically wiping its surfaces clean and dry and ensuring that the lead connections are secure. High voltages generated by the coil can easily leak to earth over its surface and prevent the spark plugs from receiving the electrical pulses. Water repellent sprays are now avail-

Chapter 4 Ignition system

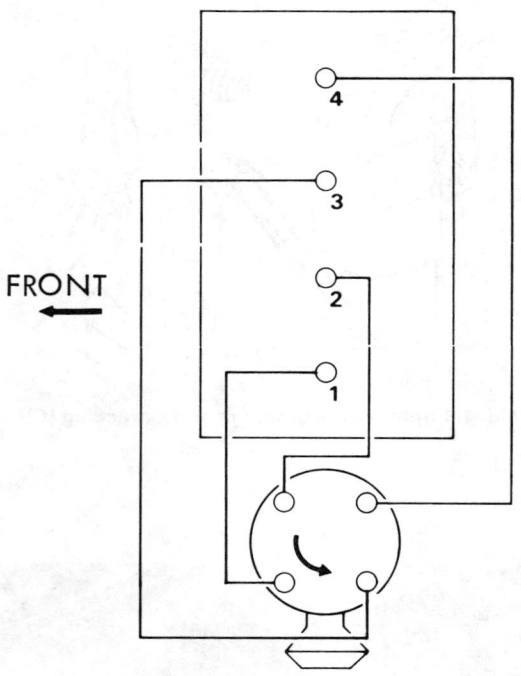

Fig. 4.4 The HT lead positions in the distributor cap – arrow indicates direction of rotation

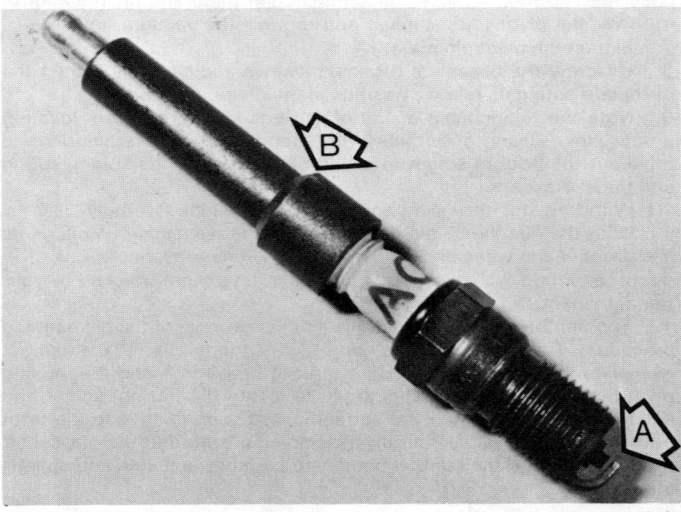

9.5 Check electrode gap clearance (A). Ensure plug and extension (B) are clean before reassembly

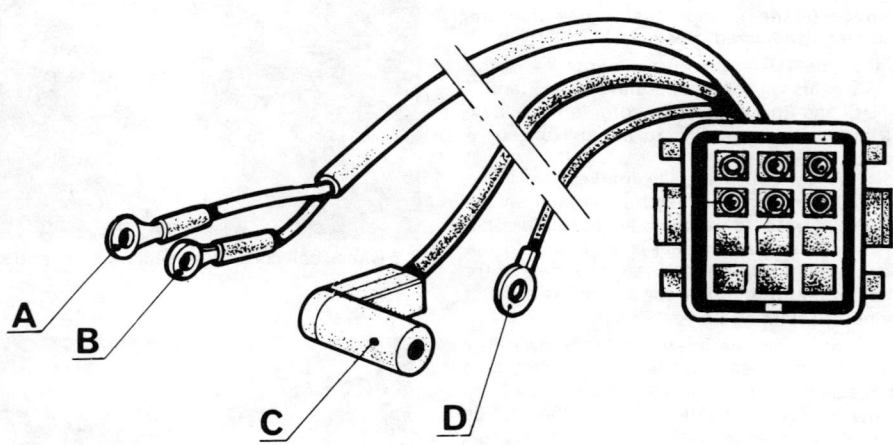

Fig. 4.5 The diagnostic socket and TDC pick up connections

A – To ignition coil negative (–) terminal (black wire/red sleeve)
B – To ignition coil positive (+) terminal (grey wire/blue sleeve)
C – TDC pick up   D – Earth wire (yellow)

able to prevent dampness causing this type of malfunction.
2   Wipe clean and spray the HT leads and distributor cap also.
3   Testing of the coil is covered in the faults Section at the end of this Chapter.

9   Spark plugs and leads

1   The spark plugs should be cleaned and their gaps reset every 3000 miles (4500 km). Use the correct spanner for removal and installation (photo).
2   Be especially careful when refitting plugs to do so without force and screw them up as far as possible by hand first. Do not overtighten. The aluminium head does not take kindly to thread crossing and extra force. The proprietary non-cranked plug extensions should always be used to ease fitting and to ensure against HT lead shorting.
3   The gaps should be set to a clearance of 0.022 to 0.026 in (0.55 to 0.65 mm). It is difficult to get a normal feeler into the gap, so ideally a wire feeler should be used. The gaps are set by bending the side electrodes. Under no circumstances must any pressure be put on the central electrode, or the porcelain insulator nose may be cracked, and later drop a chip into the engine.
4   The HT leads and their connections at both ends should always be clean and dry and, as far as possible, neatly arranged away from each other and nearby metallic parts which could cause premature shorting

Chapter 4 Ignition system

in weak insulation. The metal connections at the ends should be a firm and secure fit and free from any signs of corrosive deposits. If any lead shows signs of cracking or chafing of the insulation it should be renewed. Remember that radio interference suppression is required when renewing any leads.

10 Ignition diagnostic socket and pick up

1 The diagnostic socket is secured to the clutch housing, and is designed specifically for use by your Renault agent who has the appropriate Souriau or Sun diagnostic bay equipment for usage with the socket.

2 With the above mentioned equipment the socket can be used for the following checks:

(a) Contact points – condition and adjustment check
(b) Engine rpm reading
(c) Low tension circuit check
(d) Initial advance check
(e) Vacuum and centrifugal advance curve characteristics

3 The socket and/or pick up should not normally be interfered with but if for any reason it has to be removed or the wires disconnected, ensure that they are correctly and securely located on reassembly. The various leads are shown in Fig. 4.5

4 If the TDC pick-up is ever removed it should be re-installed so that it is about 0.040 in (1 mm) distant from the flywheel.

11 Ignition switch – removal and installation

1 Disconnect the battery earth cable connector.
2 Unscrew and remove the lower steering column panel. Note length and position of screws as they are removed to facilitate their correct refitting.
3 Disconnect the electrical connector at the switch.
4 Unscrew the cross-head screw from the lock barrel.
5 Drill out the shear bolt in the barrel and remove the switch unit.
6 Installation is a direct reversal of the removal procedure. Check the switch operation prior to refitting the lower panel.

12 Condenser – testing, removal, and refitting

1 The condenser absorbs current and reduces arcing when the contact points open. It is fitted in parallel with the points and in the event of its failure will prevent the points from interrupting the low tension circuit.

2 To test the condenser, position the engine so that the points are closed, and disconnect the HT lead from the coil. Switch on the ignition and separate the contact points by hand. If this is accompanied by a blue flash then condenser failure is indicated. Other symptoms are badly pitted or discoloured points, missing, and difficult starting.

3 To renew the condenser disconnect the lead from the terminal block on the side of the distributor, remove the retaining screw, and withdraw the condenser.

4 Refitting is a reversal of removal.

13 Ignition system – fault diagnosis

1 Engine troubles normally associated with, and usually caused by, faults in the ignition system are:

(a) Failure to start when the engine is turned
(b) Uneven running due to misfiring or mistiming
(c) Smooth running at low engine revolutions but misfiring when under load or accelerating or at high constant revolutions
(d) Smooth running at high revolutions and misfiring or cutting-out at low speeds

2 First check that all wires are properly connected and dry. If the engine fails to start when the starter is operated do not continue for more than 5 or 6 short burst attempts or the battery will start to get tired and the position made worse. Remove the spark plug lead from a plug and turn the engine again holding the lead (by the insulation) about ¼ inch from the side of the engine block. A spark should jump the gap audibly and visibly, if it does then the plugs are at fault or the static timing is very seriously adrift. If both are good, however, then there must be a fuel supply fault, so go on to that.

3 If no spark is obtained at the end of a plug lead detach the coil HT lead from the centre of the distributor cap and hold that near the block to try and find a spark. If you now get one, then there is something wrong between the centre terminal of the distributor cap and the end of the plug lead. Check the cap itself for damage or damp, the 4 terminal lugs for signs of corrosion, the centre carbon brush in the top (is it jammed?) and the rotor arm.

4 If no spark comes from the coil HT lead check next that the contact breaker points are clean and that the gap is correct. A quick check can be made by turning the engine so that the points are closed. Then switch on the ignition and open the points and, once again, if the coil HT lead is held near the block at the same time a proper HT spark should occur. If there is a big fat spark at the points but none at the HT lead then the condenser is defective and should be renewed.

5 If neither of these things happen then the next step is to see if there is any current (12 volts) reaching the coil (+ve terminal). (One could check this at the distributor, but by going back to the input side of the coil a longer length of possible fault line is bracketed and could save time).

6 With a 12 volt bulb and piece of wire suitably connected (or of course a voltmeter if you have one handy) connect between the +ve and BATT terminal of the coil and earth and switch on the ignition. No light means no volts so the fault is between the battery and the coil via the ignition switch. This is moving out of the realms of just ignition problems – the electrical system is becoming involved in general. So get a piece of wire and connect the +ve terminal of the coil to the +ve terminal on the battery and see if sparks occur at the HT leads once more.

7 If there is current reaching the coil then the coil itself or the wire from its –ve terminal to the distributor is at fault. Check the –ve or RUP terminal with a bulb with the ignition switched on. If it fails to light then the coil is faulty in its LT windings and needs renewal.

8 Uneven running and misfiring should first be checked by seeing that all leads, particularly HT are dry and connected properly. See that they are not shorting to earth through broken or cracked insulation. If they are, you should be able to see and hear it. If not, then check the plugs, contact points and condenser just as you would in a case of total failure to start.

9 If misfiring occurs at high speed check the points gap, which may be too small, and the plugs in that order. Check also that the spring tension on the points is not too light this causing them to bounce. This requires a special pull balance so if in doubt it will be cheaper to buy a new set of contacts rather than go to a garage and get them to check it. If the trouble is still not cured then the fault lies in the carburation or engine itself.

10 If misfiring or stalling occurs only at low speeds the points gap is possibly too big. If not then the slow running adjustment on the carburettor needs attention.

Chapter 5 Clutch

Contents

Clutch actuating rod/fork and release bearing – removal, inspection and installation ... 9
Clutch – adjustment ... 2
Clutch cable – removal and refitting ... 3
Clutch housing – installation ... 7
Clutch – inspection and renovation ... 5
Clutch – installation ... 6
Clutch – removal ... 4
Engine and transmission unit – relocation and final reassembly ... 8
Fault diagnosis – clutch ... 10
General description ... 1

Specifications

Type .... Diaphragm spring, single dry plate cable operation

Model No .... 180 DBR 285 – 181.5 mm (7.2 in) dia.

Operating lever clearance .... $\frac{3}{32}$ to $\frac{1}{8}$ in (2 to 3 mm)

Torque wrench settings

	lbf ft	Nm
Flywheel bolts	48 to 52.5	65 to 70
Clutch housing bolts:		
7 mm	11.25	15
8 mm	15.0	20.0
Clutch cover-to-flywheel bolts	15.0	20.0

1 General description

The clutch is a cable operated single dry plate diaphragm type.

The clutch pedal pivots on the same shaft as the brake pedal (see Chapter 9) and operates a cable to the clutch release arm. The release arm activates a thrust bearing (clutch release bearing) which bears on the diaphragm spring of the pressure plate. The diaphragm then releases or engages the clutch driven plate which floats on a splined shaft. This shaft (the engine output shaft) is part of the transfer gear assembly which is mounted on the clutch housing. The drive passes via an intermediate pinion to the gearbox input shaft.

The clutch release mechanism consists of a fork and bearing which are in contact with the release fingers on the pressure plate assembly. The fork pushes the release bearing forwards to bear against the release fingers, so moving the centre of the diaphragm spring inwards. The spring is sandwiched between two annular rings which act as fulcrum points. As the centre of the spring is pushed in, the outside of the spring is pushed out, so moving the pressue plate backwards and disengaging it from the clutch disc.

When the clutch pedal is released, the diaphragm spring forces the pressure plate into contact with the friction linings on the clutch disc and at the same time pushes the clutch disc a fraction of an inch forwards on its splines so engaging the clutch disc with the flywheel. The clutch disc is now firmly sandwiched between the pressure plate and the flywheel, so the drive is taken up.

As wear takes place on the clutch disc the clearance between the release bearing and the diaphragm decreases. This wear can be compensated for by adjusting the screws and locknut on the clutch operating lever.

2 Clutch – adjustment

1 Clutch adjustment will be necessary to compensate for the wear of the clutch lining. This will result in less free play at the rod between the clutch operating lever and the release bearing arm.

2 Loosen the locking nut and then rotate the adjusting screw in the clutch operating lever so that the clearance between the operating lever and the rod is as given in the Specifications (photo).

3 Retighten the locknut and check the clutch operation.

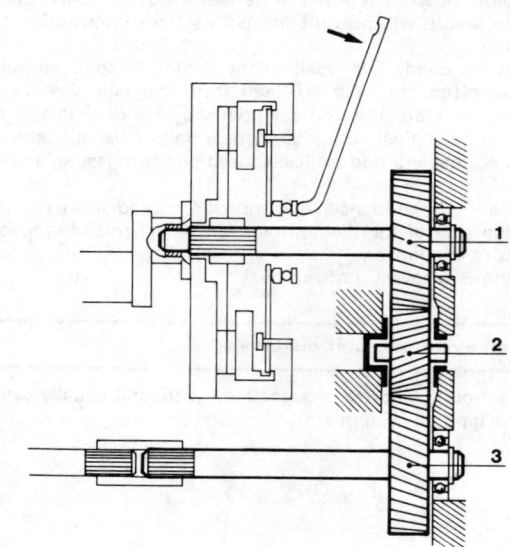

Fig. 5.1 Illustration showing the relative positions of the clutch and the transfer gear assembly

1 Output shaft from engine
2 Intermediate gear
3 Input shaft to gearbox

Chapter 5 Clutch

2.2 The clutch adjustment screw and locknut

3.3 The clutch cable connection at the operating lever

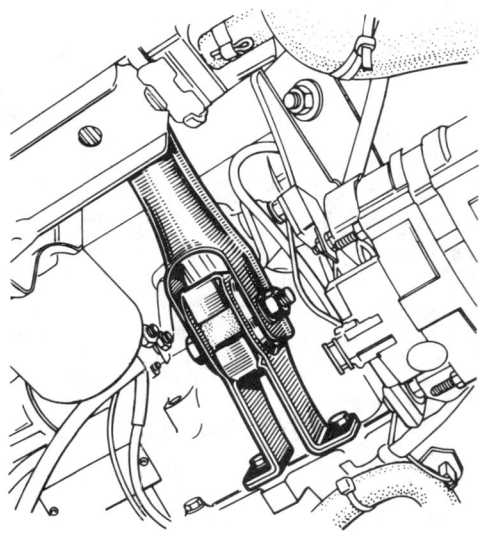

Fig. 5.1a Engine tie-rod

3 Clutch cable – removal and refitting

1 The clutch cable connects the clutch pedal to the operating lever/rod assembly. It is a simple item to replace. To remove proceed as follows:
2 Unscrew the clutch lever adjustment screw locknut and slacken the adjustment.
3 Inside the car, unhook the cable from the clutch pedal. Release the other end of the cable from the clutch operating lever (photo).
4 Carefully release the cable and pull it through the aperture in the bulkhead to remove.
5 Replacement is a direct reversal of the removal procedure. When installed readjust the operating lever clearance as given in Section 2.

4 Clutch – removal

1 Should it become necessary to renew the friction plate or examine the clutch assembly the clutch housing and transfer gear casing will have to be removed. To do this it is not necessary to entirely remove the engine and transmission units, but it will be necessary to lift the power unit clear of its mountings so that it can be swung over to the right-hand side of the engine compartment. It is not necessary to disconnect the driveshafts. The clutch housing assembly can then be removed. Proceed as follows:
2 Refer to Chapter 1, Section 5 and follow the instructions given in paragraphs 2 to 29 but note the following:

3 Drain the engine/transmission oil.
4 Remove the air cleaner, fuel pump (RHD only), the battery and battery tray.
5 Remove the spare wheel bracket the clutch control rod and disconnect the clutch operating cable from the support bracket.
6 Remove the distributor cap, the diagnostic socket mounting bracket, the engine mounting pad nuts and the engine tie-rod.
7 Without disconnecting the coolant hoses, move the expansion bottle to one side, laying the radiator on the engine.
8 Disconnect the exhaust downpipe.
9 Disconnect the selector quadrant retaining pin on the steering box.

10 When the power unit is ready for lifting connect the hoist to the sling and raise the engine/transmission unit carefully to a maximum height of 3·5 in (90 mm) from the mountings. At this height the engine can be carefully pushed to the right-hand side of the engine compartment and suitably tied to retain it in this position.
11 The clutch housing retaining bolts can now be unscrewed and the housing withdrawn together with the transfer gear housing. When withdrawing the housing, pull it straight out and in line with the engine/transmission so that no strain is placed on the input and output shafts. Note the position of the various attachments (starter motor, earth strap etc) retained by the housing bolts and also the respective bolt locations.
12 Before removing the clutch cover bolts mark the position of the cover in relation to the flywheel so that it may be put back the same way.
13 Slacken off the cover retaining bolts ½ a turn at a time in a

Fig. 5.1b Selector quadrant retaining pin

Chapter 5 Clutch

Fig. 5.2 The clutch cover retaining bolts (arrowed)

6.3 Using a special alignment tool to centre the clutch disc

diagonal fashion evenly so as to relieve the diaphragm spring pressure without distorting it.

14 When the bolts are removed the cover and the driven plate can be removed.

5 Clutch – inspection and renovation

1 The clutch driven plate should be inspected for wear and for contamination by oil. Wear is gauged by the depth of the rivet heads below the surface of the friction material. If this is less than 0·025 in (0·6 mm) the linings are worn enough to justify renewal.

2 Examine the friction faces of the flywheel and clutch pressure plate. These should be bright and smooth. If the linings have worn too much it is possible that the metal surfaces may have been scored by the rivet heads. Dust and grit can have the same effect. If the scoring is very severe it could mean that even with a new clutch driven plate, slip and juddering and other malfunctions will recur. Deep scoring on the flywheel face is serious because the flywheel will have to be removed and machined by a specialist, or renewed. This can be costly. The same applies to the pressure plate in the cover although this is a less costly affair. If the friction linings seem unworn yet are blackened and shiny then the cause is almost certainly due to oil. Such a condition also requires renewal of the plate. The source of oil must be traced also. It could be due to a leaking seal on the transmission input shaft (Chapter 6 gives details of renewal) or from a leaking rear main bearing oilseal (see Chapter 1 for details of renewal).

3 If the reason for removal of the clutch has been because of slip and the slip has been allowed to go on for any length of time it is possible that the heat generated will have adversely affected the pressure spring in the cover. It may have been affected with the result that the pressure is now uneven and/or insufficient to prevent slip, even with a new friction plate. It is recommended that under such circumstances a new assembly is fitted.

4 Clutch cover assemblies are available on an exchange basis. It will probably be necessary to order an assembly in advance as most agencies other than the large distributors carry stocks only sufficient to meet their own requirements. However, it is possible to get assemblies from reputable manufacturers other than Renault; Borg and Beck for example, but be specific as to your requirements.

6 Clutch – installation

1 Support the driven plate centrally between the flywheel and the cover. The offset side of the driven plate (the side where the boss has the larger diameter) faces outwards.

2 Align the marks made on the flywheel and cover prior to removal and retain the cover in position on the flywheel with the bolts (hand tight only).

3 It is now necessary to align the centre of the driven plate with that

7.2 Refitting the housing (engine removed)
Note gasket (1) and input shaft (2)

9.3 The clutch fork and bearing in position

Chapter 5 Clutch

of the flywheel. To do this use a special alignment tool (photo) or alternately use a suitable diameter bar inserted through the driven plate into the flywheel spigot bearing, but take care not to damage the output shaft seal. It is possible to align the driven plate by eye, but difficulty will probably be experienced when refitting the output shaft.

4 With the friction plate centralized the cover bolts should be tightened diagonally and evenly to the specified torque. Ideally new retaining bolts should be used each time a replacement clutch is fitted. When the bolts are tight remove the centralising tool.

5 Before refitting the clutch housing, check the condition of the release bearing and operating mechanism, renewing any parts as necessary.

7 Clutch housing – installation

1 Before refitting the housing, check that the mating surfaces are clean and dry. Smear the bearing surface of the withdrawal pad on the diaphragm spring with medium grease.

2 Place a new gasket over the location dowels and then carefully offer the clutch housing/transfer pinion unit to the engine and insert the output and input shafts (photo).

3 To assist the respective shaft splines to engage, rotate the flywheel and gearbox input shaft alternately until they slide home into position with the housing flush.

4 Insert the retaining bolts, remembering to replace any fittings retained by them. Tighten the bolts progressively to the specified torque settings according to size.

8 Engine/transmission unit – relocation and final assembly

1 With the clutch housing fully installed the engine/transmission unit can be relocated on its mountings and the various fittings reconnected. This procedure is a direct reversal of the removal sequence and further information is given in Chapter 1, Section 42, paragraph 8 onwards as applicable.

2 Readjust the clutch as described elsewhere in Section 2.

3 Top up the engine/transmission oil and engine coolant. Bleed the cooling system after topping up (see Chapter 2). After starting the engine check for leaks from the joints and connections.

9 Clutch actuating rod/fork and release bearing – removal, inspection and installation

1 The clutch actuating rod can be removed by slackening the operating adjustment and unhooking the return spring.

2 To remove or overhaul the withdrawal bearing and fork the clutch housing must be removed. This is described in Section 4.

3 With the housing removed the withdrawal fork and bearing can be withdrawn from the output shaft for inspection (photo).

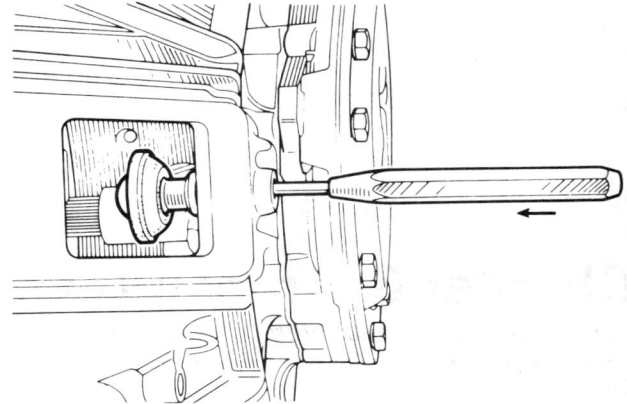

Fig. 5.3 Drift out the ball-pin

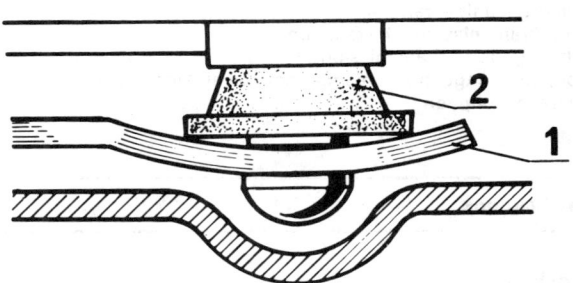

Fig. 5.4 Location of fork spring blade

1 Spring blade   *2 Rubber cover*

4 Do not clean the bearing with fluid as it will harm the bearing. Wipe it clean and check for excessive wear or play. Always renew if in doubt.

5 Inspect the fork retaining ball-pin and if obviously distorted or worn renew it. Drift the ball-pin from the housing using a suitable diameter drift (Fig. 5.3). Install the new one by driving it carefully into position using a soft faced hammer. Support the housing during this operation to prevent it being damaged.

6 To refit the fork, install the spring blade so that it is located under the rubber cover as shown in Fig. 5.4.

7 Position the release bearing over the engine output shaft and engage the retainers behind the fork fingers. The release bearing can be slid along the sleeve whilst holding the fork.

8 Check the fork and bearing for correct operation and then refit the housing – see Section 7. Readjust the clutch operating clearance on completion – see Section 2.

10 Fault diagnosis – clutch

Symptom	Reason/s
Judder when taking up drive	Loose engine/gearbox mountings or over-flexible mountings Badly worn friction surfaces or friction plate contamination with oil carbon deposit Worn splines in the friction plate hub or on the engine output shaft
Clutch spin (or failure to disengage) so that gears cannot be meshed	Clutch actuating cable clearance too great Clutch friction disc sticking because of rust on splines (usually apparent after standing idle for some length of time) Damaged or misaligned pressure plate assembly Incorrect release bearing fitted
Clutch slip (increase in engine speed does not result in increase in car speed – especially on hills)	Clutch actuating cable clearance from fork too small resulting in partially disengaged clutch at all times Clutch friction surfaces worn out (beyond further adjustment of operating cable) or clutch surfaces oil soaked

Chapter 6 Transmission

For modifications, and information applicable to later models, see Supplement at end of manual

Contents

Differential oilseal – replacement 9	Transmission components – inspection 5
Differential unit – general 4	Transmission to engine – reassembly 7
Engine/transmission – installation 8	Transmission unit – dismantling 3
Fault diagnosis – transmission 11	Transmission unit – reassembly 6
Floor gearchange mechanism – removal and installation 10	Transmission unit – removal 2
General description 1	

Specifications

Gearbox
Type number	408 – 005
Number of gears	4 forward, 1 reverse
Synchromesh	All forward gears
Ratios:	
1st	3.083 : 1
2nd	1.823 : 1
3rd	1.192 : 1
4th	0.827 : 1
Reverse	2.833 : 1

Engine-to-gearbox transfer gears
Engine output shaft	27 teeth
Gearbox input shaft	34 teeth
Reduction ratio	1.259 : 1

Lubricant and lubricant type
Lubricant and lubricant type . . . . As engine (see Chapter 1) – COMMON LUBRICATION SYSTEM

Final drive
Type	Integral with gearbox. Helical gear crownwheel and pinion
Differential end thrust	Copper face thrust washers
Differential bearing	Shell type
Ratio	3.866 : 1 (15 x 58)
Road speed per 1000 rpm in top gear	15.9 mph

Torque wrench settings
	lbf ft	Nm
Crownwheel bolts	45	60
Bottom cover bolts	7.5	10
Pinion shaft nut	20.0	22.5
Half-housing bolts (top and bottom):		
Initial tightening:		
7 mm	11.25	15
8 mm	7.5	10
10 mm	15	20
Final tightening:		
7 mm bolts	11.25	15
8 mm bolts	15	20
10 mm bolts	33.75	45
Primary shaft nut:		
Initial tightening	15.0	20
Final tightening	7.5	10
Flywheel bolts (Loctite)	48.75	65

Chapter 6 Transmission

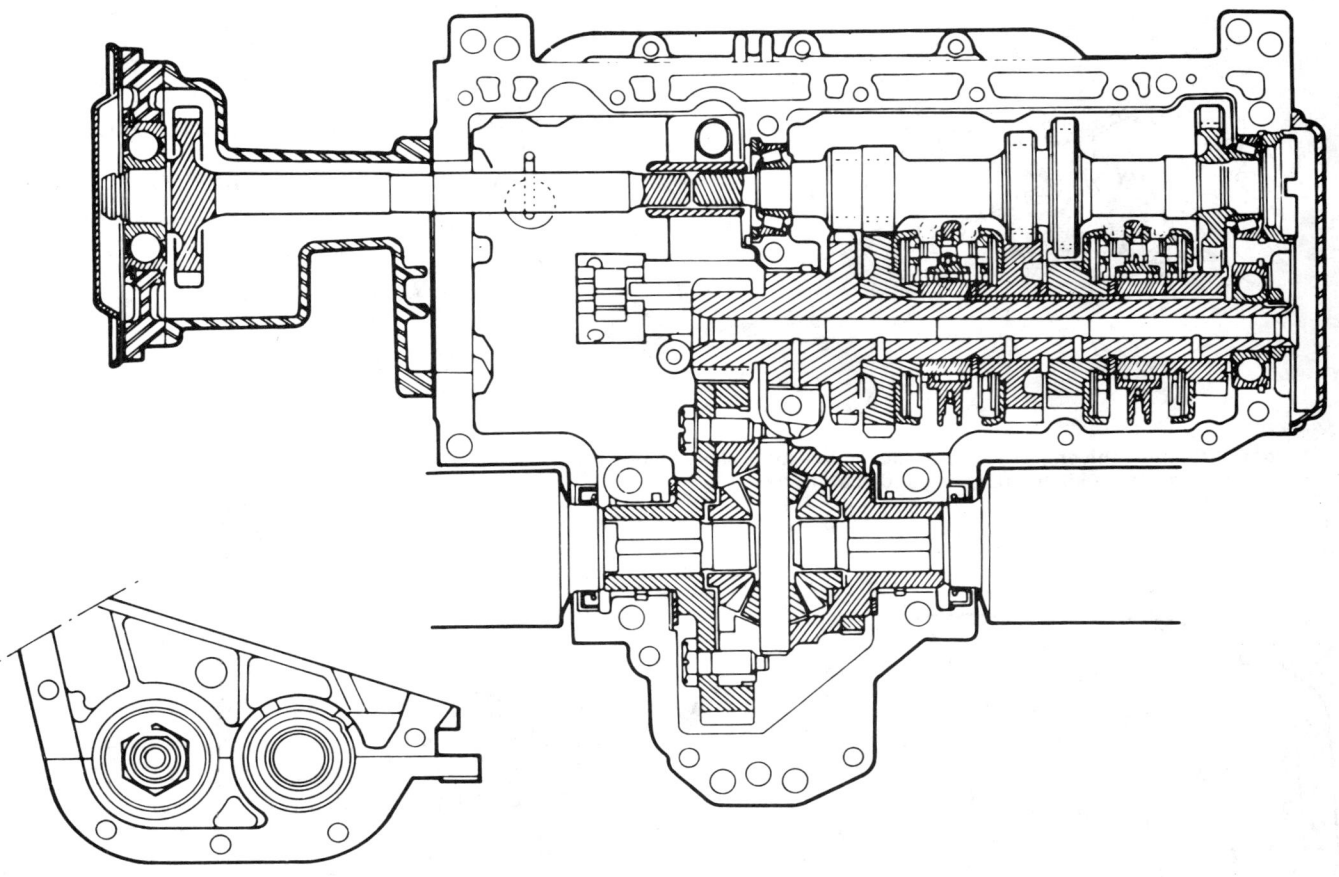

Fig. 6.1 Sectional view of the transmission system showing the input shaft from the transfer gear, the primary and secondary (pinion shaft) gear assemblies and the differential unit

1 General description

The Renault 14 is fitted with a four-speed manual gearbox mounted transversely directly beneath and in line with the engine. The transmission housing is cast in aluminium alloy and besides the gearbox, also contains the differential and final drive units. Drive to the gearbox from the engine is via transfer gears which are mounted separately on the outside of the clutch housing.

The gearbox has a conventional two shaft constant-mesh layout. There are four pairs of gears, one for each forward speed. The gears on the primary shaft are fixed to the shaft, while those on the secondary or pinion shaft float, each being locked to the shaft when engaged by the synchromesh unit. The reverse idler gear is on a third shaft. The gear selector forks engage in slots in the synchromesh unit; these slide axially along the shaft to engage the appropriate gear. The forks are mounted on selector shafts which are located in the base of the gearbox.

The gear on the end of the pinion shaft drives directly onto the crownwheel mounted on the differential unit. This differs from normal practice in that it runs in shell bearings and the end thrust is taken up by thrust washers in a similar manner to the engine crankshaft.

Although the transmission system employed is relatively simple there are nevertheless a few words of warning which must be stressed.

First of all decide whether the fault you wish to repair is worth all the time and effort involved. Secondly, if the transmission unit is in a very bad state then the cost of the necessary component parts may well exceed the cost of an exchange factory unit. Thirdly, be absolutely sure that you understand how the transmission unit works.

Special care must be taken during all dismantling and assembly operations to ensure that the housing is not overstressed or distorted in any way.

When dismantled, check the cost and availability of the parts to be renewed and compare this against the cost of a replacement unit, which may not be much more expensive and therefore a better proposition.

On reassembly, take careful note of the tightening procedure and torque settings. This is most important to prevent overtightening, distortion and oil leakage.

2 Transmission unit – removal

1 It is necessary to remove the engine and transmission assemblies as a combined unit. The procedure for this operation is given in Chapter 1, Sections 4 and 5 or Section 6 as applicable.

2 With the engine/transmission unit removed detach the following ancillary components from the engine:

(a) Alternator and tension linkage
(b) Starter motor
(c) Oil pressure switch unit
(d) Oil filter and dipstick
(e) The front engine mountings
(f) The water pump unit (note O-ring)

3 Remove the following items from the clutch housing:

(a) Diagnostic socket and TDC pick up

76

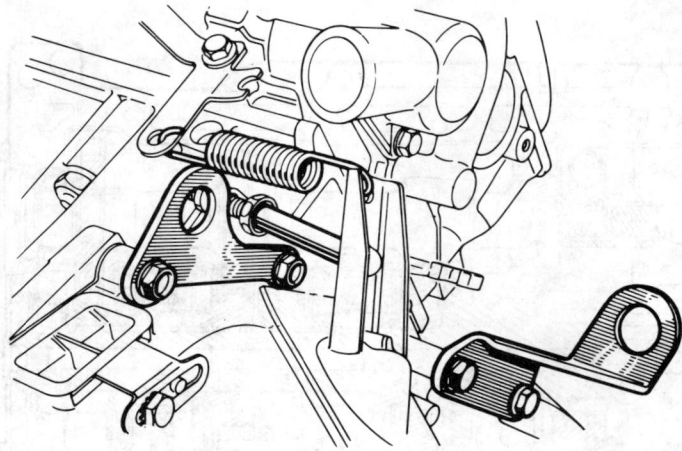

Fig. 6.2 Remove the engine lift hook, the clutch cable sleeve stop and the clutch fork spring and operating rod

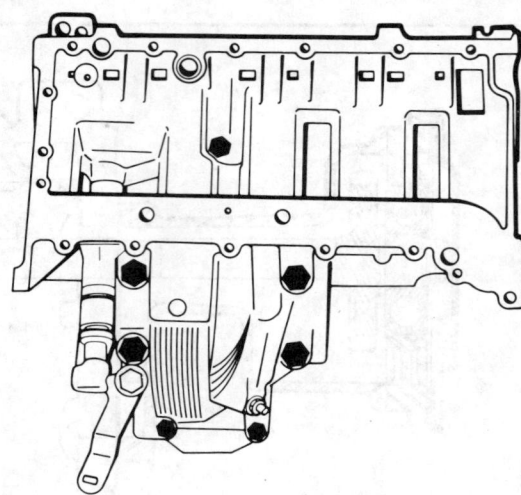

Fig. 6.3 The upper half-housing bolt positions

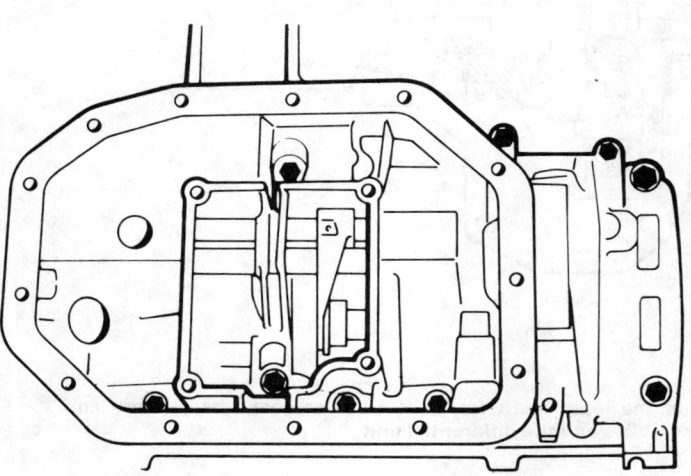

Fig. 6.4 The lower half-housing bolt positions

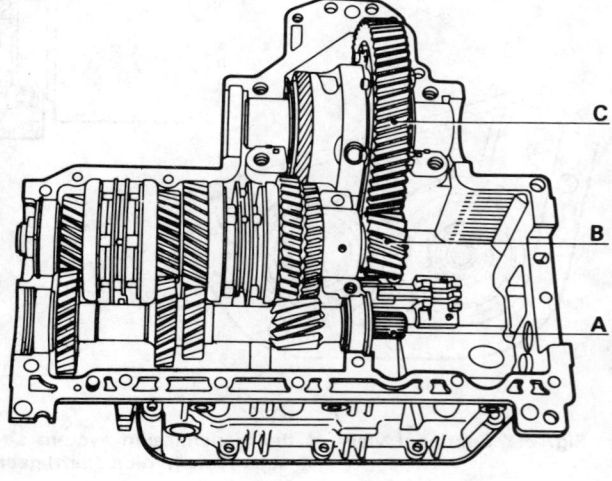

Fig. 6.5 General view of gears and differential

A Primary shaft
B Secondary (pinion) shaft
C Differential unit

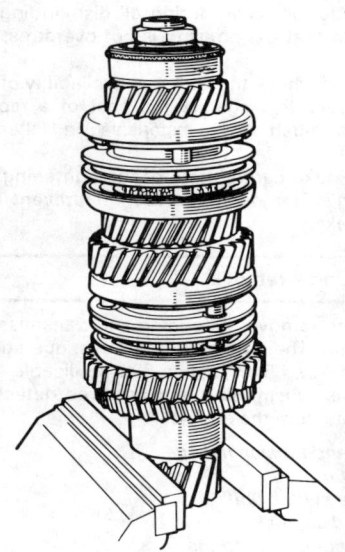

Fig. 6.6 Support the pinion shaft in soft jawed vice

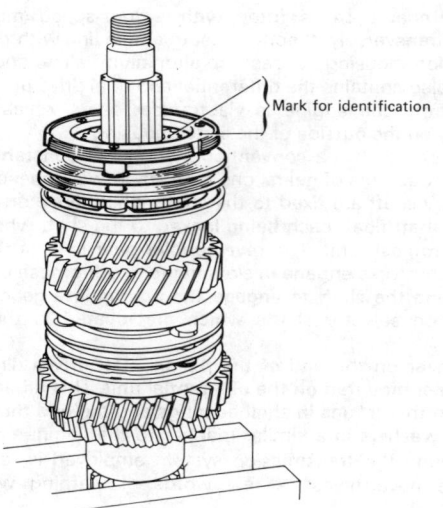

Fig. 6.7 Mark the synchromesh unit for identification

Chapter 6 Transmission

(b) Engine lifting hook
(c) Clutch cable sleeve stop
(d) Clutch fork (if necessary)

4 Refer to Chapter 1, Section 10 and remove the clutch housing, clutch unit and flywheel as detailed in paragraphs 4 to 7 inclusive.
5 Unscrew and remove the crankshaft pulley nut. To prevent the crankshaft from turning, reinsert two flywheel bolts into the end of the crankshaft and wedge a bar between them. Apply sufficient pressure to enable the pulley nut to be loosened and remove the bar and bolts.
6 Withdraw the crankshaft pulley.
7 Remove the rocker cover.
8 Unscrew the retaining bolts and remove the timing cover.
9 To separate the transmission from the engine refer to Chapter 1, Section 12.

3 Transmission unit – dismantling

1 Before proceeding according to the directions given below, first read Section 1 (General description). It is assumed that the unit is out of the vehicle and on the bench. (It is not advisable to dismantle it on the floor. It will do the kitchen table no harm as it is not heavy). Do not throw away gaskets when dismantling for they will act as a guide for the fitment of the new ones supplied in the gasket set which should have already been purchased. Always renew all gaskets, locking washers and roll pins. Clean the outside casing thoroughly, and allow to dry. Start work with clean hands and a plentiful supply of clean rag.
2 To separate the upper and lower gearbox half-housings first unscrew and remove the bolts (see Fig. 6.3) from the upper housing.
3 Now invert the gearbox and unscrew the bottom cover plate retaining bolts. Remove the cover plate and gasket and also the guard plate.
4 Unscrew the four retaining bolts and remove the oil pump suction filter. Pull the gauze strainer out carefully noting the O-ring.
5 Unscrew and remove the bottom housing bolts as shown in Fig. 6.4 and carefully separate the two half-housings.
6 Unscrew the retaining bolt and withdraw the speedometer drive sleeve. The gear train assemblies and the differential unit are now accessible for inspection and removal if further dismantling is required. Simply lift out the appropriate assembly. If the bearing shells are removed mark them with their location in case they are to be reused. Unless new bearings have been fitted recently and are in good condition it is sensible to renew them.

Secondary shaft – dismantling

7 If the secondary (or pinion) shaft is to be dismantled, first support it vertically in a soft jawed vice as shown (Fig. 6.6).
8 The end nut must now be unscrewed. This will have been (or should have been) staked on its inner protruding flange to lock it in position. Tap out the indented portion of flange and then unscrew the nut to remove.
9 The bearing must now be withdrawn. This is a press fit onto the shaft and will require the use of a suitable bearing puller to remove it. Take care not to damage the adjacent 4th speed gear. Note that the bearing is fitted with the snap-ring groove outwards.
10 With the bearing removed, withdraw the distance washer and the 4th speed gear.
11 Before removing the 3rd/4th speed synchromesh unit it should be suitably marked for identification so that it is not confused with the 1st/2nd gear synchromesh unit which is identical.
12 Withdraw the 3rd/4th synchromesh unit.
13 Remove the assembly from the vice and invert to remove the key from within the wide spline groove.
14 Rotate the distance washer to align with shaft splines and then withdraw it (photo).
15 Withdraw the 3rd speed gear, distance washer and the 2nd speed gear.
16 Rotate and remove the washer from the shaft groove and then withdraw the 1st/2nd gear synchromesh unit.
17 Remove the 1st speed gear.

Primary shaft – dismantling

18 The primary shaft runs in taper roller bearings. The outer tracks of the bearings are easily removed once the gear unit is lifted clear but keep the cups separate (photo).

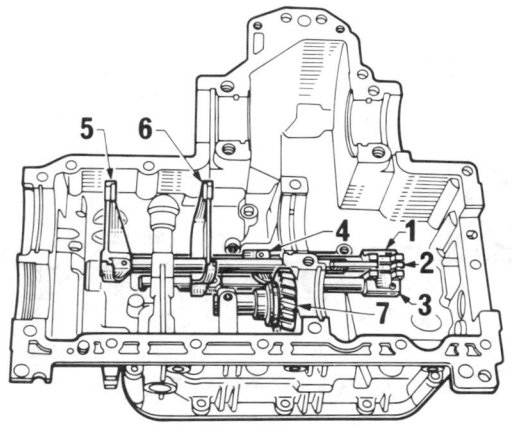

Fig. 6.8 The selector assemblies

1 Reverse shaft
2 3rd/4th selector shaft
3 1st/2nd selector shaft
4 Reverse gear selector fork
5 3rd/4th selector fork
6 1st/2nd selector fork
7 The sliding reverse idle

3.14 Twist the distance washer and withdraw it

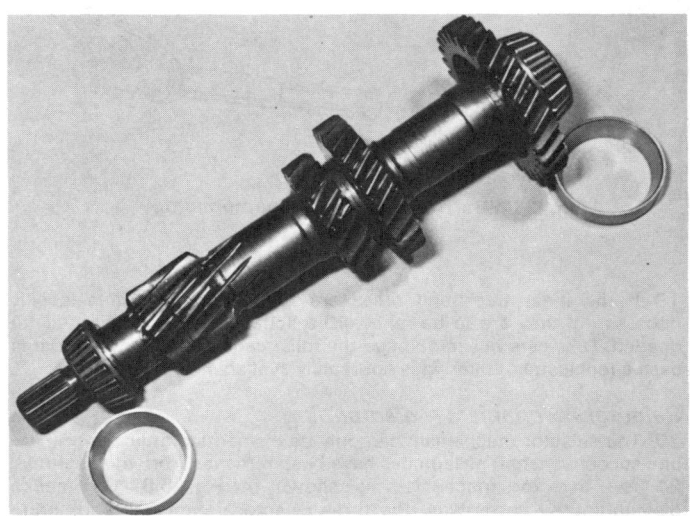

3.18 The primary shaft removed showing taper-roller bearings and outer tracks

Chapter 6 Transmission

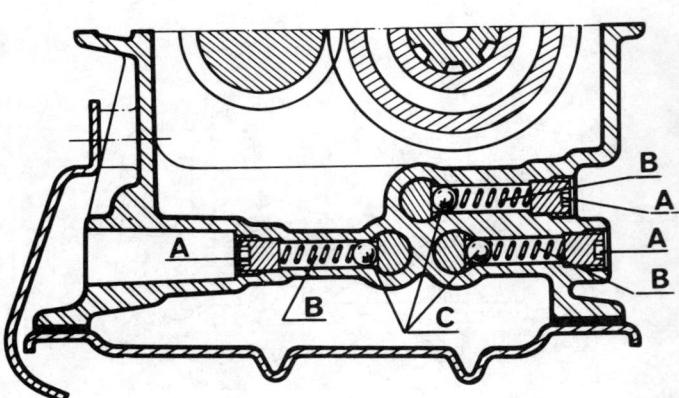

Fig. 6.9 The selector rod detent plugs (A), springs (B) and balls (C), showing their relative positions in the gearbox

5.6 Inspect the synchromesh units for wear and damage

22 Use a suitable drift and drive out the roll-pins from the reverse gear selector fork, the 1st/2nd gear selector fork and the 3rd/4th speed selector fork.
23 Extract the reverse selector shaft and its interlock disc.
24 Remove the 3rd/4th gear selector fork, then the 1st/2nd and 3rd/4th selector shafts. Withdraw the 1st/2nd gear selector fork and the reverse gear selector fork.

Reverse sliding gear – removal
25 The gear can be removed after withdrawing the shaft (located in the gearbox casting by a roll-pin). Drive the roll-pin out with a punch, extract the shaft and gear. Note which way round the gear faces. Recover the thrust washer.

Gear shift control lever
26 Drive out the roll-pin holding the control finger to the shaft using a suitable punch and then withdraw the selector shaft control finger. Withdraw the half-bushes and remove the shift control lever and shaft together with the spacer, spring and thrust bush.

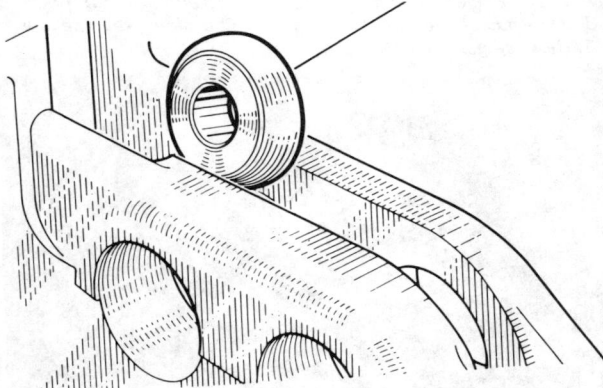

Fig. 6.10 The interlock disc

4 Differential unit – general

1 Apart from lifting it from the gearbox half-housing, cleaning it and giving it a general external inspection, the differential unit should not be dismantled or interfered with.
2 If on inspection it is found to be defective or is suspect, have your Renault dealer examine it or fit a new unit.
3 The crownwheel is mounted on the differential unit, which also incorporates the drivegear for the speedometer cable. The unit runs in shell bearings and is located by thrust washers. Make sure that the copper face of the thrust washer faces the differential (Fig. 6.11a).

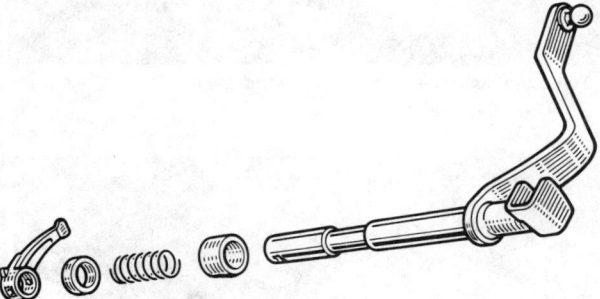

Fig. 6.11 The gearshift control components

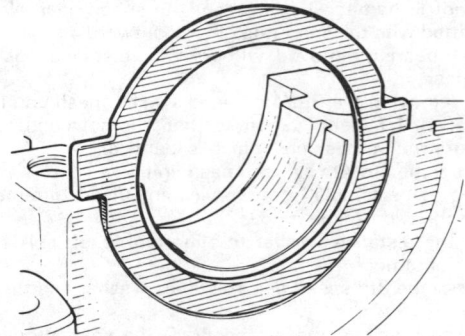

Fig. 6.11a Differential thrust washer

19 If the inner track and races are to be removed, (this is only necessary if they are to be renewed) a suitable bearing puller will be needed. Take care not to damage the roller cage during removal just in case a replacement bearing is not readily available.

Selector mechanism – dismantling
20 The selector mechanism can only be dismantled after the primary and secondary shaft assemblies have been removed from the gearbox.
21 The selector mechanism is shown in Fig. 6.8. Commence dismantling by unscrewing the three selector detent ball and spring retaining plugs (Fig. 6.9). Extract the springs and balls and keep them with their respective plug.

5 Transmission components – inspection

1 Having removed and dismantled the transmission, the various components should be throughly washed with a suitable solvent, or with petrol and paraffin, and then wiped dry. Take great care not to

Chapter 6 Transmission

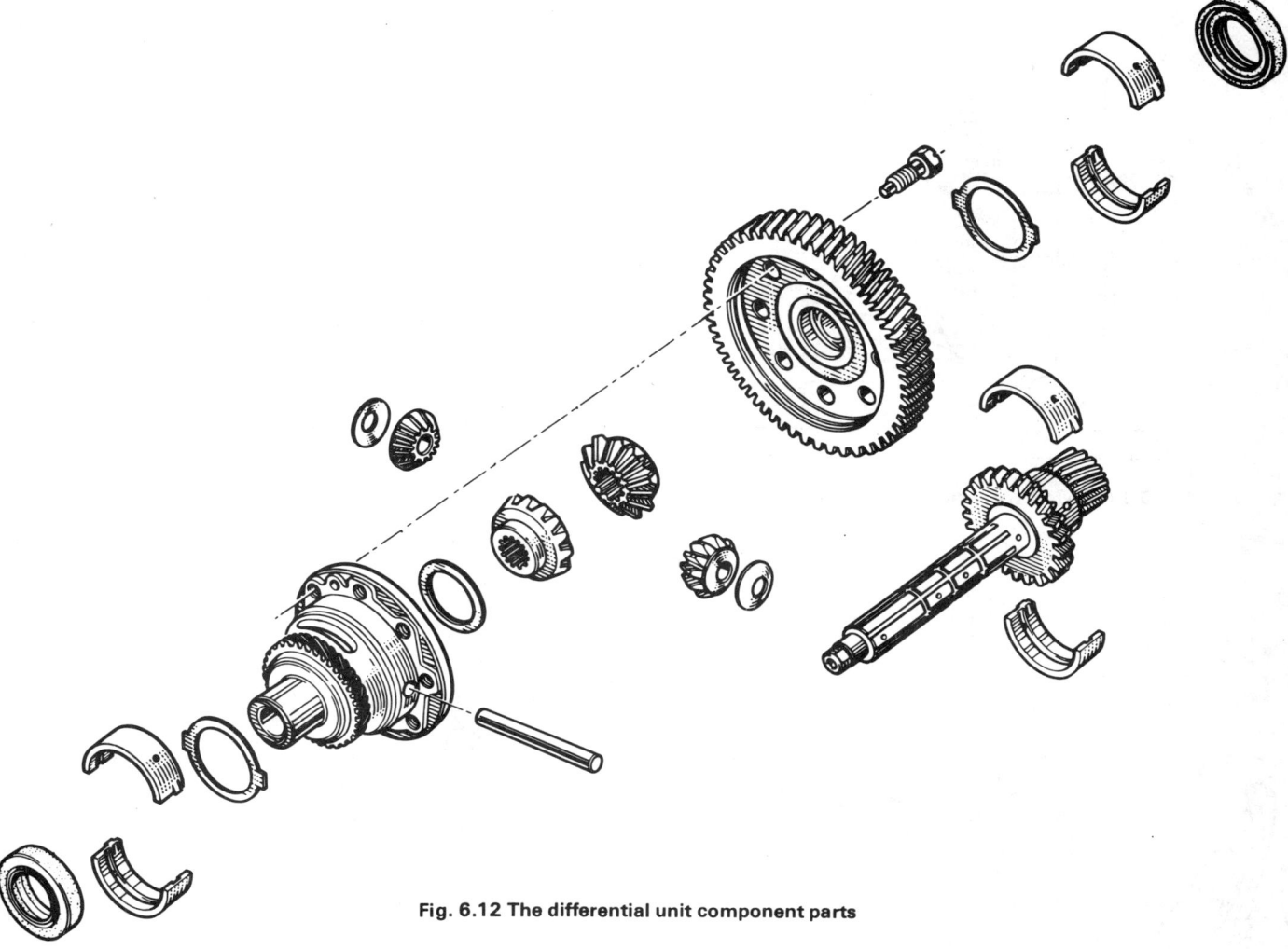

Fig. 6.12 The differential unit component parts

damage the mating faces of the half-housings.

2 Check the transmission half-housing for cracks or damage, particularly near the bearings or selector rod bushes. The housings are a matched pair and cannot be renewed individually.

3 Components requiring particular attention will have been noted as a result of the performance of the transmission prior to dismantling.

4 Check the gears for chips or uneven wear. The pinion shaft gears should be a good sliding fit on the shaft.

5 Check the pinion shaft for wear on the splines or bearing surfaces.

6 Carefully inspect the synchromesh units (photo). If weak synchromesh has been experienced, renew the synchromesh units and carefully inspect the synchronising friction face on the pinion shaft gears.

7 Examine the gearbox bearings. The primary shaft taper roller bearings are normally reliable and hard wearing. Check them for wear, scoring and freedom of rotation which should be perfectly smooth. Check the pinion shaft ball bearing for excessive play between the inner and outer races. Hold the bearing by its inner race and spin the outer race – the movement should be smooth with no signs of harshness or binding. Check the shell bearings at the pinion end of the mainshaft and either side of the differential, taking care not to get them mixed up. They must be in good condition showing no signs of scoring or excessive wear. In view of their low cost and ease of fitting it is suggested that they be renewed as a matter of course. The differential thrust washers should also be renewed at this stage – again the replacement cost being relatively low.

8 Check that the nylon speedometer drive gearwheel is in good condition and running easily in its bush.

9 Check the selector forks for wear. Measure them with a pair of calipers and compare their ends with the thickest point; if in doubt replace. They should be only fractionally worn.

10 Check the gear shift mechanism. The tongue which also slots into the top of the selectors wears quite rapidly often resulting in non-selected gears and sloppy action.

11 Blow through the oilways to ensure that they are clear before reassembly.

6 Transmission unit – reassembly

1 All components must be spotlessly clean prior to assembly, as must the upper and lower half-housings. Lubricate the respective sub-assemblies as they are installed, particularly the bearings and moving parts.

2 As with the dismantling process, the various sub-assemblies are dealt with in turn:

Reverse sliding gear assembly

3 Insert the shaft into the housing and as it is pressed in fit the reverse sliding gear and thrust collar as shown (Fig. 6.13).

4 Turn the shaft and position to align the roll-pin holes through the shaft and housing. Tap a new roll-pin into position to locate the shaft. Check that the gear slides and rotates freely.

Selector mechanism assembly

5 To reassemble the gear selection control mechanism commence by lubricating and fitting a new O-ring into position on the selector shaft.

6 Fit the shaft bush and then carefully insert the shaft into position. Do not mix up the shaft thrust bush with the spacer bush. They are

Chapter 6 Transmission

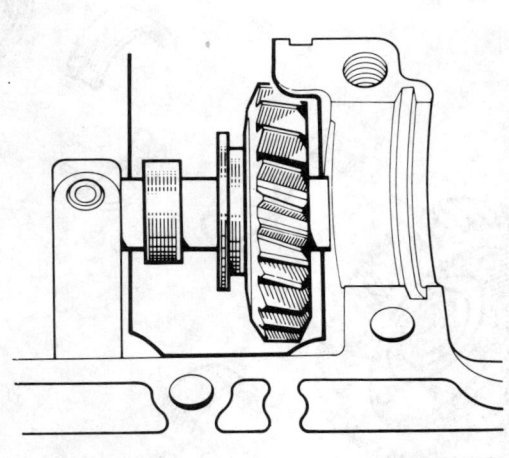

Fig. 6.13 The reverse sliding gear in position

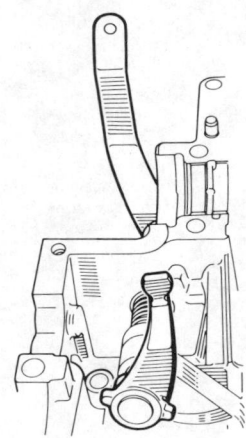

Fig. 6.14 Align the selector finger and shift lever

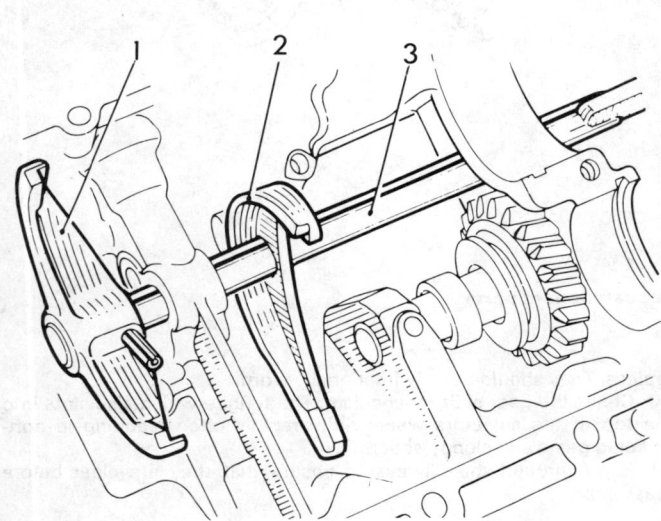

Fig. 6.15 The 3rd/4th selector shaft in position showing

1 3rd/4th selector fork and roll-pin
2 1st/2nd selector fork
3 Selector shaft (3rd/4th)

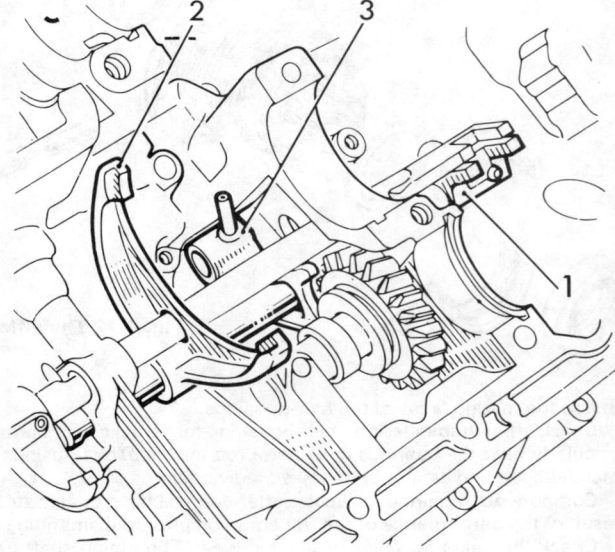

Fig. 6.16 The 1st/2nd selector shaft (1) and selector fork (2) Also shown is the reverse selector fork (3)

easily identifiable by their differing lengths:

Thrust bush length 0·875 in (22 mm)
Spacer bush length 0·437 in (11 mm)

7 As the shaft is slid into position fit the spring and spacer bush.
8 Compress the spring at each end in turn and insert the four half-bushes to locate the spring.
9 The selector control finger can now be refitted to the shaft, and positioned as shown in relation to the gearshift control finger, aligning the roll-pin holes in the shaft. Drift the new roll-pins into position (photo). The large roll-pin (7 mm diameter) is fitted first and then the smaller (4 mm dia) is drifted into its internal bore to secure.
10 Check the shaft for operation.
11 Slide the 3rd/4th selector shaft into position with its slot upwards. As it is pushed through the housing, locate the 1st/2nd selector fork on the shaft. Then locate the 3rd/4th selector fork onto the shaft and with the holes carefully aligned drift the new roll-pin into position (photo).
12 Slide the 1st/2nd selector shaft into position. As it is pushed through, locate the reverse selector fork against the groove of the sliding gear and pass the 1st/2nd shaft through the aperture (or cutaway section) in the reverse fork. Position the 1st/2nd selection fork onto its shaft and tap the new roll-pin into position to secure.
13 The reverse selector shaft can now be fitted. As it is passed through do not forget to insert the interlock disc. Align the shaft and selector fork holes and tap the new roll-pin into position (photo).

Primary shaft assembly
14 The gears and shaft are a combined unit and therefore the only things likely to be fitted will be new bearings. The bearing inner race must be pressed or drifted into position using a suitable diameter tube.

6.9 The selector mechanism reassembled

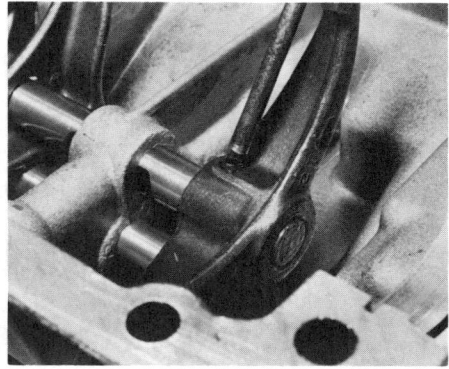

6.11 3rd/4th selector fork roll-pin installation

6.13 Reverse selector fork in position and located with roll-pin

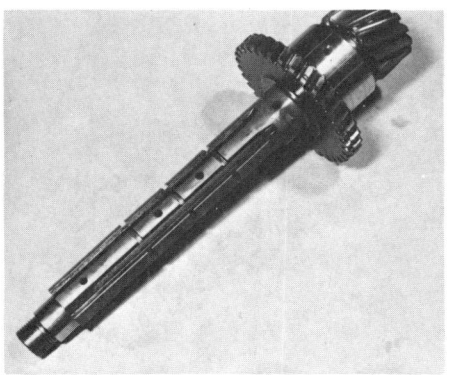

6.15a The pinion shaft cleaned and ready for reassembly

6.15b Fit the 1st gear onto the shaft ...

6.16 ... followed by the 1st/2nd synchromesh unit

6.17 Locate the distance washer and ...

6.18a ... assemble 2nd gear ...

6.18b ... followed by its distance washer

6.19a Fit 3rd speed gear and ...

6.19b ... distance washer

6.20 Slide key down wide groove – tapered end first

6.21 Fit the 3rd/4th synchromesh unit. Note spacer line marks (arrowed)

6.22a Fit 4th gear ...

6.22b ... and its distance washer

6.23 Locate the bearing with snap-ring offset outwards

6.24a Stake nut flange to secure

6.24b The pinion shaft assembly complete and ready for installation

6.25 Insert the bearing shells

6.26 Reinstall the speedometer drive pinion

6.27a Thrust washer location

6.27b Differential unit installed

6.28 Insert the primary shaft – note thrust washer (A)

6.29a Insert the pinion shaft – check snap-ring is in groove (A)

The important thing is to line up the bearing with the shaft as accurately as possible and maintain the alignment while it is being pressed or drifted into position.

Pinion shaft assembly

15 First check that the shaft and its bearing surfaces are perfectly clean. Commence by fitting the 1st speed gear (photos).

16 Slide the synchromesh unit down the shaft to fit against the 1st speed gear (photo). If the original synchromesh unit is being used it must be fitted facing the same direction as before. The spacer pins of the synchromesh unit are marked on one side with 2 lines (Fig. 6.17) and when the hub is installed these lines must face the 1st speed gear.

17 Slide the distance washer down the shaft and align with the groove around the splines (photo). Adjust the washer position to allow the key to be installed in the wide groove between the shaft splines. Check that the key can be fitted.

18 Slide the 2nd speed gear into position as shown (photo). Fit the next distance washer (photo) and again position to allow the key to pass through between the shaft splines.

19 Slide the 3rd speed gear into position (photo), and as before fit a distance washer (photo).

20 Now slide the key down the wide groove of the shaft with its tapered end inwards (photo). Check that when the key is fully installed it does not protrude beyond the last distance washer, fitted against 3rd gear. If the key does not fit fully home the chances are that a distance washer has moved and is obstructing the key.

21 Slide the 3rd/4th synchromesh unit into position on the shaft. If using the old unit it must be fitted facing the same direction as before, with the two lines on the spacer pins facing the 3rd speed gear (photo).

22 Fit the 4th speed pinion and its distance washer (photos).

23 The ball bearing is now pressed into position on the shaft. The snap-ring groove must be offset outwards as shown (photo). Support the shaft and press or drift the bearing carefully into position using a suitable diameter tube against the bearing inner race flange. If the snap-ring was removed this should now be refitted.

24 Refit the nut and tighten to the specified torque. The nut is locked in position by staking the flange against the flat section of the shaft, using a punch (photos).

General transmission reassembly

25 The sub-assemblies can now be refitted into the bottom half-housing. Start by installing the lower half differential and pinion shaft bearings. The bearings must be perfectly clean as must their respective recesses. If the old shells are being re-installed they must be fitted in their original positions (photo).

26 Re-install the speedometer drive pinion if it was removed (photo), using a new retaining pin. Check that it rotates freely (Refer to Chapter 9 – Section 27).

27 Insert the differential thrust washers. Each washer has a coppered face and this must be positioned to face the differential. It will also be apparent that the lugs of the washer are of different sizes and are asymmetrically positioned so that they sit on the mating face of the half-housing when the washer is correctly installed (photo). Lubricate the washers and fit them into position on the differential, then lower the differential unit into position in the lower half-housing (photo). Check that it rotates freely and that the washers are fully seated.

28 Lubricate the taper-roller bearings of the primary shaft and fit the outer races. Position the thrust washer and carefully lower the shaft into the lower half-housing (photo).

29 The pinion shaft assembly is now lowered into position as shown (photo). The shell bearing must be well lubricated and the selector forks must engage with their respective grooves as the gear assembly is lowered into position. Check that the snap-ring in the end bearing is seated in the location groove of the housing (photo).

30 Before refitting the top half-housing, check that the gears rotate freely and that the case joint flanges are perfectly clean. Generously lubricate the gear assemblies with clean engine oil. Install the upper differential and pinion shaft bearing shells in their respective recesses.

31 Smear an even amount of sealant (Rhodorsil CAF33 or equivalent) over the joint faces. The oilseals can also be inserted into position at this stage if desired but ensure that they are facing the correct way and are correctly seated.

32 Check that you have not left any tools or loose articles inside the housing and carefully lower the upper half-housing into position (photo).

33 Insert and hand tighten the top housing fixing bolts, then refer to

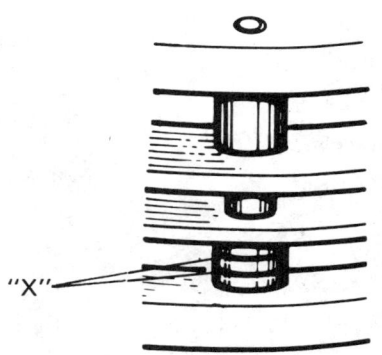

Fig. 6.17 The synchro hub spacer markings (X)

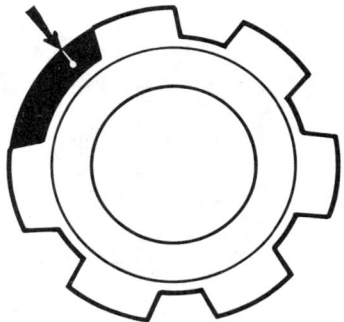

Fig. 6.18 End view of the pinion shaft showing wide groove (arrowed)

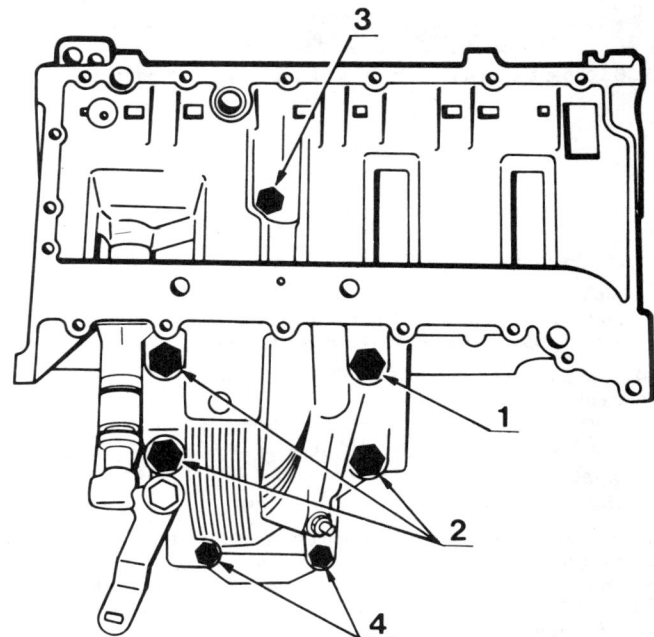

Fig. 6.19 Top half-housing bolts must be tightened to individual torque settings – see text

6.29b Check that bearings are fully seated in housing

6.32 Carefully refit the top half-housing

6.36 Locate the special primary shaft bearing adjustment nut

6.42 Refit the bottom plate – note gasket and gauze filter in position

Fig. 6.19 and tighten the bolts as follows:

Bolts numbered 1 and 2 (10 mm) 15 lbf ft (20 Nm)
Bolt number 3 (8 mm) 7·5 lbf ft (10 Nm)
Bolt number 4 (7 mm) 11·25 lbf ft (15 Nm)

34 The above bolts vary in length, their lengths and respective locations are shown in Fig. 6.20.
35 Invert the gearbox and insert the bottom retaining bolts as shown. Hand tighten them first, then tighten as follows:

Bolts 5 and 6 (8 mm) 7·5 lbf ft (10 Nm)
Bolts 7 and 8 (7 mm) 11·25 lbf ft (15 Nm)

36 Screw the special nut into position at the end of the gearcase in the primary shaft aperture (photo). This will adjust the endfloat of the primary shaft and give the correct preload to the taper-roller bearing. When the nut is in position recheck the torque settings of the number 3 and 5 housing bolts and if satisfactory select a gear. Molegrips clamped onto the selector rod will assist here.
37 Rotate the gear train by turning the pinion shaft nut (clockwise), and at the same time tighten the primary shaft nut. An assistant will be required here to support the gearbox. Renault dealers use a special peg socket (Fig. 6.23) to tighten the bearing adjustment nut and unless this can be borrowed or hired you will have to fabricate one that is suitable for use with your torque wrench. Initially tighten the nut to a torque figure of 15 lbf ft (20 Nm). Now slacken the nut and tighten it a second time to a pressure of 7·5 lbf ft (10 Nm).
38 Now reselect neutral and check the torque necessary to rotate the pinion shaft. A figure of 3·75 to 6·0 lbf ft (5·0 to 8·0 Nm) should be necessary to start turning it. Bend over the nut flange to lock in position.
39 When the above is achieved, retighten the upper housing retaining bolts to the following torque settings:

10 mm bolts	33·75 lbf ft (45 Nm)
8 mm bolts	15·0 lbf ft (20 Nm)
7 mm bolts	11·25 lbf ft (15 Nm)

40 Invert the gearbox again and retighten the bottom housing retaining bolts to the following torque settings:

8 mm bolts	15·0 lbf ft (20 Nm)
7 mm bolts	11·25 lbf ft (15 Nm)

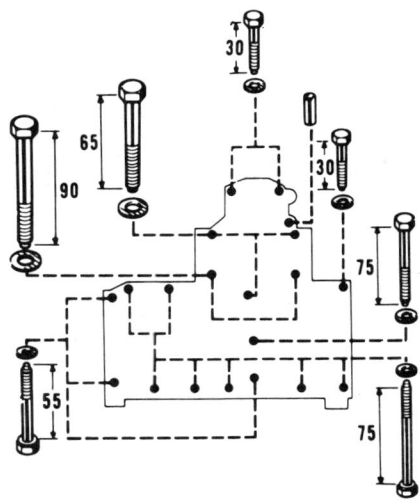

Fig. 6.20 The top half-housing bolt positions and lengths (mm)

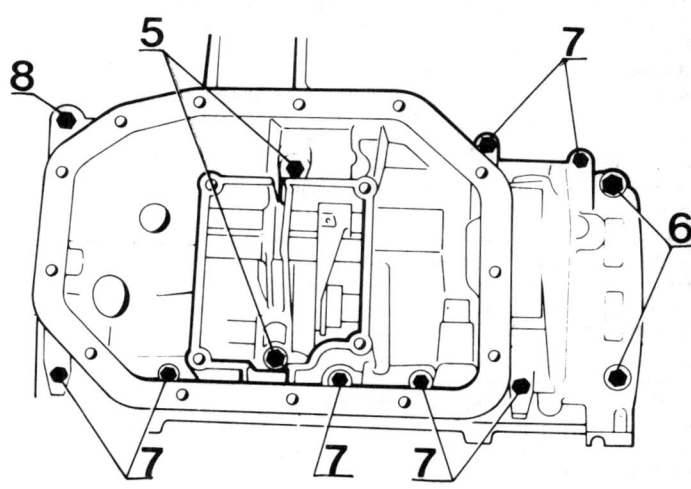

Fig. 6.21 The bottom half-housing bolts must be tightened to individual torque settings – see text

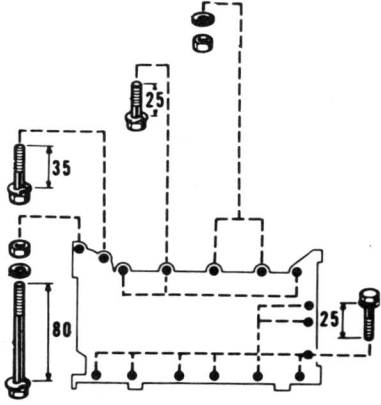

Fig. 6.22 The bottom half-housing bolt positions and lengths (mm)

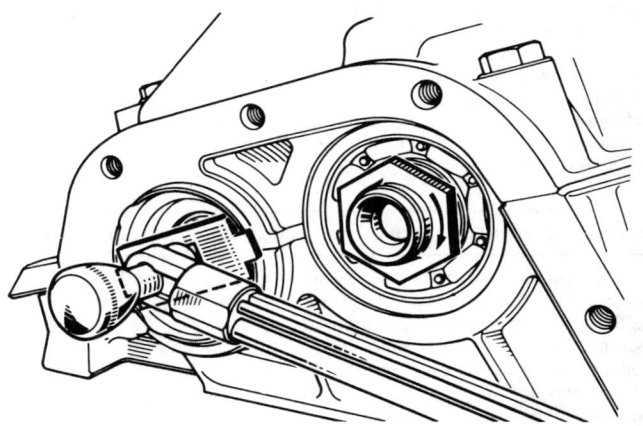

Fig. 6.23 This shows the special Renault tool used for tightening the primary shaft bearing adjustment nut. Turn the pinion shaft in direction of arrow when checking torque (see text)

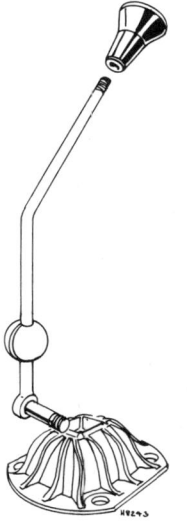

Fig. 6.24 Gear lever and base

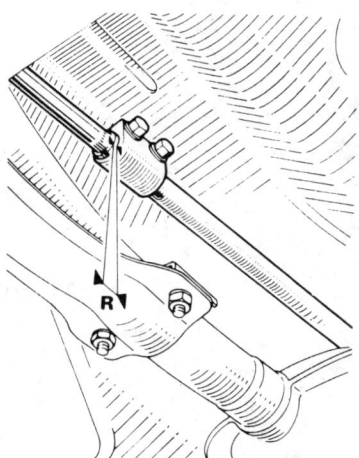

Fig. 6.25 Realign the clamp marks and adjust to give distance of $\frac{9}{64}$ in (3.5 mm) at R

41 To complete the gearbox assembly, refit the oil pump suction filter unit and tighten the four bolts to the specified torque.
42 Refit the bottom plate, using a new gasket. Fit the flat washers under the bolts and tighten to the specified torque (photo).
43 Refit the sump guard plate.
44 Use a new O-ring seal and re-insert the speedometer cable sleeve if not already fitted.
45 Carefully tap the oilseals into position in the differential housing if not already fitted.
46 The gearbox is now ready for refitting to the engine.

7 Transmission to engine – reassembly

Refer to Chapter 1, Section 36, where full instructions are given for refitting the gearbox to the engine.

8 Engine/transmission – installation

Refer to Chapter 1, Section 42 or 43 accordingly, where full installation instructions are given.

9 Differential oilseal – replacement

1 The differential oilseals can be removed and replaced with the engine/transmission unit in position in the car, but the driveshafts will obviously have to be removed. This operation is covered in Chapter 7.
2 With the driveshafts withdrawn the old oilseals can be extracted from the differential housing using a suitable screwdriver.
3 Clean out the seating before installing a new seal. Lubricate the seal to assist assembly and drift carefully into position.
4 Always take care not to damage the oilseals when removing or installing the driveshafts.

10 Floor gearchange mechanism – removal and installation

1 This mechanism does not normally give problems but if for any reason it is to be removed the car will have to be raised on ramps or positioned over a pit to gain suitable access to the gear change linkage. Check that the handbrake is applied and the wheels are chocked before working underneath the car.
2 Position the gear lever in neutral or take note which gear is engaged, then from underneath the car clean off and mark the relative position of the two portions of the selector rod where they join, then loosen the clamp (photo) to separate. Never disconnect the swivel lever balljoints.
3 Unhook the spring (photo) and unclip the rubber shroud.
4 Disconnect the gear lever from the rear portion of the selector rod, then remove the bolts securing the base of the lever (working inside the car). The gear lever can now be withdrawn.
5 Refit in the reverse order, but check the alignment of the selector shaft before tightening the clamp bolt (Fig. 6.25). Check the adjustment of the gears to ensure that they all select in a satisfactory manner. Further adjustment may be required (See Chapter 1, Section 42).

11 Fault diagnosis – transmission

1 Transmission faults can be divided into two main groups. The first being a definite failure preventing the transmission from operating. The second may be partial failure or unusual noises caused by a worn or damaged component.
2 In the first instance the problem is almost certain to be an internal gearbox fault in which case it will be necessary to remove and dismantle the transmission for inspection and rectification.
3 The only possible external fault could be the gear lever control linkage. The selector rod between the lever and transmission may have been distorted or damaged. Adjustment may also be lost due to the joint clamp coming loose. This is most unlikely but is worth a check.
4 In the second instance a partial failure such as difficulty in engaging a gear, noises and/or vibrations, it should first be confirmed that the problem is actually within the transmission. Strange noises caused by a component malfunction may carry through to the car and mislead the unsuspecting operator into an incorrect diagnosis. Check the basics first, items such as the selector rod adjustment end to clutch operating clearance. Difficulty in changing gear may be caused by a worn or incorrectly adjusted clutch.
5 Noises can be traced to a certain extent by doing the test sequence as follows:
6 Find the speed and type of driving that makes the noise. If the noise occurs with engine running, car stationary, clutch disengaged, gear engaged: the noise is not in the transmission. If it goes after the clutch is engaged in neutral, halted, it is the clutch.
7 If the noise can be heard faintly in neutral, clutch engaged, it is in the gearbox or transfer gears. It will presumably get worse on the move, especially in some particular gear.
8 Final drive noises are only heard on the move. They will only vary with speed and load, whatever gear is engaged.
9 Noise when pulling is likely to be a worn crownwheel and pinion.
10 Gear noises when free-wheeling may to a lesser amount be due to a worn crownwheel and pinion.
11 Noise on corners implies excessive tightness or excessive play of the bevel side gears or idler pinions in the differential, but first suspect

10.2 Loosen the clamp to separate the selector rod

10.3 Unclip the spring

the front wheel bearings.

12 In general, whining is gear teeth at the incorrect distance apart. Roaring or rushing or moaning is bearings. Thumping or grating noises suggest a chip out of a gear tooth.

13 If subdued whining comes on gradually, there is a good chance the transmission will last a long time to come.

14 Whining or moaning appearing suddenly, or becoming loud should be examined quickly.

15 If thumping, or grating noises appear stop at once. If bits of metal are loose inside, the whole transmission, including the casing, could quickly be wrecked.

16 Synchromesh wear is obvious. You just beat the gears and crashing occurs.

17 If uncertain about a problem, get a second qualified opinion before dismantling — you may save yourself some time, effort and possibly money.

Chapter 7 Driveshafts, hubs, wheels and tyres

Contents

Driveshaft joints 3	General description 1
Driveshafts – removal and refitting 2	Rear hub bearings – removal and refitting 5
Fault diagnosis – see Chapter 10 8	Tyres – general 7
Front wheel hub bearings – removal and refitting ... 4	Wheels – general 6

Specifications

Driveshafts
Type .... Removable with constant velocity universal joints on inner and outer ends, inner joints permit axial movement

Wheel hub bearings
Type:
    Front .... Twin track ball bearings
    Rear .... Two taper-roller bearings
Lubricant type/specification .... Multi-purpose lithium-based grease (Duckhams LB 10)

Wheels
Type .... Pressed steel disc, 3 stud fitting
Size .... $4\frac{1}{2} \times 13$
Maximum rim run out .... 0.040 in (1 mm)

Tyres
Type .... Radial ply, tubeless
Size .... 145 SR 13
Tyre pressures:
    Front .... 24.5 lbf/in^2 (1.7 bars)
    Rear .... 27.5 lbf/in^2 (1.9 bars)
For motorway use or when fully laden increase above pressures by 1.5 lbf/in^2 (0.1 bar)

Torque wrench settings

	lbf ft	Nm
Stub axle nut – rear	22.0	30
Stub axle nut – front	187.0	250
Lower suspension arm to strut	24.0	35
Lower suspension arm-to-anti-roll bar nut	40.0	55
Steering arm-to-strut nut	24.0	35
Steering arm adjustment clamp nuts	11.0	15
Wheel nuts	43.4	60

1 General description

The drive to the front wheels of the Renault 14 is transmitted directly from the final drive unit to the front hubs by the driveshafts. Constant velocity universal joints are fitted at each end and accommodate the steering and suspension angular movements.

Little maintenance is possible, even changing the rubber bellows is a specialised operation best entrusted to your Renault dealer.

The driveshafts are splined to the front hubs. These run on double row ball-races located in the hub carrier at the base of the MacPherson strut.

The rear hubs run on conventional taper-roll bearings on individual stub axles.

2 Driveshafts – removal and refitting

1 Raise the car at the front and support with chassis stands or blocks. Check that the handbrake is fully applied and chock the rear wheels.
2 Remove the front roadwheel on the side concerned.
3 Refer to Chapter 8 and remove the brake caliper unit but do not detach the hydraulic line. When the caliper is removed from the disc tie it to a suitable body member where it will not be in the way. Take care not to stretch or put stress on the brake hose.

Chapter 7 Driveshafts, hubs, wheels and tyres

4 Unscrew and remove the stub axle retaining nut. To prevent the hub from turning when undoing the nut, wedge a suitably padded bar between two wheel studs, taking care not to damage the threads.
5 Detach the steering arm from the MacPherson strut. Use a ball-joint separating tool or a suitable wedge.
6 Before removing the securing nut from the anti-roll bar take up the tension in the bar. Renault use an adjustable chain tensioner to hold the two ends of the bar together (Fig. 7.2) but the same result can be achieved by improvising a tourniquet from wire cable or strong rope. Loop this around the bar close to the lower suspension arm. Place a rod between the two ropes and twist them together until the required tension is achieved.
7 Now remove the retaining pin, nut and washer from the end of the anti-roll bar.
8 Remove the nut and washer from the lower suspension arm inner pivot pin and unscrew the pin to detach the arm from the subframe.
9 The tension can now be carefully released from the anti-roll bar and the bar extracted from its location in the lower suspension arm.
10 Pull the MacPherson strut to the rear and tap the bar out using a soft-faced hammer.
11 The driveshaft can now be separated from the hub by sliding it inwards while the hub is pivoted outwards. If the shaft is not being removed from the transmission tie it up to prevent it from being inadvertently withdrawn from the transmission unit.
12 To remove the driveshaft from the transmission, withdraw it carefully keeping it horizontal until clear of the differential housing (photo).
13 Installation is a direct reversal of the removal procedure but note the following:

   (a) Clean the pivot pins and stub axle splines prior to reassembly. Lubricate them with Duckhams LBM 10 grease and Duckhams LB 10 grease respectively
   (b) Take special care when inserting the driveshaft into the transmission not to damage the oil seal
   (c) Check that the lower suspension arm is accurately located before inserting the pivot pin. Grip the arm with a self-gripping wrench or similar to twist it as necessary
   (d) Do not torque tighten the suspension fastenings until the weight of the car is on the suspension
   (e) On completion check the hydraulic fluid reservoir level and top up if necessary. If there has been a leakage of fluid when the caliper was removed or refitted the brakes will require bleeding – see Chapter 8

3 Driveshaft joints

1 As has been stated in General description, there is very little maintenance which can be carried out on the driveshaft joints. The

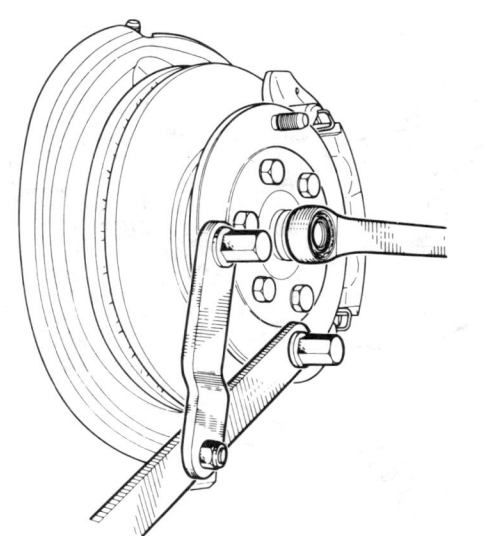

Fig. 7.1 The special tool used by Renault dealers to prevent the wheel from rotating

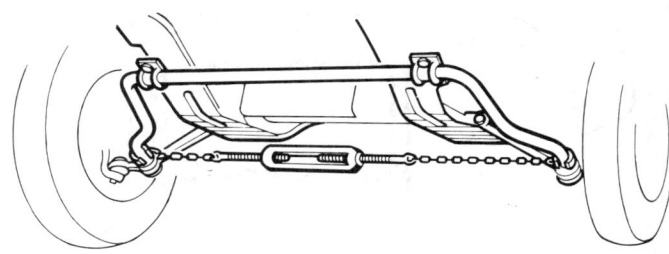

Fig. 7.2 Tensioning the anti-roll bar

2.12 The inner ends of the driveshafts (engine removed)

Fig. 7.3 Detach the suspension arm from the subframe (arrowed)

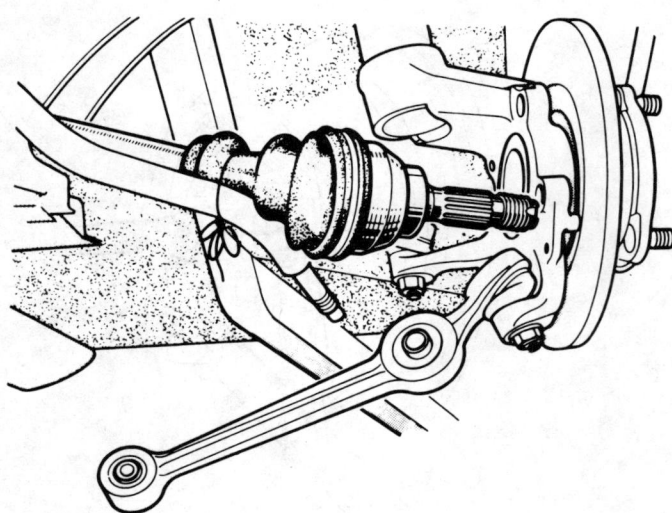

Fig. 7.4 Remove the driveshaft from the hub

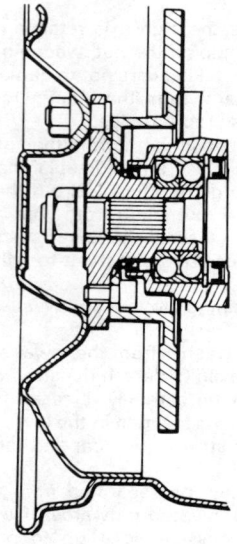

Fig. 7.5 Sectional view of the front hub unit

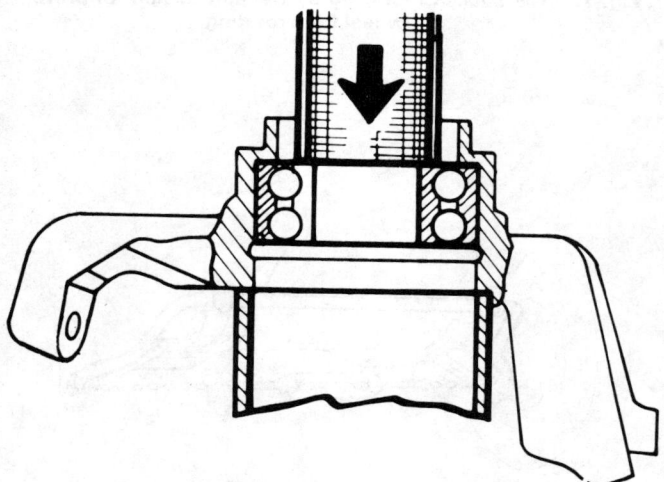

Fig. 7.6 Apply pressure to bearing inner race to remove

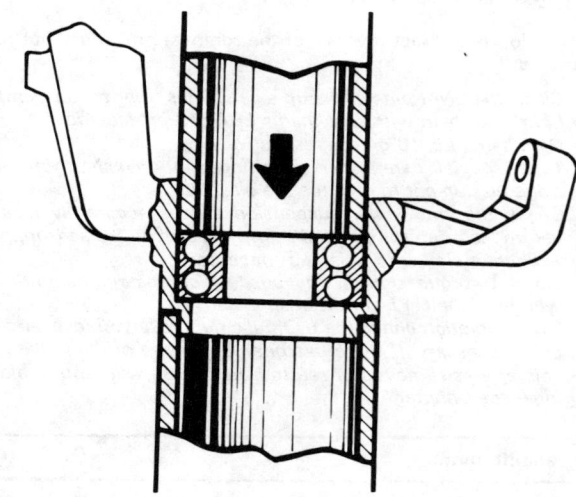

Fig. 7.7 Apply pressure to bearing outer race to install

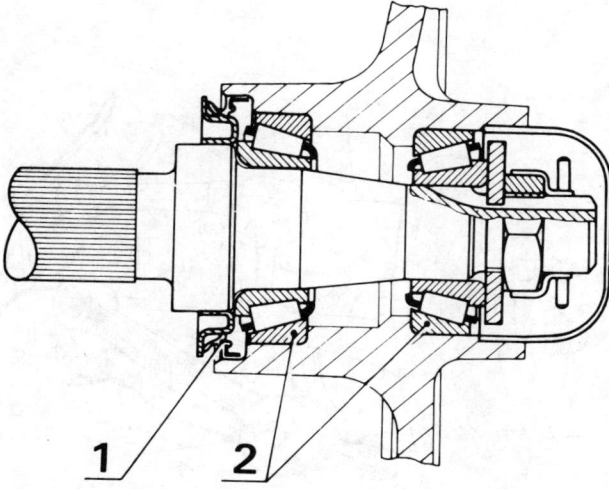

Fig. 7.8 The rear hub cross-section

1 Seal   2 Inner and outer bearings

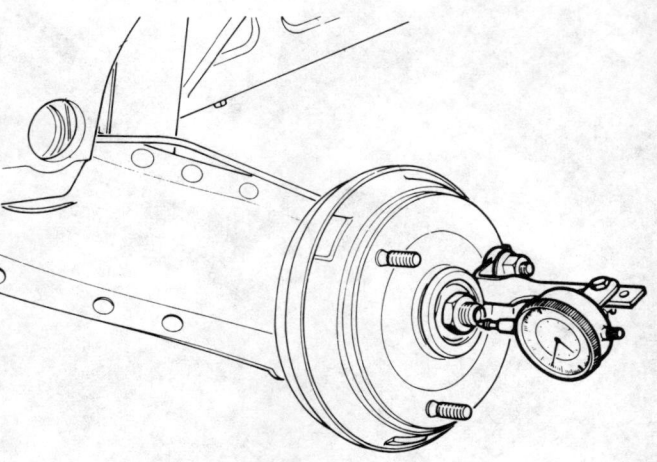

Fig. 7.9 Use a dial gauge to check hub endfloat

Chapter 7 Driveshafts, hubs, wheels and tyres

joints can be dismantled by the do-it-yourself mechanic but he will find that he is unable to reassemble them.

2 Special tools are required to effect an efficient repair. Even replacing the rubber bellows and relubricating the joint is beyond the use of ordinary tools. Under all circumstances it is more efficient to remove the driveshaft and then take it to a Renault garage (only) to have them effect any repair or maintenance. With the specific special tools available to them all repairs to the joint can be carried out very quickly and safer than attempting it yourself.

3 With the spider joints it is possible to have the bellows, yoke and spider itself replaced, together or separately. However experience shows that unless the bellows is punctured and lubricant allowed to escape and the joint to become dry, the outer universal joint wears at a far greater rate than the inner, consequently the shaft is nearly always replaced before the total life of the inner joint is reached.

4 Front wheel hub bearings – removal and refitting

1 The front hub bearings are twin track ball-races and removal for cleaning, inspection or replacement necessitates removing the wheel hub and brake disc.
2 Raise the car at the front and support with axle stands or blocks. Check that the handbrake is applied.
3 Remove the roadwheel.
4 Refer to Chapter 8 and remove the brake caliper unit, but do not detach the hydraulic line. Suspend the caliper unit from a suitable body member to prevent straining or distortion of the brake line.
5 Place a suitable bar between the wheel hub studs to prevent the hub from turning and then unscrew and remove the stub axle nut.
6 The hub and disc unit will have to be withdrawn using a three legged puller. Do not try to remove with hammers and drifts or serious damage may result – get the proper puller. If you are unable to borrow one, hire one.
7 Having removed the hub and disc assembly the MacPherson strut must now be removed as described in Chapter 10.
8 To extract the bearing from the strut first remove the retaining circlip. The bearing can now be pressed or drifted out of its location using a mandrel or tube drift of 2 in (50 mm) diameter which will apply pressure to the bearing inner ring.
9 When the bearing is removed it is an easy matter to withdraw the outer oilseal from the housing.
10 The bearing can now be cleaned and checked for wear and damage. Spin the outer race whilst holding the inner race. If a harshness is felt then it is almost certain that the bearing is in need of replacement. If in any doubt regarding the condition of the bearing have it checked by your Renault dealer or a competent auto engineer. Always renew the oilseal as a matter of course.
11 Reassembly is a direct reversal of the removal procedure but note the following:

 (a) Lubricate the oilseal to assist assembly and take care not to distort it
 (b) When fitting the bearing, the mandrel or tube drift diameter must suit that of the bearing outer race (as opposed to the inner during removal) a tube of 2.687 in (68 mm) diameter being required
 (c) Use a new retaining circlip if the old one was distorted or damaged on removal. Make sure that it is properly located
 (d) Tighten all nuts and bolts to the specified torque settings, but do not tighten the suspension fastenings until the weight of the car is on the suspension

5 Rear hub bearings – removal and refitting

1 Only ever replace the inner and outer rear hub bearings as a pair. Raise the rear of the car and support with axle stands or blocks. Chock the front wheels.
2 Refer to Chapter 8 and loosen the tension on the handbrake cable (to ease removal of brake drum).
3 Remove the grease cap from the drum and have a new one ready for fitting. A screwdriver will easily remove this.
4 Extract the split-pin from the hub nut and withdraw the locking cap (photo).
5 Unscrew and remove the hub nut and washer (photo). Withdraw the brake drum taking care not to let the inner track and roller cage of the outer bearing fall out – it should be removed and set aside for inspection.

5.4 Extract the split-pin

5.5 Removing the washer and outer bearing

5.7 Using two leg puller to remove inner bearing

6 Extract the grease seal from the inside face of the brake drum and then remove the outer bearing tracks using a suitable drift or extractor.
7 The inside bearing track and roller cage can be removed from the hub using a two leg puller as shown in the photo.
8 Clean the respective components for inspection and renew where necessary. Always renew the grease seal as a matter of course.
9 Reassembly is a reversal of the removal sequence, the inner bearing being drifted into position, using a suitable diameter tube drift, butting against the inner track — not against the roller cage! When fitting the bearing tracks into position ensure that they are pressed/drifted into position squarely and are fully located.
10 Brake drum replacement is a reversal of the removal procedure, but it will be necessary to adjust the hub bearings. To do this tighten the hub nut to a torque of 22·5 lbf ft (30 Nm), then unscrew the nut by $\frac{1}{8}$th turn. This should provide the specified bearing endfloat of 0 to 0·0012 in (0 to 0·03 mm). This is best checked using a dial gauge if available.
11 Relocate the locking cap over the nut and insert a new split-pin. Fill the grease cap with approximately $\frac{1}{3}$ oz (10 gm) of general purpose grease and then carefully tap it into position on the hub.
12 Readjust the brakes and handbrake as described in Chapter 8 and finally refit the wheel.

6 Wheels – general

1 Because of the design of the suspension of the car the strength and the trueness of the roadwheels is critical, particularly at the front. Excessively fast wear on the wheel bearings and universal joints can often be attributed to buckled and deformed wheels. Check every 3000 miles or when there is a sudden difference of feeling at the steering wheel that the wheels are not buckled or dented. Check also that the front wheels are balanced.
2 If it is suspected that the wheels are out of balance have your Renault dealer rebalance them.
3 If a wheel is badly rusted or damaged in any way do not attempt to repair it – get a new replacement.
4 Do not overtighten the wheel nuts for this can deform the wheel. Always check that the inner side of the wheel is free from mud and grit for the accumulation of these can create imbalance.

7 Tyres – general

In the same way that the condition and suitability of the wheels fitted is critical so it is with the tyres. Because of the long suspension travel and fully independent suspension it is always wise to fit radial tyres on all wheels of these cars. Tyre wear is not great under any circumstances but the front tyres wear faster than the rear. Do not fit oversize tyres. The wheel rims are not readily able to take a larger section tyre. See Specifications for suitability of tyres. Tyre pressures are also critical.

8 Fault diagnosis – see Chapter 10

Chapter 8 Braking system

For modifications, and information applicable to later models, see Supplement at end of manual

Contents

Bleeding the hydraulic system	3
Brake adjustment	4
Brake disc – examination, removal and refitting	7
Brake limiter – general	14
Disc brake caliper – removal, overhaul and refitting (ATE type)	6
Disc pads – inspection and refitting (ATE type)	5
Fault diagnosis	16
General description	1
Handbrake system	15
Hydraulic fluid pipes – inspection and renewal	10
Master cylinder (ATE type) – removal, overhaul and refitting	11
Rear drum brakes (Girling type) – removal, inspection and refitting	8
Rear wheel cylinder (Girling type) – removal, overhauling and refitting	9
Routine maintenance	2
Servo unit – general and maintenance	12
Servo unit – removal and installation	13

Specifications

Type .... Front disc brakes, rear drum brakes. Servo-assisted (except basic models), self-adjusting. Cable operated handbrake to rear wheels

Hydraulic fluid
Type/specification .... Hydraulic fluid to SAE J1703F, DOT 3 or DOT 4 (Duckhams Universal Brake and Clutch Fluid)

Front disc brakes
Disc diameter	9.488 in (241 mm)
Disc thickness	0.394 in (10 mm)
Minimum allowable disc thickness	0.354 in (9 mm)
Pad thickness including backing strip	0.630 in (16 mm)
Minimum pad thickness allowable (with backing)	0.276 in (7 mm)
Allowable disc run out	0.004 in (0.1 mm)
Caliper cylinder bore	1.890 in (48 mm)

Rear drum brakes
Drum diameter	7.096 in (180.25 mm)
Maximum allowable drum diameter	7.136 in (181.25 mm)
Lining width	1.575 in (40 mm)
Lining thickness with shoe	0.275 in (7 mm)
Minimum allowable thickness	0.020 in (0.5 mm) above rivet head
Wheel cylinder bore diameter	0.866 in (22 mm)

Master cylinder
Type	Tandem type dual-circuit, with fluid level warning device
Bore	0.748 in (19 mm)
Pushrod operating clearance (servo equipped models)	0.354 in (9 mm)

Servo unit
Type	Master-Vac, 6 in (152 mm) diameter
Pedal adjustment – clevis pin to bulkhead	3.812 in (97 mm)

Torque wrench settings
	lbf ft	Nm
Limiter inlet union	20.0	25
Bleed screws	4.0 to 5.0	5.0 to 7.0
All other hose unions	15.0	20
Caliper bolts	60	80
Rear stub axle nut	22.5	30
Front stub axle nut	187.0	250
Brake disc to hub bolts	40.5	55

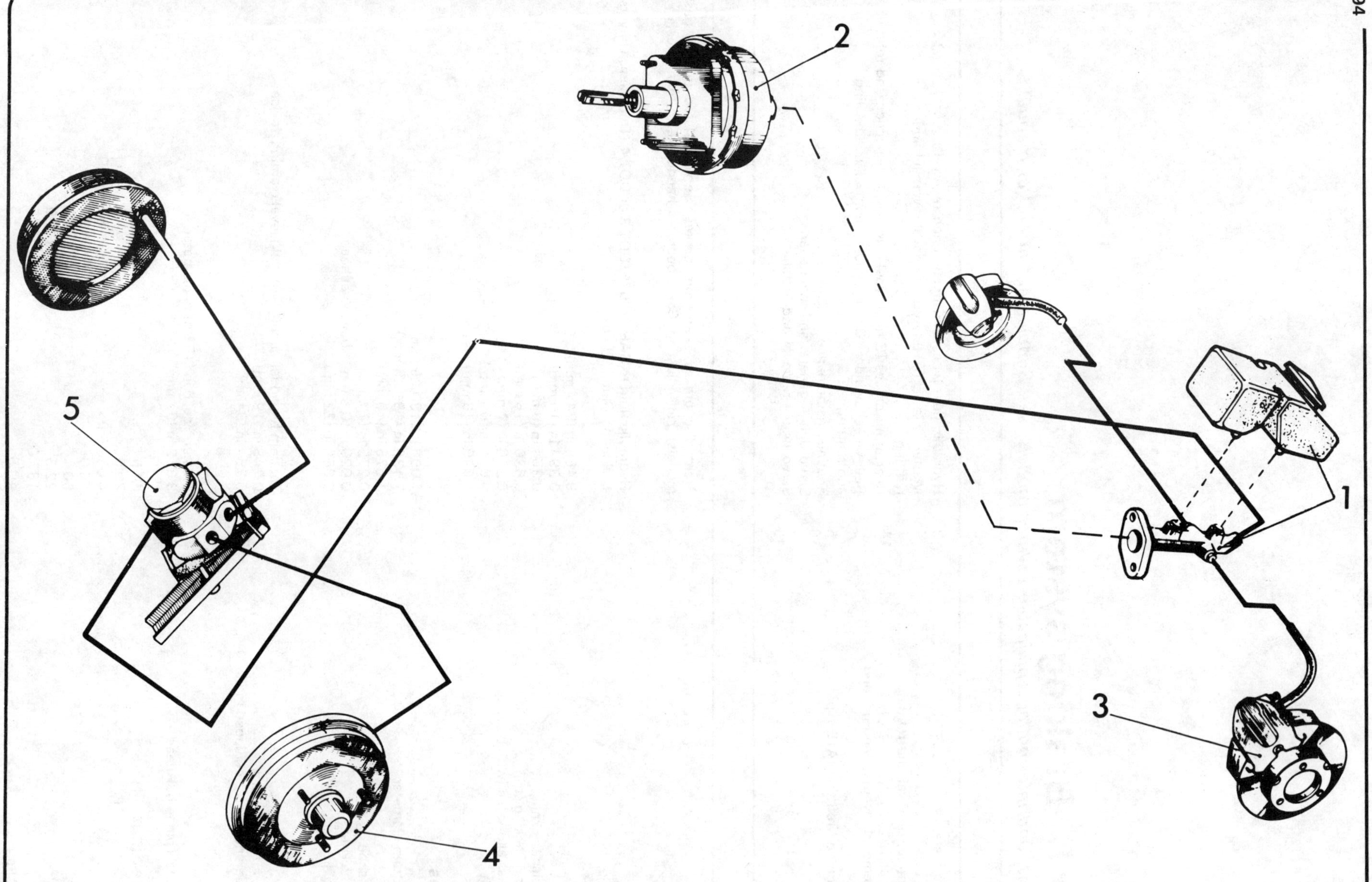

Fig. 8.1 The brake system layout

1 Master cylinder and fluid reservoir
2 Servo unit (where fitted)
3 Front disc brakes
4 Rear drum brakes
5 Limiter

Chapter 8 Braking system

1 General description

A conventional modern brake system is employed on the Renault 14. The front disc brakes and rear drum brakes are hydraulically operated with servo-assistance on all models except the basic. A tandem master cylinder is fitted to provide a dual-circuit braking system. A pressure limiting valve is incorporated in the system to prevent the rear wheels from locking under hard braking.

2 Routine maintenance

Every week
1 Remove the hydraulic fluid reservoir cap, having made sure that it is clean, and check the level of the fluid which should be just below the bottom of the filler neck. Check also that the vent hole in the cap is clear. Any need for regular topping-up, regardless of quantity, should be viewed with suspicion and the whole hydraulic system carefully checked for signs of leakage.

Every 9000 miles
2 Remove the brake drums and examine the shoe linings. They should be renewed when the friction material has very nearly reached the level of the rivet heads or to the specified minimum allowance. If either the rivets or the shoes themselves come into contact with the brake drum they will cause scoring and greatly reduced braking efficiency. Never interchange worn shoes to even-out wear.
3 For the disc brakes — inspect the disc pad wear, the total pad and backing should be not less than 9 mm. If less they must be replaced in full sets.
4 Make sure the handbrake functions at all times. At the same time as friction material is examined the hydraulic pipes and unions should be examined for any signs of damage or corrosion. Brake lining pad wear varies according to driving style but no set of brake shoes should be expected to last more than 20 000 miles. The front pads are likely to wear at a faster rate than the rear linings.

Every 18 months to 2 years
5 Every 18 months to 2 years, depending on usage, it is good policy to renew all hydraulic cylinder seals and disc caliper piston seals as a matter of routine, together with the fluid and flexible hoses. Any repair work in the interim period should also, of course, be taken into account.
6 If you have just acquired a secondhand car it is strongly recommended that all brake drums and shoes and/or disc and pads are thoroughly examined for condition and wear immediately. Even though braking efficiency may be excellent the friction materials could be nearing the end of their useful life and it is as well to know this without delay. Similarly, the hydraulic cylinders, pipes and connections should be carefully examined for leaks or chafing. Faults should be rectified immediately. It should be remembered that three year old cars will be subject to safety tests and that apart from safety, which is paramount, defects in the system even though they may not yet affect stopping power cause the vehicle to fail the test.
7 **Warning**: *Always ensure absolute cleanliness when dismantling brake hydraulic components and never bring anything except clean hydraulic fluid or methylated spirit into contact with internal components of hydraulic assemblies.*

3 Bleeding the hydraulic system

1 The system should need bleeding only when some part of it has been dismantled which would allow air into the fluid circuit; or if the reservoir level(s) has been allowed to drop so far that air has entered the master cylinder.
2 Ensure that a supply of clean non-aerated fluid of the correct specifications is to hand in order to replenish the reservoir(s) during the bleeding process. It is advisable, if not essential, to have someone available to help, as one person has to pump the brake pedal while the other attends to each wheel. The reservoir level has also to be continuously watched and replenished. Fluid bled out should not be re-used. A clean glass jar and a 9 – 12 inch length of $\frac{1}{8}$ inch internal diameter rubber tube which will fit tightly over the bleed nipples is also required (photo).
3 Bleed the rear brakes first as these are furthest from the master

3.2 The bleed nipple with tube attached to the caliper

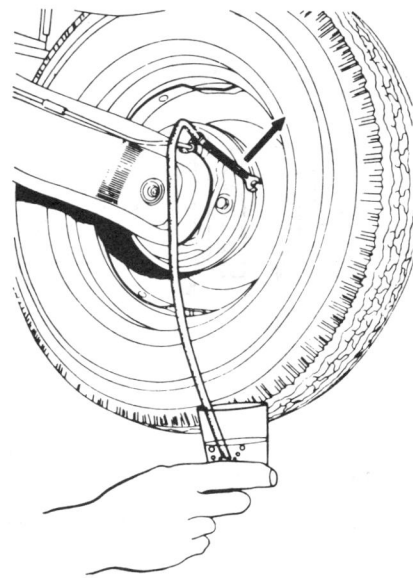

Fig. 8.2 Brake bleed nipple – rear

cylinder. Bleed one system first keeping its reservoir topped up. Then bleed the by-pass circuit from the bleed nipple on the top of the unit.
4 Make sure the bleed nipple is clean and put a small quantity of fluid in the bottom of the jar. Fit the tubes onto the nipple and place the other end in the jar under the surface of the liquid. Keep it under the surface throughout the bleeding operation.
5 Unscrew the bleed screw $\frac{1}{2}$ turn and get the assistant to depress and release the brake pedal in short sharp bursts when you direct him. Short sharp jabs are better than long slow ones because they will force any air bubbles along the line ahead of the fluid rather than pump the fluid past them. It is not essential to remove all the air the first time. If the whole system is being bled, attend to each wheel for three or four complete pedal strokes and then repeat the process. On the second time around operate the pedal sharply in the same way until no more bubbles are apparent. The bleed screw should be tightened and closed with the brake pedal fully depressed which ensures that no aerated fluid can get back into the system. Do not forget to keep the reservoir topped up throughout.
6 When all four wheels have been satisfactorily bled depress the foot pedal which should offer a firm resistance and give no trace of sponginess. The pedal must not go down under sustained pressure and if it does the master cylinder seals are likely to require replacement.
7 To bleed a servo braking system, refer to Supplement.

Chapter 8 Braking system

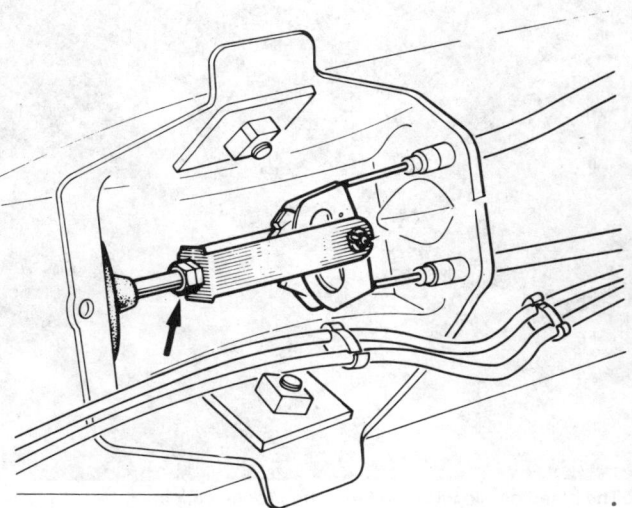

Fig. 8.3 Handbrake clevis locknut and adjustment nut

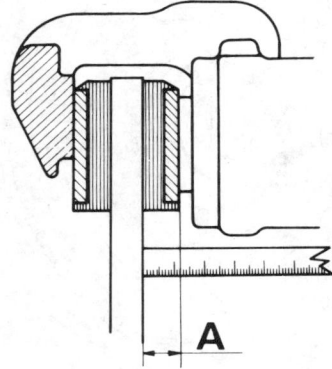

Fig. 8.4 Measure the pad wear as shown using a rule

Minimum thickness allowable at A = 0.275 (7 mm)

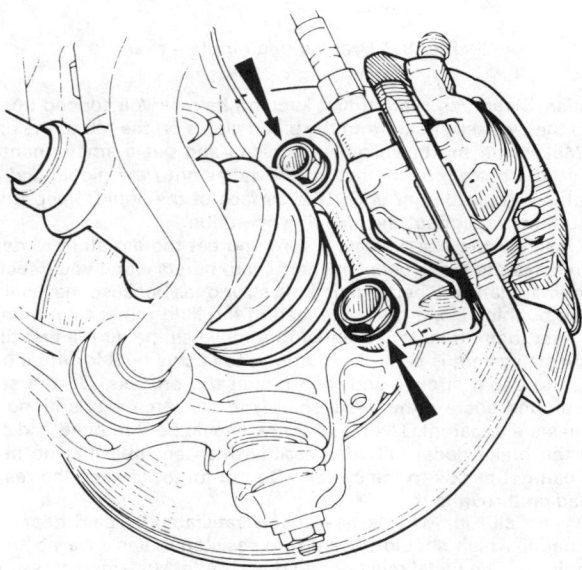

Fig. 8.5 Caliper retaining bolts

4 Brake adjustment

Front disc brakes
1 The front disc brakes are fully self-adjusting and therefore apart from the normal service checks they do not require any attention. The disc brake pads must be checked for wear periodically and if necessary, they must be replaced (Fig. 8.4), as described in Section 5.

Rear drum brakes
2 As with the front disc brakes the rear drum brakes are self-adjusting and do not require any attention apart from the occasional service check to inspect the linings. This involves removing the drums and is fully covered in Section 8.
3 If the linings have been renewed or the handbrake cables have been detached or replaced, the rear brakes can be readily adjusted by simply applying pressure to the brake pedal several times. The handbrake cable adjustment should then be checked, as described below:

Handbrake adjustment
Note: *The handbrake should only be adjusted when the rear brake linings have been renewed or the cable/s replaced, otherwise the operation of the self-adjusting mechanism may be affected.*
4 Raise the rear of the vehicle and support on axle stands. Place chocks each side of the front wheels to prevent the vehicle rolling and release the handbrake.
5 Slacken off the handbrake clevis locknut and turn the primary rod nut until the shoes are just touching the brake drums when the wheel is rotated. Unscrew the nut slightly so the wheels turn freely. Adjustment should be made so that a minimum of twelve notches of lever travel remains between the off and on positions. Retighten the locknut when adjustment is complete and lower the car.

5 Disc pads – inspection and refitting (ATE type)

Refer to Supplement for Bendix type brakes
1 Before dismantling any part of the brakes they should be thoroughly cleaned. The best cleaning agent is hot water and a mild detergent. Do not use petrol, paraffin or any other solvents which could cause deterioration to the friction pads or piston seals.
2 Jack-up the car and remove the wheel.
3 Inspection does not necessitate the removal of the disc pads themselves. The pads abut the disc surface at all times. Therefore it is possible to put the end of a steel rule (the measure must start at the end of the rule) into the recess above the caliper and the tops of the pads. It should be be evident where the outer edge of the pad comes to, on the rule, from the surface of the disc outwards. The total thickness of the pad, and its backing must not be less than 7 mm (0.275 in). If less than this figure the pads must be renewed.

5.4 Remove the retaining pins

Chapter 8 Braking system

4 To remove the pads, extract the retaining clips from the pad pins and drift the pins out (photo).
5 Extract the anti-rattle spring and then withdraw the inner disc pad first (photo).
6 Press the caliper unit towards the outside of the hub and extract the outer disc pad. Keep the two pads from each wheel separate.
7 Inspect the disc friction surface area. If it is badly scored, cracked or exceeds the specified wear limits it will have to be renewed. The brake discs cannot be re-surfaced.
8 If the pads are not worn out but have a black and shiny surface it is helpful to roughen them up a little on some emery cloth before refitting them. Disc pads last normally about 12 000 miles.
9 Visually check the condition of the caliper, the hydraulic fluid pipe and their connection.
10 Behind the pads on the carrier bracket are two pad anchor springs. Remove these and clean them up after ensuring that they are intact. Renew them otherwise.
11 Check the disc and caliper before replacing the original or new pads.
12 Replacement of the pads and calipers is a reversal of the removal procedure, but when fitting new pads certain additional matters must be attended to.
13 Make sure that the new pads are exactly similar to the ones taken off. Push back the piston in the caliper with a suitable blunt instrument to provide the necessary clearance for the new thicker pads.
14 With ATE type disc brakes ensure that the pistons are correctly positioned (Fig. 8.8) – see next section.
15 Check that the outside pad is fitted over the boss.
16 Use new pad pin retaining clips if the old ones are rusty and/or distorted.
17 When the pads are in position and fully located pump the brake pedal several times to bring the caliper piston up to the pads.
18 Refit the roadwheel lower the car and pump the brake pedal.
Note: *Disc pads must be renewed in sets. Always renew pads on both front wheels – never just one.*

6 Disc brake caliper – removal, overhaul and refitting (ATE type)

Refer to Supplement for Bendix type brakes

1 Note and follow the instructions given in paragraphs 1 to 6 in Section 5.
2 Disconnect the brake hose connection to the caliper and plug the hose to prevent spillage of fluid and ingress of dirt. Tie the brake line back out of the way but do not distort it or damage may result.
3 Unscrew the caliper retaining bolts and withdraw the caliper unit from the stub axle carrier.
4 Unclip the spring from the caliper bracket and then detach the bracket and caliper. Remove the cylinder from the caliper.
5 Prise the rubber dust cover free and then extract the piston from its cylinder. This is best done by carefully blowing the piston out using an air line or tyre pump applied to the brake line hose aperture (Fig. 8.7). If however the piston has seized within the bore, you will have difficulty in removing it without damaging the caliper and piston. In this case try soaking the assembly in methylated spirit. If this fails to soften things up the caliper assembly will have to be replaced.
6 Assuming the pistons have been removed without difficulty, clean them thoroughly with methylated spirit and remove the seal from the annular groove in the cylinder bore. Any hard residue deposits may be removed with careful use of some 600 grit wet and dry paper. If there are any ridges or scores in the cylinder or on the piston the parts must be renewed. Fit new seals in the cylinder groove, lubricate the cylinders and pistons with hydraulic fluid and replace the pistons. Fit the dust seals so that they fit in the caliper groove and on to the piston.
7 Installation of the calipers is a direct reversal of the removal procedure but note the following:
   (a) The bleed screw should be removed and the caliper tilted in each direction whilst it is being topped up with fluid, to expel any air in the unit and ease bleeding later, when the system is reconnected
   (b) Check that the hoses are correctly located with no distortions and their connections are secure. Remember they must not rub against surrounding components when fitted
   (c) Renew the copper washer when refitting the flexible hose to

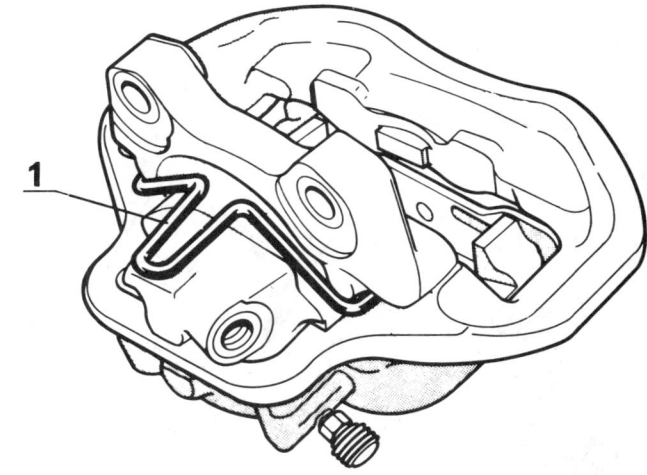

Fig. 8.6 Remove the caliper retaining clip (1)

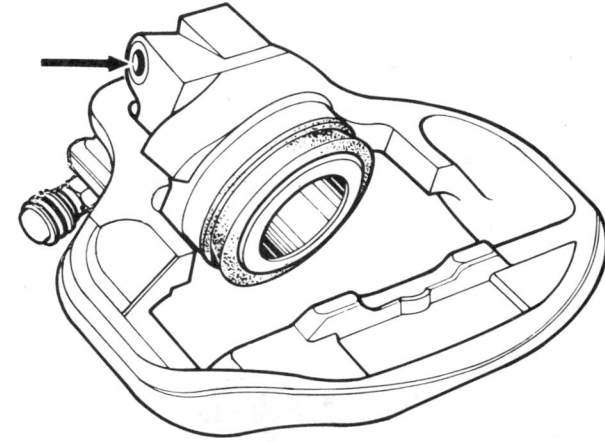

Fig. 8.7 Inject air into aperture indicated to remove piston

5.5 Removing the inner pad

Chapter 8 Braking system

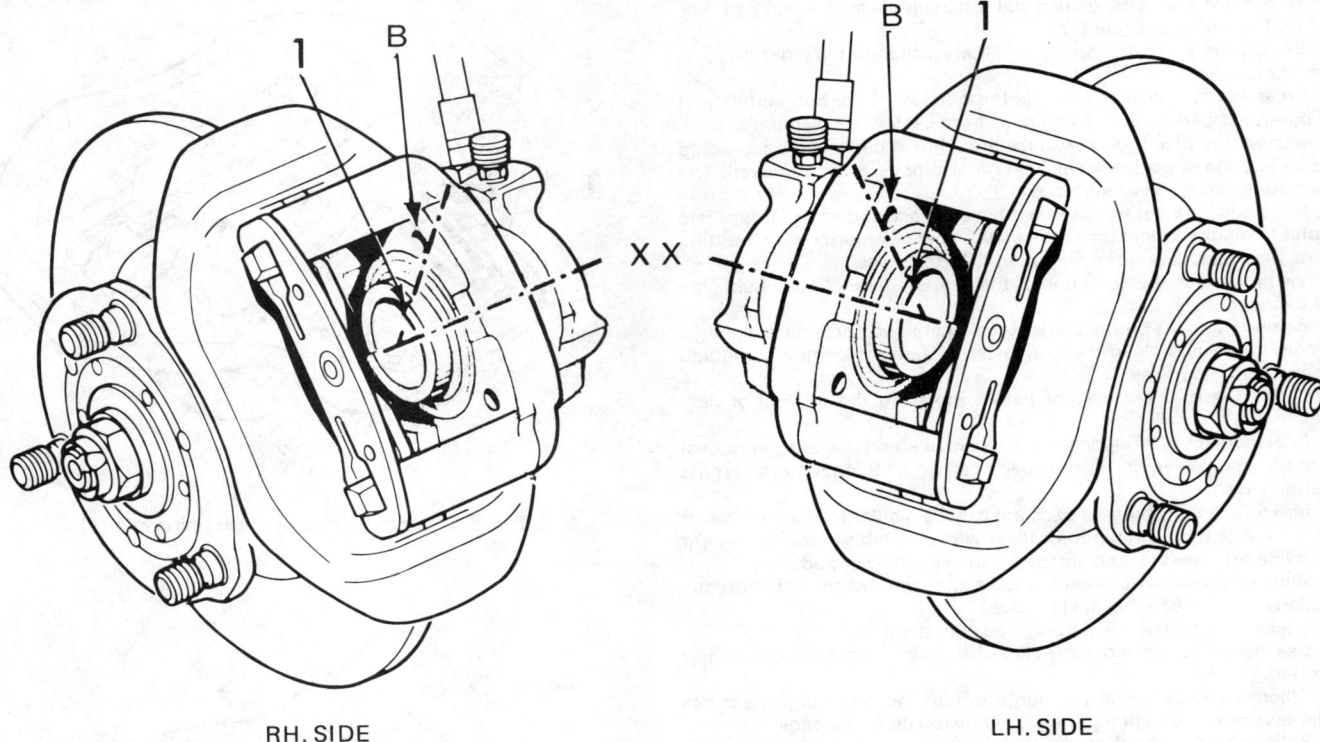

RH. SIDE LH. SIDE

Fig. 8.8 Views showing the ATE type cutaway pistons. Position right and left-hand accordingly with the step (1) in line with hole B as shown. The opposing step is to be aligned with centre line of caliper (X)

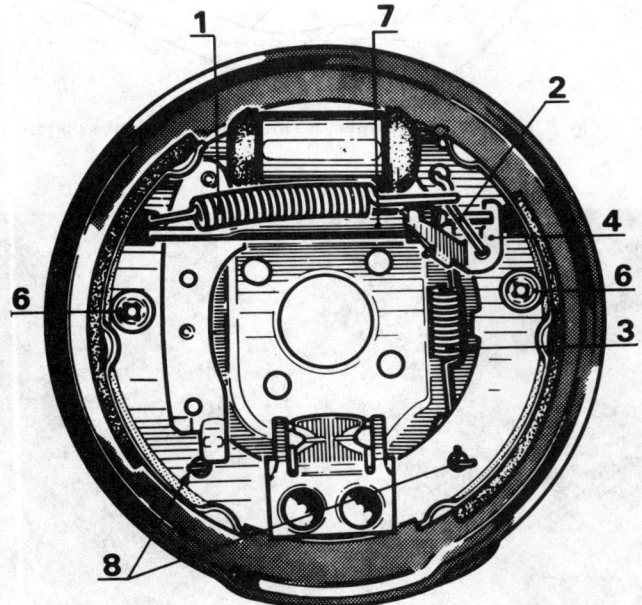

Fig. 8.9 General view of Girling brake assembly with drum removed

1 Top return spring
2 Clip
3 Self-adjuster spring
4 Self-adjuster lever
6 Brake shoe retainer
7 Thrust link and ratchet
8 Bottom return spring

the caliper
(d) Bleed the brake system to complete the operation (Section 3)

Note: Where ATE type front disc brake units have been fitted the piston face to the pad is stepped as shown in Fig. 8.8. The cutaway section reduces the thrust area on the pad and spreads the brake loading to lengthen the life of the pads. It is important with this type of brake that the piston is correctly positioned. The correct alignment is shown for the right and left-hand sides.

7 Brake disc – examination, removal and refitting

Refer to Supplement for Bendix type brakes

1 Refer to Section 6 and remove the brake caliper without disconnecting the hydraulic hose; tie the caliper out of the way.
2 Inspect the disc for deep scoring. Light scoring is normal but severe scoring will necessitate the renewal of the disc, or alternatively machining within the specified limits.
3 Using a dial gauge or feeler blades and a fixed block, check that the run-out of the disc is within the specified limit.
4 To remove the disc first withdraw the hub as described in Chapter 7. Unscrew the three nuts, remove the bolts, and separate the disc from the hub.
5 Refitting is a reversal of removal but make sure that the disc to hub mating surfaces are clean and tighten the bolts to the specified torque. Refit the hub as described in Chapter 7.

Chapter 8 Braking system

8 Rear drum brakes (Girling type) – removal, inspection and refitting

For Bendix type rear brakes refer to Supplement

1 Jack up the car at the rear and support with axle stands or blocks. Remove the roadwheels. Chock the front wheels at front and rear and release the handbrake.
2 Release the handbrake tension by slackening the cable locknut and unscrewing the adjustment nut.
3 Prise the plug from the backplate concerned (photo) and insert a suitable screwdriver through the backplate hole and apply pressure to the handbrake operating lever within the drum, freeing the peg from the brake shoe.
4 Now push the lever rearwards to free the brake drum.
5 The brake drum is retained in position by the stub axle nut. The drum also houses the wheel hub bearings and therefore to remove the drum (and bearing assemblies as required) refer to Chapter 7, Section 5.
6 First examine the drum, in particular the friction surface area for any signs of scoring or excessive wear marks. The surface should be smooth and bright but minor hairline scores are of no consequence and could have been caused by grit or brake shoes with linings just worn to the rivets. A drum that is obviously badly worn should be renewed. Although it is possible to have a scored or badly pitted brake drum re-surfaced they must not exceed the maximum diameter specified. Furthermore the drums must either be machined equal amounts or both renewed as a pair, whichever is the case. If in doubt check with your Renault dealer.
7 The brake shoes should be examined next. There should be no signs of contamination by oil and the linings should be above the heads of the rivets. If the level is close (less than 0.020 in) it is worth changing them. If there are signs of oil contamination they should be renewed also and the source of oil leakage found before it ruins the new ones as well!
8 To remove the shoes, first unhook the handbrake cable from it's operating lever (photo).
9 Now unhook the return spring at the top using a suitable screwdriver to release the spring from the shoes.
10 The vertical spring is now unhooked from the self-adjustment mechanism. Withdraw the ratchet and the spring noting how they are located. The thrust washer must be retained on the ratchet spindle.
11 Use a pair of pliers and twist the shoe retainer pin to align the slot in the washer then withdraw the washer and spring from each shoe (photo).
12 Remove the adjustment link between the shoes.
13 Remove the brake shoes and then tie a length of string around the end of the wheel cylinder to prevent the pistons from coming out whilst the shoes are removed.
14 Before installing new brake shoes check the operating components for signs of damage and distortion, and renew as necessary.

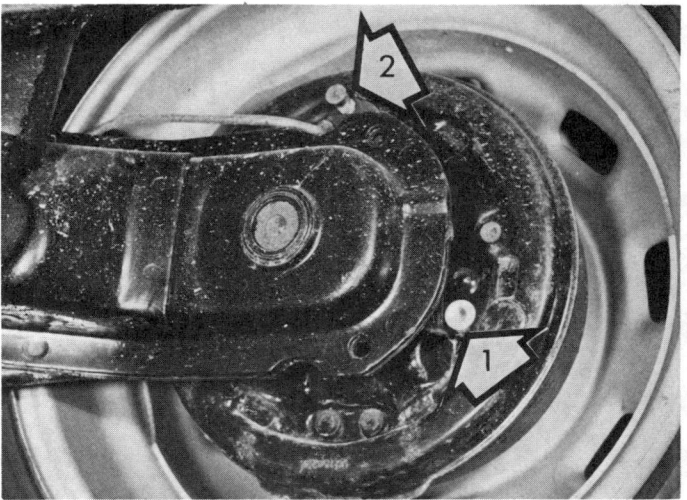
8.3 View of backplate showing plastic plug (1) and bleed nipple (2)

8.8 Unhook the handbrake cable

8.11 Shoe retainer

8.18 The reassembled brake assembly prior to fitting drum and bearings

Chapter 8 Braking system

9.12 The reassembled brake wheel cylinder in position

Note that the left-hand side adjustable link has a right-hand thread whilst the right-hand side link has a left-hand thread. Do not get the parts from each drum assembly mixed up. Never renew the brake linings on one side only, this is false economy and can promote unequal braking on the rear wheels.
15 Check the wheel cylinder for signs of leakage and ensure that it is secure but do not actuate the brake pedal whilst the shoes are removed!
16 Refitting the brake shoes and the associated components is a direct reversal of the removal process. Position the shoes under the anti-rattle retainers and fit the bottom return spring.
17 Prior to installing the upper return spring, the thrust link length must be adjusted so that the lining diameter is approximately 7 inches (178 mm) to permit the brake drum to be fitted over the shoes.
18 Before refitting the drum and bearings, check that the return springs are properly located, and make a final check for correct assembly of the ratchet and handbrake mechanisms (photo).
19 Refit the brake drum and lubricate the hub bearings as described in Chapter 7, Section 5.
20 When the drum is refitted insert a new plastic plug into the backplate. Centralise and adjust the brakes by operating the brake pedal several times.
Note: *If the wheel cylinders were overhauled the brakes will need to be bled as described in Section 3.*

9 Rear wheel cylinder (Girling type) – removal, overhaul and refitting

Refer to Supplement for Bendix type brakes

1 Raise the rear of the car and remove the brake drum as described in Chapter 7, Section 5.
2 Unclip the upper shoe return spring.
3 Carefully unscrew the hydraulic brake line from the wheel cylinder.
4 Unscrew and remove the two wheel cylinder retaining bolts, prise the brake shoes apart at the top and remove the wheel cylinder.
5 If the cylinder has been leaking and there is fluid spillage on the brake linings then they must be removed and renewed, (on both sides). This is described in the previous Section.
6 Clean off the outside of the cylinder using methylated spirit and remove to a clean work area for dismantling.
7 Pull the dust cover from each end of the cylinder and extract the pistons, seals and spring.
8 Inspect the cylinder bore for signs of scoring or rust pittings, if present the cylinder must be renewed.
9 Assuming that the cylinder is in a serviceable condition, clean thoroughly in methylated spirit or hydraulic fluid and wipe dry with a clean non-fluffy cloth.
10 The old rubber seals will have deteriorated and must be automatically renewed on reassembly.
11 Smear the new seals and the pistons in hydraulic fluid and reassemble the cylinder in the reverse order of removal. Fit the two dust covers onto the cylinder.
12 Installing the cylinder to the brake backplate (photo) is a reversal

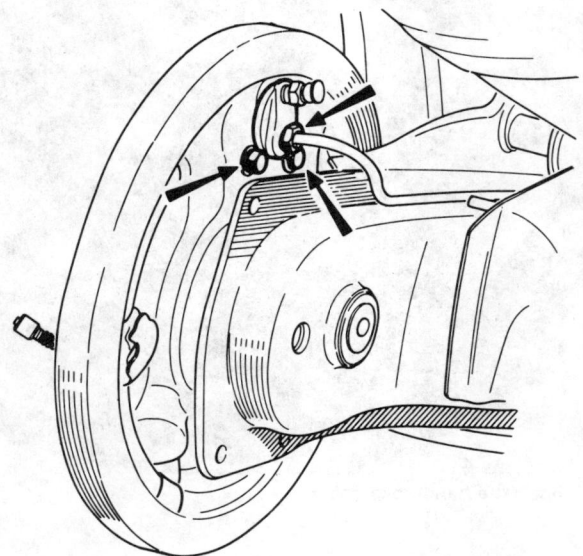

Fig. 8.10 Unscrew the cylinder hydraulic line connection and the retaining bolts

Fig. 8.11a Metric hydraulic component identification
A Pipe flares   C Pipeline unions
B Cylinder union sockets

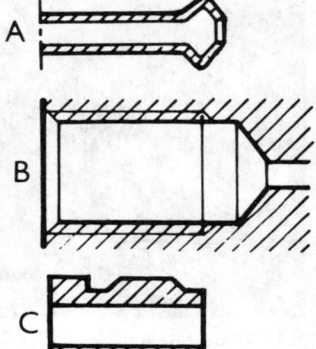

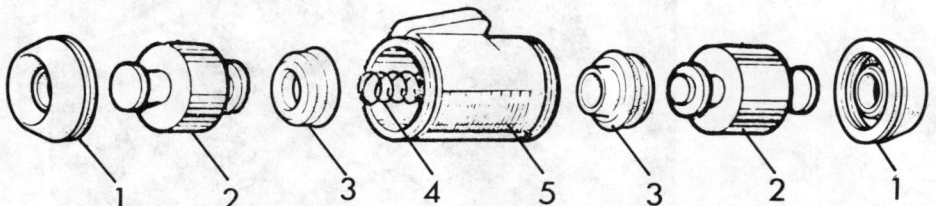

Fig. 8.11 The Girling rear wheel cylinder components

1 Dust cover   3 Seal   4 Spring   5 Cylinder
2 Piston

Chapter 8 Braking system

of the removal sequence but note the following:

(a) Check that the wheel hub endfloat is adjusted as described in Chapter 7 Section 5
(b) Take care not to cross-thread the brake pipe when reconnecting the wheel cylinder
(c) Bleed the brake system and top up the hydraulic fluid reservoir, see Section 3
(d) Centralise and adjust the brakes by operating the brake pedal several times

10 Hydraulic fluid pipes – inspection and overhaul

1 Periodically, and certainly well in advance of the DoE (MoT) test, if due, all brake pipes, connections and unions should be carefully examined.
2 Examine first all the unions for signs of leaks. Then look at the flexible hoses for signs of fraying and chafing (as well as for leaks). This is only a preliminary inspection of the flexible hoses as exterior condition does not necessarily indicate interior condition which will be considered later.
3 The steel pipes must be examined equally carefully. They must be thoroughly cleaned and examined for signs of dents or other percussive damage, rust and corrosion. Rust and corrosion should be scraped off and, if the depth of pitting in the pipes is significant, they will need replacement. This is most likely in those areas underneath the chassis and along the rear suspension arms where the pipes are exposed to the full force of road and weather conditions.
4 If any section of pipe is to be removed, first of all take off the fluid reservoir cap, and place some polythene film over the filler neck aperture and secure with an elastic band. Sealing the system in this manner will minimise the amount of fluid dripping out of the system when the pipes are removed.
5 Rigid pipe removal is usually quite straightforward. The unions at each end are undone and the pipe drawn out of the connection. The clips which may hold it to the car body are bent back and it is then removed. Underneath the car exposed unions can be particularly stubborn, defying the efforts of an open ended spanner. As few people will have the special split ring spanner required, a self-grip wrench (Mole) is the only answer. If the pipe is being renewed new unions will be provided. If not then one will have to put up with the possibility of burring over the flats on the union and use a self-grip wrench for replacement also.
6 Flexible hoses are always fitted to a rigid support bracket where they join a rigid pipe, the bracket being fixed to the chassis or rear suspension arm. The rigid pipe unions must first be removed from the flexible union. Then the locknut securing the flexible pipe to the bracket must be unscrewed, releasing the end of the pipe from the bracket. As these connections are usually exposed they are more often than not rusted up and a penetrating fluid is virtually essential to aid removal. When undoing them, both halves must be supported as the bracket is not strong enough to support the torque required to undo the nut and can easily be snapped off.
7 Once the flexible hose is removed examine the internal bore. If clear of fluid it should be possible to see through it. Any specks of rubber which come out, or signs of restriction in the bore, mean that the inner lining is breaking up and the pipe must be replaced.
8 Rigid pipes which need replacement can usually be purchased at any local garage where they have the pipe, unions and special tools to make them up. They will need to know the pipe length required and the type of flare used at the ends of the pipe. These may be different at each end of the same pipe. It is very important to note that all pipe flares and union and cylinder bores are to metric standards. Always test the compatibility of threads initially by screwing them together by hand.
9 Installation of the pipes is a reversal of the removal procedure. The pipe profile must be pre-set before fitting. Any acute bends must be put in by the garage on a bending machine otherwise there is the possibility of kinking them and restricting the fluid flow.
10 With the pipes refitted, remove the polythene from the reservoir, top up and bleed the system as described in Section 3.

11 Master cylinder (ATE type) – removal, overhaul and refitting

For Bendix type master cylinder, refer to Supplement

1 If the wheel hydraulic cylinders are in order and there are no leaks elsewhere, yet the brake pedal still does not hold under sustained pressure then the master cylinder seals may be presumed to be ineffective. To renew them the master cylinder must be removed.
2 Remove the filler cap and place out of the way together with the fluid level indicator (photo).
3 Hydraulic fluid is harmful to paintwork. Try to avoid spillage and take steps to protect the paintwork – place rags under the master cylinder prior to removal.
4 Unscrew the hydraulic pipe unions and push the pipes carefully to one side, just enough to allow removal of the master cylinder.
5 Remove the two master cylinder securing nuts.
6 Lift the unit away wrapped in a piece of rag to stop fluid dripping onto the paintwork. Empty the reservoir contents into a clean container.
7 To dismantle the master cylinder first pull off the reservoir using a rocking motion to release it from the rubber sleeves.
8 Push the piston inwards using a thin wood dowel and unscrew the stop screw.
9 Extract the snap-ring from the end of the cylinder and remove the primary and secondary piston assemblies, noting the order of assembly. The piston assemblies are best removed by blowing compressed air into the cylinder.
10 Clean the respective parts in methylated spirit ready for inspection.
11 New seals will have to be fitted to the piston assemblies so discard the old seals. Check the cylinder bore for signs of scoring or excessive wear. Renew any defective or suspect components.
12 Reassembly is a reversal of the dismantling process. Lubricate all components in hydraulic fluid prior to assembly, the utmost cleanliness is essential at all times. When the piston assemblies are inserted into the cylinder retain them with the stop washer and snap-ring.
13 When the cylinder is refitted to the car top up with new hydraulic fluid and bleed the system as described in Section 3.
14 Adjust the brake pedal rod as described in Section 12 for models equipped with a servo unit or as indicated in Fig. 8.12 for non-servo models.

12 Servo unit – general and maintenance

1 The servo unit (when fitted) is located between the bulkhead and the brake master cylinder. It provides vacuum assistance to the brake pedal and reduces the effort required by the driver to operate the brakes.
2 The unit operates from vacuum obtained from the inlet manifold and consists of a booster diaphragm and control valve assembly.
3 Maintenance and adjustments are minimal. There is a filter round the pushrod from the brake pedal and its replacement is one of two service operations that are permitted, the other being the one-way valve replacement. The servo unit is normally a reliable component and will give few if any problems. Should a fault develop in the unit this will be noticed by the additional effort needed to operate the brakes, although if the fault is progressive this may not be the case.

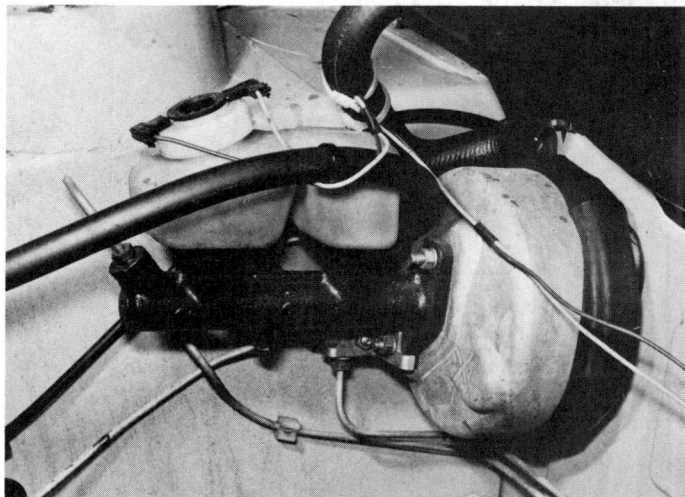

11.2 General view of the master cylinder and servo unit

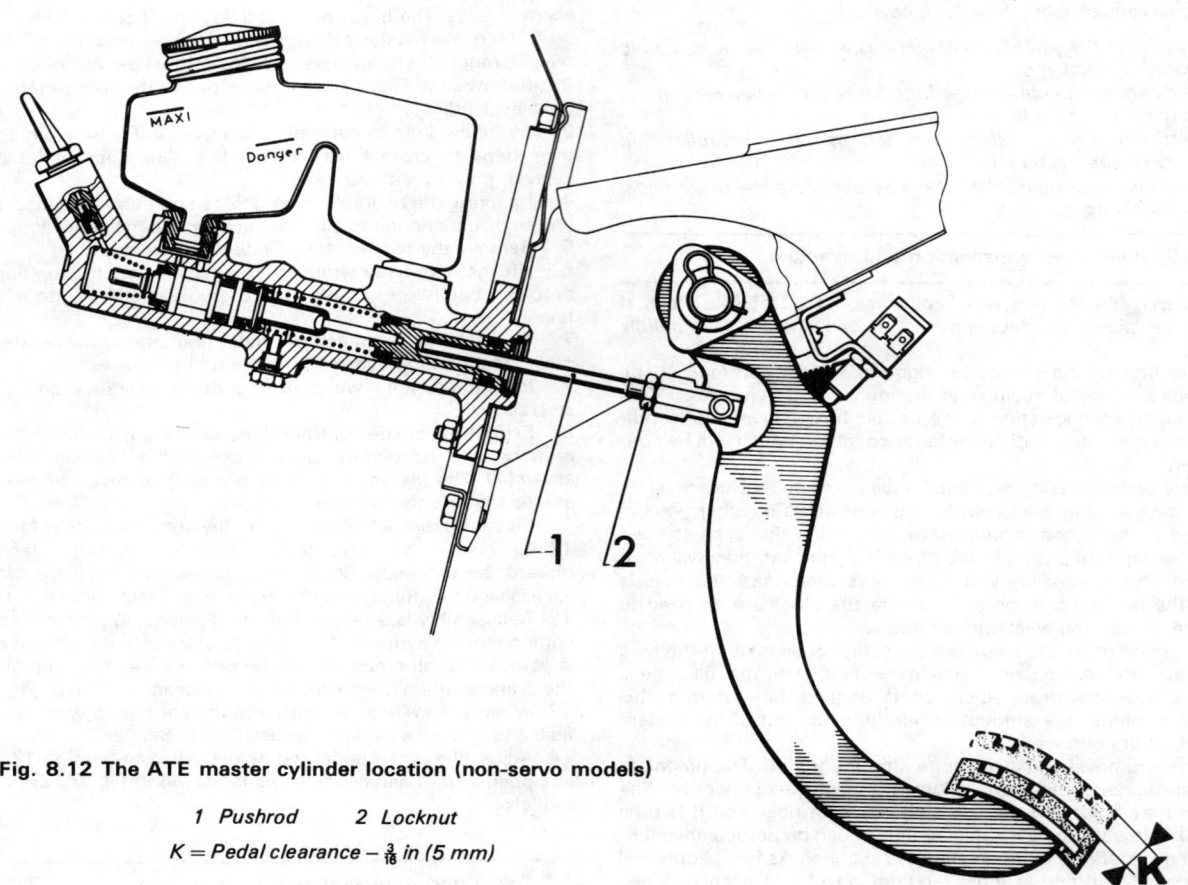

Fig. 8.12 The ATE master cylinder location (non-servo models)

1 Pushrod 2 Locknut
K = Pedal clearance – 3/16 in (5 mm)

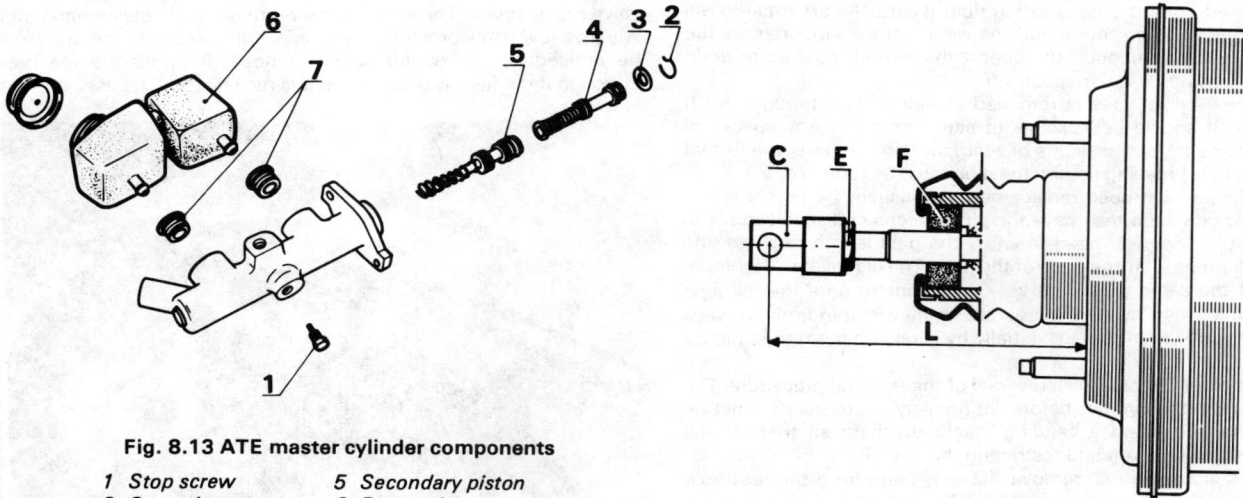

Fig. 8.13 ATE master cylinder components

1 Stop screw
2 Snap-ring
3 Stop washer
4 Primary piston
5 Secondary piston
6 Reservoir
7 Rubber sleeves

Fig. 8.14 The pushrod and clevis (C) showing locknut (E) and filter (F). Check adjustment measurement (L)

Chapter 8 Braking system

4   If a fault is suspected have the unit checked by your Renault dealer who has the necessary vacuum testing equipment. Never dismantle and attempt a repair at home — its doubtful whether you will get the replacement parts required.

Changing the air filter

5   The servo unit can stay in position for filter replacement. Refer to Fig. 8.14. Disconnect the clevis from the brake pedal and unscrew the clevis locknut. Unscrew the clevis and remove from the pushrod together with the locknut.

6   The filter retaining spring can now be extracted from the housing and the filter withdrawn along the pushrod using a hooked scriber or similar instrument.

7   Install the new filter and retaining clip. The clevis must be screwed in to adjust the pedal position and this should be measured between the center of the clevis pin hole to the servo unit (Fig. 8.14). When the correct length is obtained (3.812 in — 97 mm), secure in position by tightening the locknut.

Replacing the one-way valve

8   The one-way valve can be easily replaced if defective. Detach the vacuum intake pipe to the servo unit. The valve can then be extracted by pulling and twisting from the sealing washer. If the rubber seal washer is perished or cracked renew this as well. Replace in the reverse order.

9   The only other maintenance tasks on the servo unit are to check that the vacuum hoses and connections are secure and in good condition.

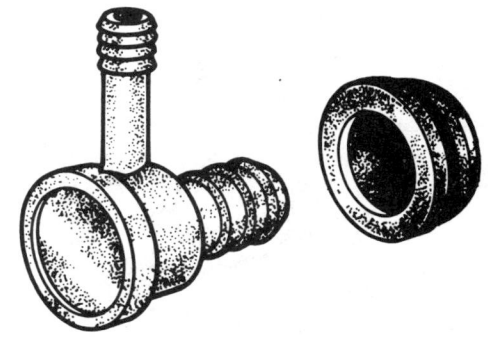

Fig. 8.15 The one way valve and sealing washer

13 Servo unit – removal and installation

1   Remove the master cylinder – see Section 11.
2   Disconnect the pushrod clevis from the brake pedal.
3   Detach the vacuum hose from the intake manifold.
4   Unscrew the servo unit retaining nuts and withdraw the unit.
5   Install in the reverse order but note the following:

  (a) Check the dimension (X) between the end of the pushrod and the master cylinder mounting face of the servo unit before bolting the master cylinder into position. This should be 9.0 mm (0.35 in) to ensure the correct clearance between servo push-rod and the end of the primary piston. Adjust if necessary by turning the domed nut (P)
  (b) Top up the hydraulic fluid and bleed the system (see Section 3)
  (c) Check and adjust the pedal clevis adjustment (see Section 12, para 7)

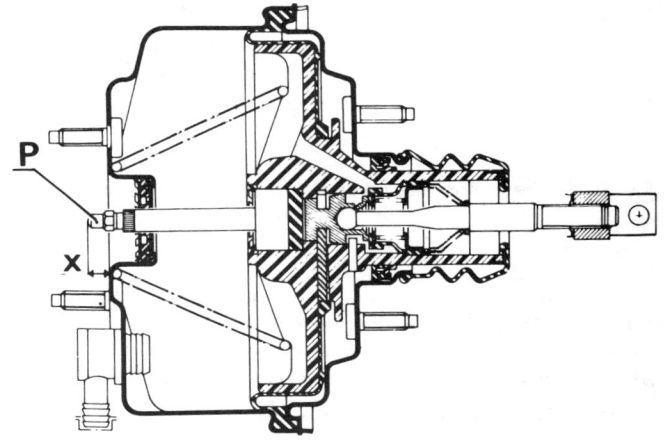

Fig. 8.16 Cross section view of the servo unit

P pushrod and adjustment nut
X Adjustment = 0.354 in (9 mm)

14 Brake limiter – general

1   A hydraulic fluid pressure limiting device is incorporated into the

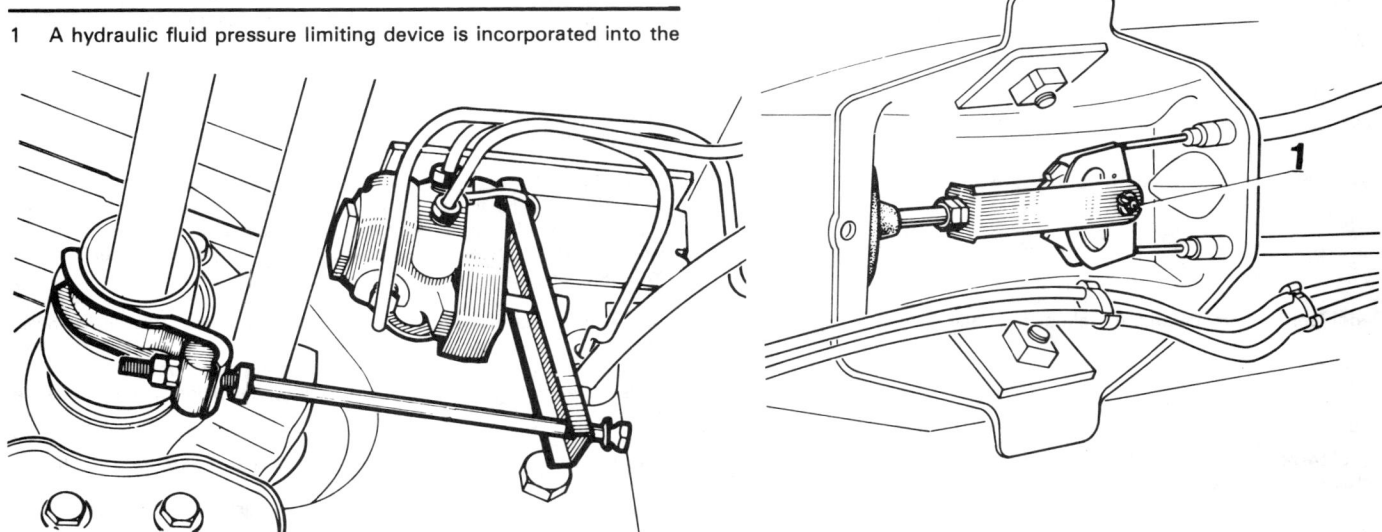

Fig. 8.17 The brake limiter

Fig. 8.18 Remove the clevis pin (1)

Chapter 8 Braking system

system and is located underneath the car just forward of the rear torsion bars.

2 It is designed to prevent the rear wheels from locking up under extreme braking conditions. Although it is adjustable, a brake fluid pressure gauge is required and therefore any adjustments which are made must be undertaken by your Renault dealer.

3 If the limiter unit is thought to be defective have your Renault dealer check it. It cannot be repaired and if defective a new replacement will have to be fitted. As the new limiter unit will have to be adjusted to the required operating pressure when installed, this again should be entrusted to your Renault agent.

15 Handbrake system

Primary rod and handbrake lever – removal and refitting

1 Unscrew the clevis rod locknut and adjuster nut. Extract the clevis pin (Fig. 8.18).
2 From inside the car prise the handbrake lever rubber shroud from the floor. Unscrew and remove the two lever retaining bolts and lift the lever clear together with the primary rod.
3 Installation is a direct reversal of the removal procedure but check the adjustment (see Section 4).

Handbrake secondary cable – renewal

4 Remove the clevis pin from the cable holder. Remove the holder and release the cable. Remove the guide sleeve. Refer to Section 8 and remove the brake drums and disconnect the brake cable from the shoe operating levers. Remove the cable with backplate cable sleeves.
5 Install in the reverse sequence and re-adjust the brakes (see Section 4).

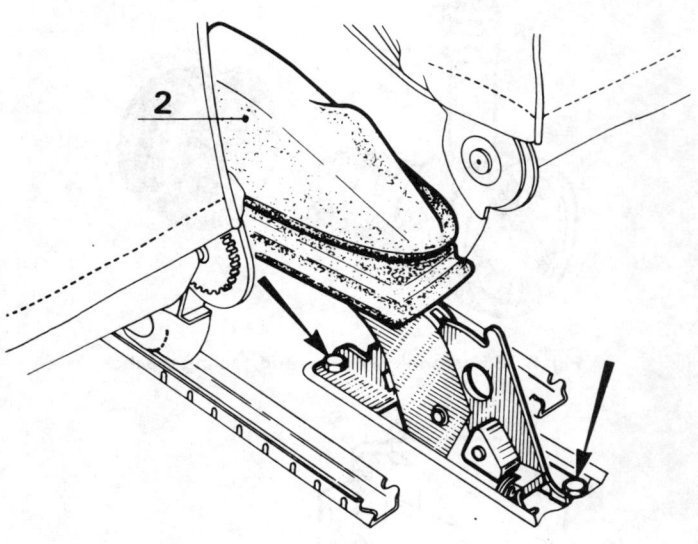

Fig. 8.19 The handbrake lever showing shroud (2) and retaining bolt positions

16 Fault diagnosis – braking system

Before diagnosing faults from the following chart, check that any braking irregularities are not caused by:

(a) Uneven and incorrect tyre pressures
(b) Incorrect mix of radial and crossply tyres
(c) Wear in the steering mechanism
(d) Defects in the suspension
(e) Misalignment of the chassis geometry

Symptom	Reason/s
Pedal travels a long way before the brakes operate	Automatic adjusters inoperative
Stopping ability poor, even though pedal pressure is firm	Linings/pads and/or drums/disc badly worn or scored One or more wheel hydraulic cylinders or caliper pistons seized, resulting in some brake shoes/pads not pressing against the drums/discs Brake linings/pads contaminated with oil Wrong type of linings/pads fitted (too hard) Brake shoes/pads wrongly assembled
Car veers to one side when the brakes are applied	Brake linings/pads on one side are contaminated with oil Hydraulic wheel cylinder(s)/caliper on one side partially or fully seized A mixture of lining materials fitted between sides Unequal wear between sides caused by partially seized wheel cylinders/calipers
Pedal feels springy when the brakes are applied	Brake linings/pads not bedded into the drums/discs (after fitting new ones) Master cylinder or brake backplate mounting bolts loose Severe wear in brake drums/discs causing distortion when brakes are applied
Pedal travels right down with little or no resistance and brakes are virtually non-operative	Leak in hydraulic systems resulting in lack of pressure for operating wheel cylinders/caliper pistons If no signs of leakage are apparent all the master cylinder internal seals are failing to sustain pressure
Binding, juddering, overheating	One or a combination of causes given in the foregoing sections

Chapter 9 Electrical system

For modifications, and information applicable to later models, see Supplement at end of manual

Contents

Alternator – description and precautions ... 4	Horn – general ... 21
Alternator – removal and refitting ... 5	Instrument panel – removal and refitting ... 16
Battery charging ... 3	Interior light bulb – replacement ... 15
Battery – maintenance, removal and refitting ... 2	Rear number plate light bulb – replacement ... 14
Combination lights (front) – removal and bulb replacement ... 11	Reversing light switch – removal and refitting ... 24
Combination lights (rear) – removal and bulb replacement ... 12	Speedometer cable – removal and refitting ... 27
Combination light switch – removal and refitting ... 17	Starter motor – dismantling and reassembly ... 8
Direction indicator switch – removal and refitting ... 18	Starter motor drive pinion – inspection and repair ... 9
Engine oil pressure warning switch – removal and refitting ... 25	Starter motor – removal and refitting ... 7
Fault diagnosis – electrical system general ... 28	Starter motor – testing ... 6
Fuel tank sender unit – removal and refitting ... 22	Windscreen washer unit – general ... 23
Fuses – general ... 10	Windscreen wiper motor and mechanism – removal and refitting ... 20
General description ... 1	Windscreen wipers and drive motor – fault diagnosis ... 19
Hazard warning light switch – general ... 26	
Headlights and bulbs – removal, refitting and adjustment ... 13	

Specifications

Battery
Voltage ... 12 volt
Polarity ... Negative (-) earth

Alternator
Type ... Paris-Rhone A12R8 or Ducellier 7591
Voltage ... 12 volt
Rating ... 35A
Rotor resistance – between rings ... 5.4 ohms

Regulator ... 8371A or 72710212

Starter motor ... Paris Rhone D8E137 or Ducellier 6247

1 General description

The electrical system operates at 12 volts DC and the major components, excluding the ignition circuits are:

(a) Battery 12 volt negative earth
(b) Alternator, driven by a V-belt from the crankshaft
(c) Regulator
(d) Starter motor, pre-engaged, mounted at the front of the engine

The battery supplies current for the ignition, lighting and other circuits and provides a reserve of power when the current consumed by the equipment exceeds that produced by the alternator.

The starter motor places very heavy demands on the power reserve. The alternator uses engine power to produce electricity to recharge the battery and the rate of charge is automatically controlled by a regulator. This regulator keeps the power output of the alternator within its capacity (an uncontrolled alternator can burn itself out) and adjusts the voltage and current output depending on the state of the battery charge and the electrical demands being made on the system at any one time.

2 Battery – maintenance, removal and refitting

1 Any new battery, if properly looked after, will last for two years at least (provided also that the alternator and regulator are in correct order).
2 The principal maintenance requirements are cleanliness and regular topping up of the electrolyte level with distilled water. Each week the battery cell cover or caps should be removed and just enough water added, if needed, to cover the tops of the separators (photo). Do not overfill with the idea of topping up lasting longer – it will only dilute the electrolyte and with the level high the likelihood of it gassing out is increased. This is the moisture one can see on the top of a battery. Little and often is the rule.
3 Wipe the top of the battery carefully at the same time removing all traces of moisture. Paper handkerchiefs are ideal for the job.
4 Every three months disconnect the battery terminals and wash

Chapter 9 Electrical system

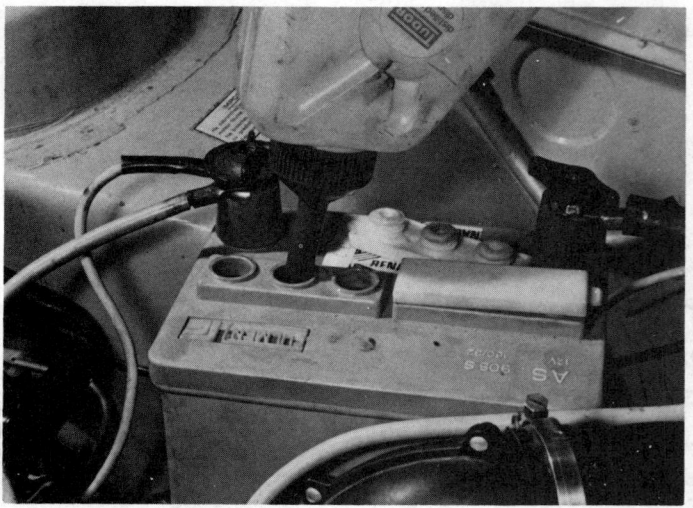

2.2 Topping up the battery

both the posts and lead connectors with a washing soda solution. This will remove any deposits. Dry them off and smear liberally with petroleum jelly – not grease, before reconnection.

5   Battery removal is simply a matter of disconnecting the terminal fittings, slackening the battery retaining clamp and lifting it out. Always undo the earth terminal clamp first and when reconnecting replace it last. In this way there is no danger of short circuiting the other terminal to earth. Always carry and place the battery in an upright position so as to prevent spillage of the electrolyte.

6   If a significant quantity of electrolyte is lost through spillage it will not suffice to merely refill with distilled water. Empty out all the electrolyte into a glass container and measure the specific gravity. Electrolyte is a mixture of sulphuric acid and water and the ready made solution should be obtainable from battery specialists or large garages. The normal solution can be added if the battery is in a fully charged state. If the battery is in a low state of charge, use the normal solution, then charge the battery, empty out the electrolyte, swill the battery out with clean water and then refill with a new charge of normal electrolyte.

3   Battery – charging

1   In winter certain conditions may result in the battery being used in excess of the alternators ability to recharge it in the running time available. This situation does not often occur on cars fitted with alternators which have output at low revolutions.

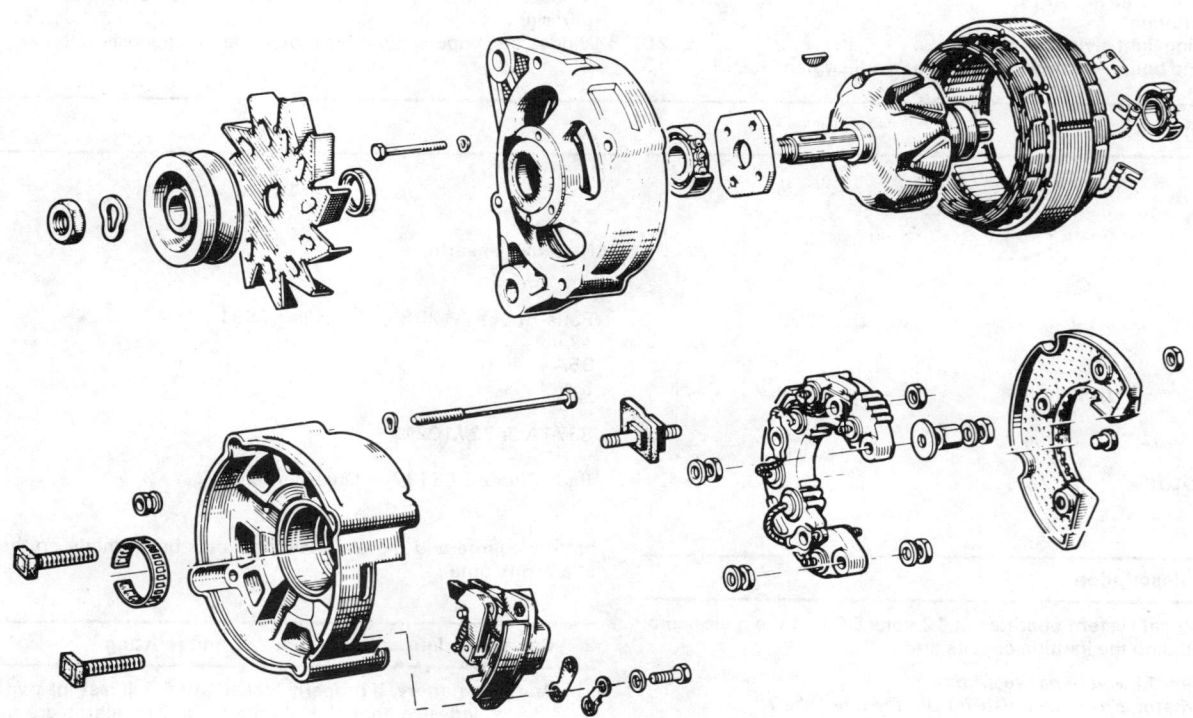

Fig. 9.1 The A12 R8 alternator components

Chapter 9 Electrical system

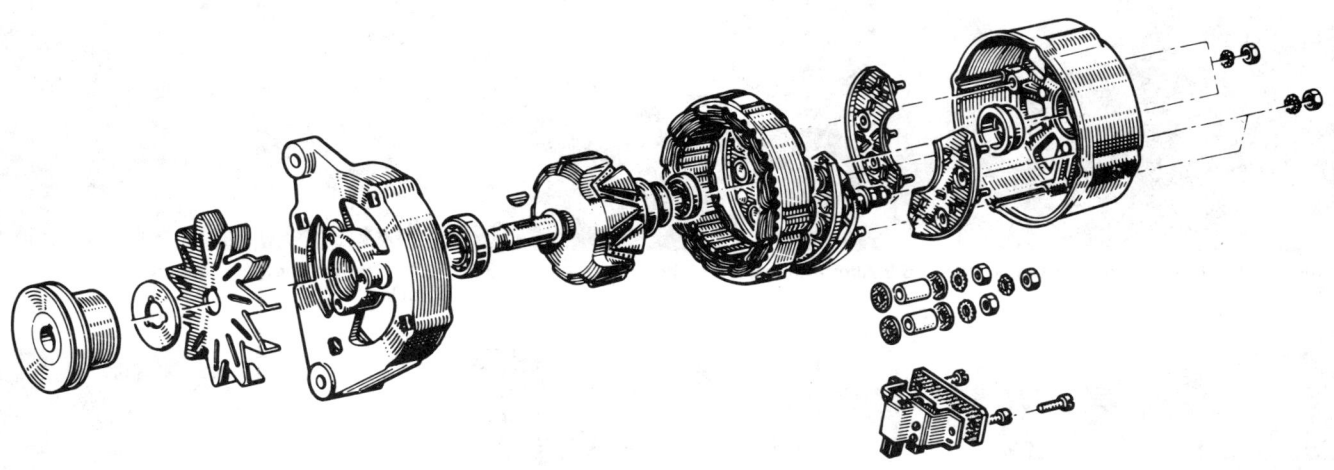

Fig. 9.2 The Type 7591 alternator components

2 An external charging source may be needed to keep the battery power reserve at the proper level. If batteries are being charged from an external source a hydrometer is used to check the electrolyte specific gravity. Once the fully charged reading is obtained charging should not continue for a period in excess of four hours. Most battery chargers are set to charge at 3-4 amps initially and as the battery charge builds up this reduces automatically to 1-2 amps. The table below gives details of the specific gravity readings at 21°C/70°F. Do not take readings just after topping up, just after using the starter motor, or when the electrolyte is too cold or too warm. The variation in SG readings is 0.004 for every 6°C/10°F change – the higher readings being for the lower temperatures.

Specific gravity	Battery state of charge
1·28	100%
1·25	75%
1·22	50%
1·19	25%
1·16	Very low
1·11	Discharged completely

4 Alternator – description and precautions

1 All models imported to the United Kingdom are fitted with alternators in place of the well known DC dynamos. The alternator generates alternating current (AC) which is rectified by diodes into DC and is the current needed for battery storage.

2 The regulator is a transistorized unit which is permanently sealed and requires no attention. It will last indefinitely provided no mistakes are made in wiring connections. It is mounted on the left-hand inner wing. On later models, the alternator has an integral regulator (see Supplement).

3 Apart from the renewal of the rotor slip ring brushes there are no other parts which need periodic inspection. All other items are sealed assemblies and must be replaced if indications are that they are faulty.

4 If there are indications that the charging system is malfunctioning in any way, care must be taken to diagnose faults properly, otherwise damage of a serious and expensive nature may occur to parts which are in fact quite serviceable.

5 The following basic requirements must be observed at all times, therefore, if damage is to be prevented:

(a) ALL alternator systems use a NEGATIVE earth. Even the simple mistake of connecting a battery the wrong way round could burn out the alternator diodes in a few seconds

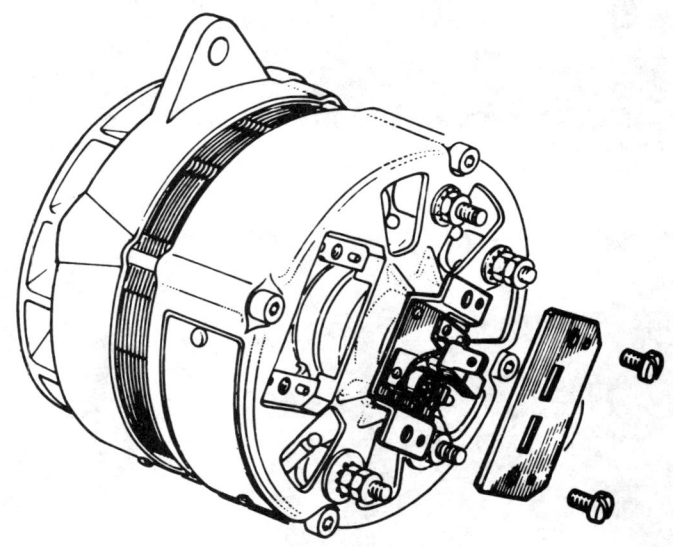

Fig. 9.3 The brush holder unit on the Type 7591

(b) Before disconnecting any wires in the system the engine and ignition circuits should be switched off. This will minimise accidental short circuits

(c) The alternator must NEVER be run with the output wire disconnected

(d) Always disconnect the battery from the car's electrical system if an outside charging source is being used

(e) Do not use test wire connections that could move accidentally and short circuit against nearby terminals. Short circuits will not blow fuses – they will blow diodes or transistors

(f) Always disconnect the battery cable and alternator output wires before any electric welding work is done on the car body

6 Fault diagnosis on alternator charging systems requires sophisticated test equipment and even with this the action required to rectify any fault is limited to the renewal of one or two components. Knowing what the fault is is only of academic interest in these circumstances.

4.9 Removing the brush holder on the A12R8 alternator

5.3 Alternator mounting bolts

7.4 Starter motor rear mounting location bracket

7.7 Refitting the starter motor

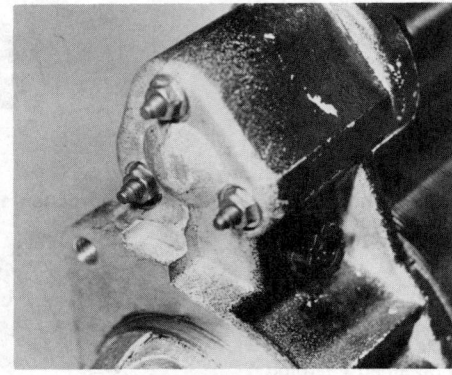
8.2a Remove solenoid retaining nuts ...

8.2b ... detach wire connection ...

8.2c ... and withdraw solenoid

8.3 Remove the end cover

8.4 Withdraw the through-bolts

8.5 Detach the brush holder

8.7 Pinion/starter gear lever pivot pin

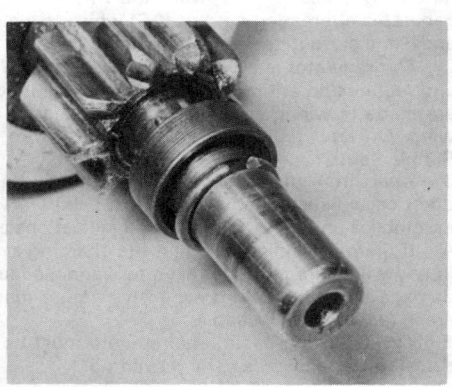
8.8 Tap back the collar washer and remove the snap-ring

Chapter 9 Electrical system

7 Rotor slip ring brushes are easily removed and replaced, however the rotor shaft bearings should be left to an auto electrician.
8 To change the brushes on the Type 7591 alternator, unscrew and remove the brush holder retaining screws and detach the insulating cover. The brush holder can then be withdrawn. Replace in the reverse order checking that all connections are secure.
9 To change the brushes on the Type A12R8 alternator, simply unscrew the brush carrier retaining bolts and remove the carrier (photo). Refit the replacement unit in reverse and reconnect the terminals.

5 Alternator – removal and refitting

1 Disconnect the battery earth terminal.
2 Detach the wire connector to the alternator.
3 Loosen the alternator mounting bolts (photo) and the drivebelt adjustment strap. Detach the drivebelt from the alternator pulley by pivoting the alternator towards the engine.
4 Now remove the retaining bolts and lift the alternator clear.
5 Install in the reverse order and readjust the drivebelt tension as described in Chapter 2.

6 Starter motor – testing

1 If the starter motor fails to turn the engine when the switch is operated there are four possible reasons why:

(a) The battery is no good
(b) The electrical connections between switch, solenoid battery and starter motor are somewhere failing to pass the necessary current from the battery through the starter to earth
(c) The solenoid switch is no good
(d) The starter motor is either jammed or electrically defective

2 To check the battery, switch on the headlights. If they go dim after a few seconds, the battery is definitely unwell. If the lamps glow brightly, next operate the starter switch and see what happens to the lights. If they go dim then you know that power is reaching the starter motor but failing to turn it. Therefore, check that it is not jammed. The starter will have to come out for examination.
3 If when the starter switch is operated, the lights stay bright, then power is not reaching the starter. Check all connections from battery to solenoid switch or cable to starter for perfect cleanliness and tightness. With a good battery installed this is the most usual cause of starter motor problems. Check that the earth link cable between the engine and frame is also intact and cleanly connected. This can sometimes be overlooked when the engine is taken out.
4 If no results have yet been achieved turn off the headlights, otherwise the battery will go flat. The solenoid contact can be checked by putting a voltmeter or bulb across the main cable connection on the starter side of the solenoid and earth. When the switch is operated, there should be a reading or lighted bulb. If not, the solenoid switch is no good. If finally, it is established that the solenoid is not faulty and 12 volts are getting to the starter then the starter motor must be the culprit.

7 Starter motor – removal and refitting

1 Disconnect the battery earth terminal.
2 The starter motor will have to be removed from underneath the vehicle. Raise the front and support with axle stands or blocks. Alternatively run it onto ramps or over an inspection pit.
3 Disconnect the starter leads from the solenoid.
4 Unscrew and remove the two starter motor mounting bracket-to-engine bolts (photo).
5 Now unscrew the three mounting bolts from the clutch housing. Use an extension wrench inserted through the aperture in the left-hand side inner wing panel.
6 Withdraw the starter motor from the clutch housing.
7 The starter motor is refitted in the reverse order (photo).

8 Starter motor – dismantling and reassembly

1 Such is the inherent reliability and strength of the starter motors fitted that it is very unlikely that a motor will ever need dismantling until it is totally worn out and in need of replacement. It is not a task for the home mechanic because although reasonably easy to undertake, the reassembly and adjustment before refitting is beyond his scope because of the need of specialist equipment. It would under all circumstances be realistic for the work to be undertaken by the specialist auto-electrician. It is possible to replace solenoids and brushes on starter motors.
2 Remove the solenoid by undoing the holding nuts and unhooking the solenoid plunger from the lever. Mark the relative positions of the solenoid body to pinion housing (photos).
3 Unscrew the two endplate nuts and remove the cover at the opposite end to the pinion (photo).
4 Now unscrew the two through-bolts. These may be tight due to being treated with thread locking fluid when assembled, so take care and use the correct size spanner (photo).
5 Withdraw the brush holder plate from the body (photo).
6 The pinion housing and armature can be withdrawn from the main body.
7 To dismantle and inspect the pinion/starter gear unit the special lever pivot pin must be tapped out (photo).
8 To withdraw the pinion from the armature, tap back the special collar washer (photo) and extract the snap-ring from the groove in the shaft. The pinion assembly can then be slid from the shaft.
9 With the starter motor dismantled the various components can be cleaned and inspected for general wear and/or signs of damage. The principal components likely to need attention will be the brushes, the solenoid or possibly the drive pinion unit (photo).
10 The brushes can be removed by unsoldering the connecting wires to the holder and to the field coil unit. Take care not to damage the latter during removal and assembly of the brushes.

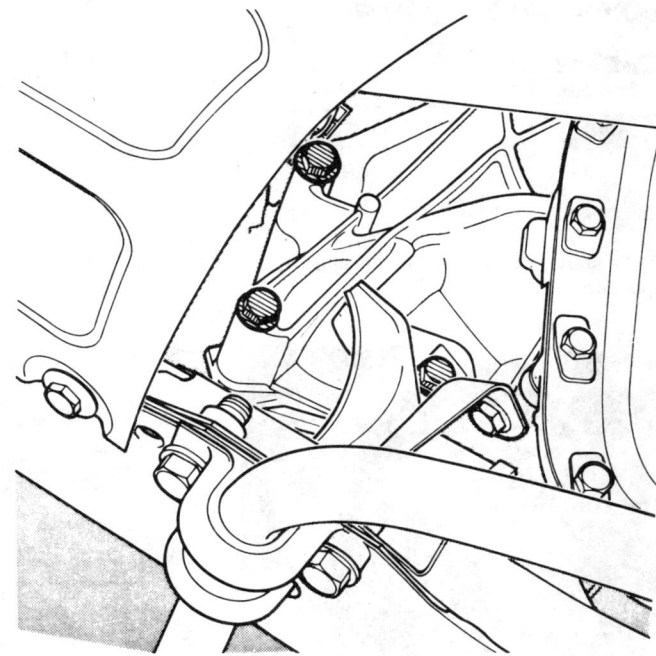

Fig. 9.4 The three starter motor mounting bolt positions (shaded)

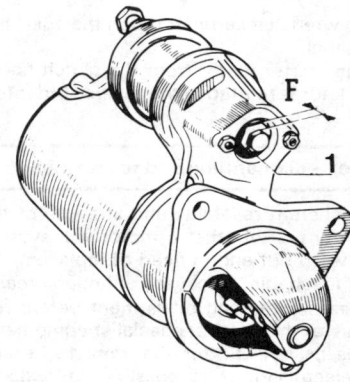

Fig. 9.4a Ducellier adjuster rod (1) and bolt to nut clearance (F)

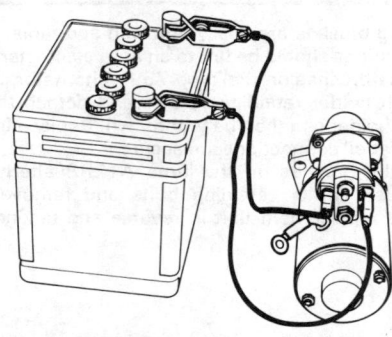

Fig. 9.4c Voltage applied to Paris-Rhone solenoid terminals

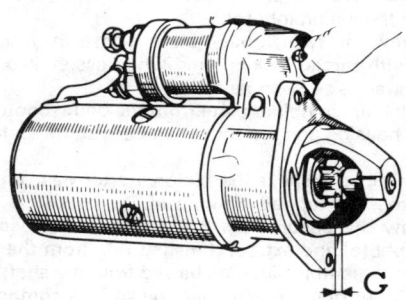

Fig. 9.4b Ducellier pinion clearance (G)

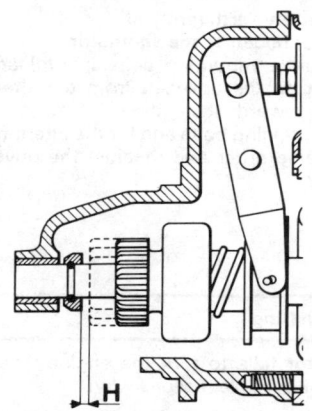

Fig. 9.4d Pinion clearance (H) on Paris-Rhone starter

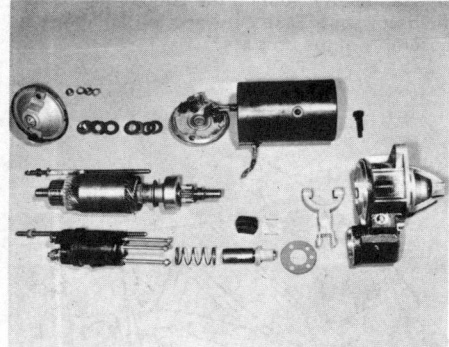

8.9 The starter motor components laid out for inspection

8.12 Undercutting the commutator separators

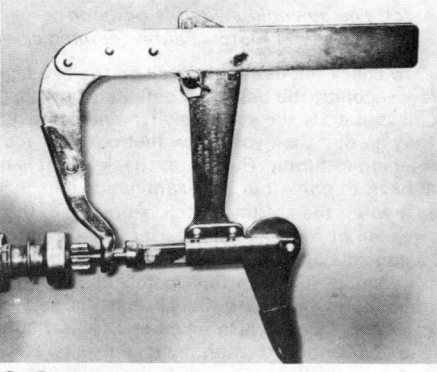

8.16a Using a valve compressor to reassemble the snap-ring and collar

8.16b Refit the thrust washers onto commutator shaft

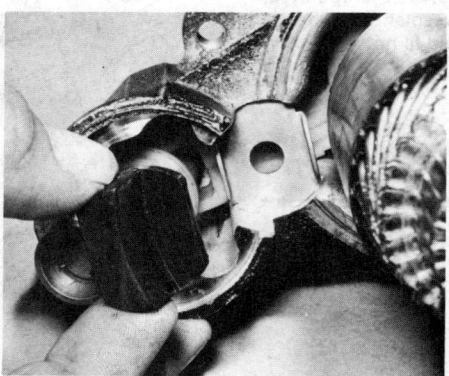

8.16c Install the separator and special washer as shown

8.16d Brush holder and body alignment notches

Chapter 9 Electrical system

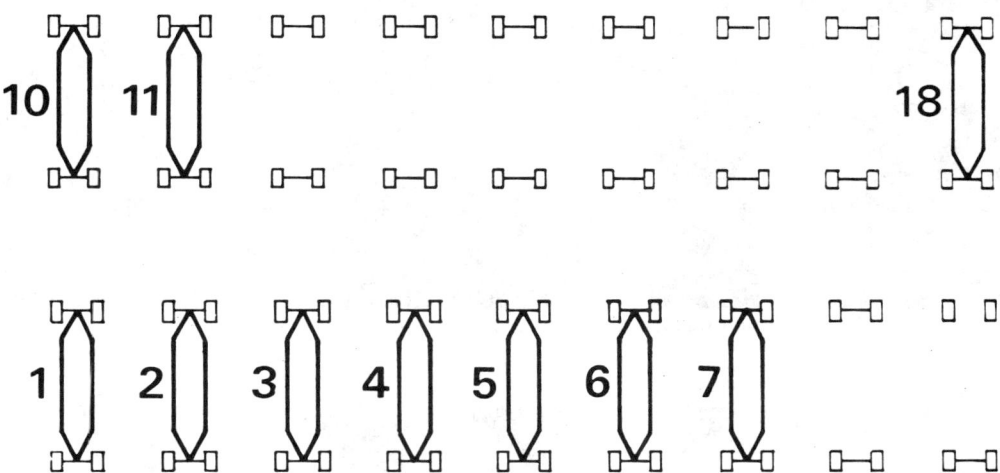

Fig. 9.5 The fuse locations for the various circuits

Fuse No	Circuit	Rating
1 and 2	Spares	
3	Instrument panel	5 amp
4	Spare	
5	Reversing light rear screen demist	8 amp
6	Wiper motor and switch	8 amp
7	Cigar lighter and interior light	8 amp
8 and 9	Spares	
10	Flasher unit	5 amp
11	Blower motor/stop light switch/radio	8 amp
12 to 18	Spares (or lights on some models)	

11 If the starter motor has shown a tendency to jam or a possible reluctance to disengage then the starter pinion is almost certainly the culprit. Dirt around the pinion and shaft could cause this and when cleaned check that the pinion can move freely in a spiral movement along the shaft. If the pinion tends to bind or is defective in any way renew it.

12 Undercut the separators of the commutator using an old hacksaw blade as shown (photo), to a depth of about 0·02 to 0·03 in (0·5 to 0·8 mm). The commutator may be further surface cleaned using a strip of very fine glass paper. Do not use emery cloth for this purpose as the carborundum particles will become embedded in the copper surfaces.

13 Testing of the armature is best left to an auto-electrician but if an ohmmeter is available it can be done by placing one probe on the armature shaft and the other on each of the commutator segments in turn. If continuity is indicated at any time during the test, then the armature is defective and must be renewed.

14 The field coil can also be tested using an ohmmeter. Connect one probe to the field coil positive terminal and the other to the positive brush holder. If there is no continuity then the field coil circuit has a break in it.

15 Connect one lead of the meter to the field coil positive lead and the other one to the yoke. If there is a low resistance then the field coil is earthed due to a breakdown in the insulation. If this proves to be the case the field coils must be renewed. As field coil replacement requires special tools and equipment it is a job that should be entrusted to your auto-electrician. In fact it will probably prove more economical and beneficial to exchange the starter motor for a reconditioned unit.

16 Reassembly of the rest of the starter motor is a direct reversal of the removal procedure, but note the following:

(a) The snap-ring and collar can be difficult to relocate and to assist in this we used a valve compressor as shown (photo)
(b) Reassemble the correct number of thrust washers (photo)
(c) Check that the brushes slide freely in their holders
(d) Lubricate the commutator shaft sparingly, using a medium grease
(e) Do not forget to install the separator and washer (photo)
(f) Realign the brush holder and body notches (photo)
(g) Realign the marks made on dismantling when assembling the solenoid, and fit the coil spring and washer as shown
(h) Check the pinion clearance. To do this on Ducellier starters, remove the plug from the front of the solenoid and check clearance (F) is as small as possible (nut to bolt head). The drive pinion should be up against the end of the armature. Now push the solenoid bolt and check clearance (G). This should be between 0.05 and 1.5 mm (0.002 and 0.059 in). Turn adjusting nut (1) to obtain the specified clearance.

With Paris-Rhone starter motors, apply battery voltage to the solenoid terminals as shown. Check that clearance (H) drive pinion to stop is 1.5 mm (0.059 in). Turn the eccentric on the fork spindle as necessary to adjust.

9 Starter motor drive pinion – inspection and repair

1 Persistent jamming or reluctance to disengage may mean that the starter pinion assembly needs attention. The starter motor should be removed first of all for inspection.

2 With the starter motor removed thoroughly remove all grime and grease with a petrol soaked rag, taking care to stop any liquid running into the motor itself. If there is a lot of dirt this could be the trouble and all will now be well.

3 If the preceding cleaning does not actually remove the fault the starter motor will need to be stripped down to its component parts and a further check made. This has been explained in the preceding Section.

10 Fuses – general

All vehicles are fitted with an adequate fuse system. The fuse box is mounted under the instrument panel (photo). The top can be lifted off and inspected easily. Always replace fuses with the correct type, see the end of this Section. They are a straight pull-out/push-in fit. Never think that you can leave them out, by-pass them or substitute a piece

10.1 The fuse holder

11.1 Remove the combination light lens (front)

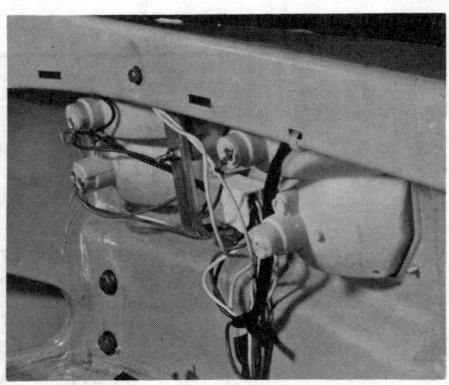
12.3 Wiring to rear combination light

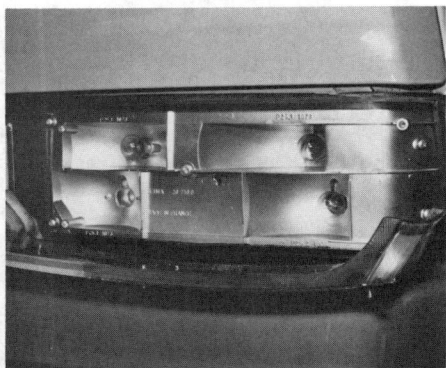
12.4 Rear combination light unit with lens removed

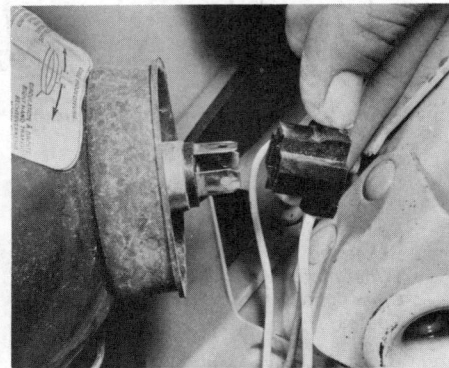
13.1a Detach the connector

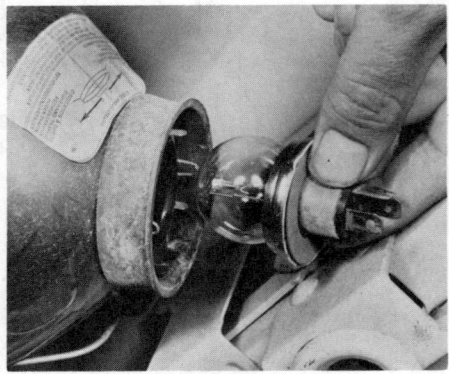
13.1b Extract the bulb

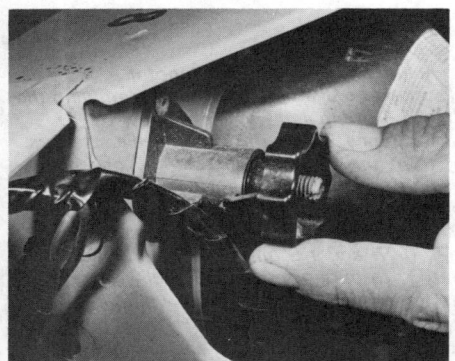
13.5a Height adjuster

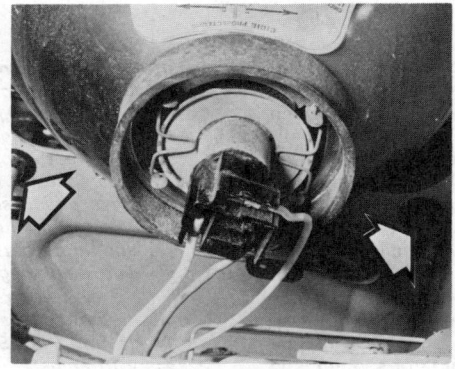
13.5b Horizontal plane adjustment screws

14.1 Number plate lens and bulb removal

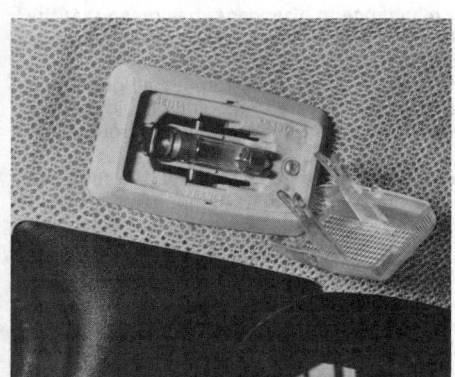
15.1 Interior light with lens removed

16.3 Withdraw the panel surround and switch panel

16.5 Panel side retaining clip

of tin foil or other conductive material unless in a real emergency. Replace the make-shift fuse as soon as possible. The layout of the fuse box is shown in Fig. 9.5 and the respective circuit and fuses are indicated accordingly.

11 Combination lights (front) – removal and bulb replacement

1 To gain access to the side light and indicator bulbs unscrew the lens retaining screws and detach the lens (photo).
2 Bayonet fixing bulbs are fitted and these are pushed and twisted anti-clockwise to remove.
3 If the light unit is to be removed, disconnect the battery earth terminal, then detach the wires to the rear of the combination light.
4 Remove the lens and unscrew the retaining screws. Withdraw the combination light unit.
5 Install in the reverse sequence and check the lights and indicators for correct operation.

12 Combination lights (rear) – removal and bulb replacement

1 Access to the rear combination light bulbs is obtained on removing the rear light lens, which is retained by screws.
2 Bayonet fixing bulbs are fitted and these are removed by pushing and twisting anti-clockwise.
3 To gain access to the wiring connections at the rear of the light units, the inside rear panel must be removed (see photo).
4 To remove the combination light unit first disconnect the battery earth terminal. Remove the lens cover (photo) and then the inside panel to enable the wiring to be detached. Unscrew the five retaining screws and withdraw the unit.
5 Install in the reverse order and check the operation of the respective lights on completion.

13 Headlights and bulbs – removal, refitting and adjustment

Bulb replacement
1 To change the headlight bulb, pull the connector from the rear of the unit (photo), hinge the bulb retaining clips away from the bulb and extract the bulb (photo). If renewing the bulb fit the sealing washer of the old bulb into place on the replacement.
2 When installing the new bulb note the offset boss to facilitate correct fitting. Close the retaining clips and re-insert the wires connector. Check the bulb for operation and adjustment.
3 Some models are fitted with a bulb rotator which enables the bulb to be set for driving on the right or left. This is shown in Fig. 9.6.

Headlight beam adjustment
4 The headlight beam adjustment is most important not only for your safety but for that of other road users. Accurate beam alignment can only be obtained using optical beam setting equipment as used by garages.
5 When the car is loaded a temporary beam height adjustment can be made using the setting knob shown in the accompanying photo. The horizontal plane of the beam is adjusted via the two screws, one each side of the headlight (photos). If the setting knob does not provide enough vertical adjustment, turn the screw in the centre of the knob as necessary.
6 When re-adjustment is made have the settings checked by your local garage or Renault dealer at the earliest opportunity.

Headlight unit removal
7 Disconnect the battery earth terminal. Detach the terminal at the rear of the bulb unit.
8 Unclip the spring retainer that secures the headlight unit at the rear (Fig. 9.7). Note that on later models, the spring retainer is replaced by a spring clip to one side of the headlight unit, which secures the direction indicator light. This clip should be released before removing the headlight unit.
9 The unit can now be withdrawn from the front of the vehicle. Give it a sharp pull to release it from the adjustment screws.
10 Install the replacement unit in the reverse order and check operation on completion.

14 Rear number plate light bulb – replacement

1 To replace a defective bulb in the number plate light unscrew the lens cover and unit retaining screws. Separate the lens from the light unit (photo).
2 To remove the bulb push it and twist it anti-clockwise to release the bayonet connection.
3 Install in the reverse order and check the light for operation on completion.
4 If the unit is to be removed repeat the above procedure and disconnect the wiring. Refit in the reverse order.

15 Interior light bulb – replacement

1 To replace the interior light bulb simply pull the lens from the light unit. The festoon bulb can now be extracted from between the two holders (photo).

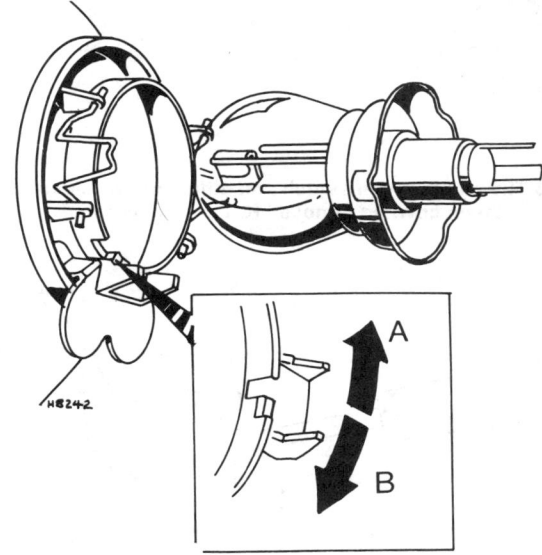

Fig. 9.6 The rotator bulb assembly for driving on the left (A) or driving on the right (B)

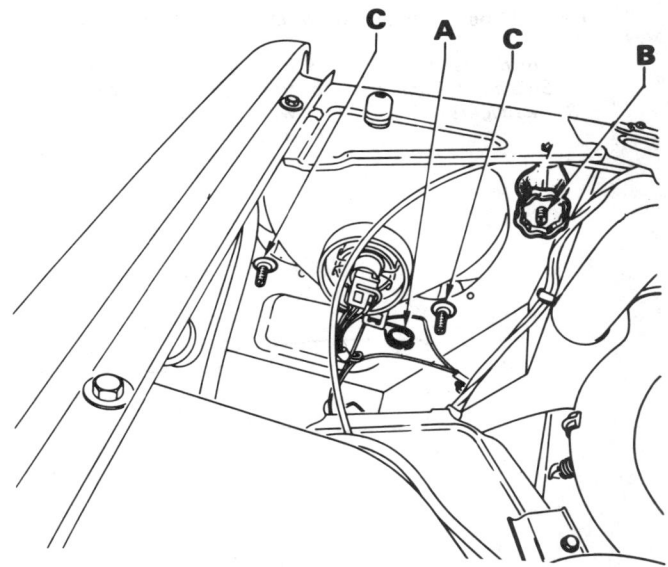

Fig. 9.7 Rear view of headlight unit

A Unit retaining spring
B Vertical adjuster
C Horizontal adjuster screws

Chapter 9 Electrical system

2 Install the new bulb and replace the lens by pushing into position.
3 If the light unit is to be removed, disconnect the battery earth terminal, remove the lens and unscrew the unit retaining screws and withdraw it. Detach the wires.
4 Refit in the reverse order and check operation.

16 Instrument panel – removal and refitting

1 Disconnect the battery earth terminal.
2 Unscrew and remove the steering column cowling noting the various screw positions.
3 Withdraw the combined instrument panel surround and switch panel by pulling outwards, then disconnect the switch wire connectors (photo).
4 Unscrew and remove the heater control panel.
5 The instrument panel can now be withdrawn. It is retained in position by a spring loaded clip on each side (photo). Press the retainers at the side and manoeuvre the panel outwards, then disconnect the junction blocks and the speedometer cable to remove completely.
6 To dismantle the panel once removed, detach the trip mileage zero knob and unclip the bezel. The instruments concerned can now be removed by unscrewing their retaining nuts.
7 Refit in the reverse sequence but ensure that the connections are secure and check the various instrument and warning lights for correct functioning on completion.

17 Combination light switch – removal and refitting

1 Disconnect the battery earth terminal.
2 Unscrew and remove the steering column cowling noting the various screw positions.
3 Detach the wiring connector from the harness to the combination switch.
4 Unscrew the retaining screws and remove the combination switch.
5 Installation is a direct reversal of the removal procedure, but check the lights for correct operation on completion.

18 Direction indicator switch – removal and refitting

1 Disconnect the battery earth terminal.
2 Unscrew and remove the steering column cowling noting the various screw positions.
3 Detach the wiring connector from the main harness to the indicator switch.
4 Unscrew and remove the indicator switch from the column.
5 Installation is a direct reversal of the removal procedure but check the indicators for correct operation on completion.

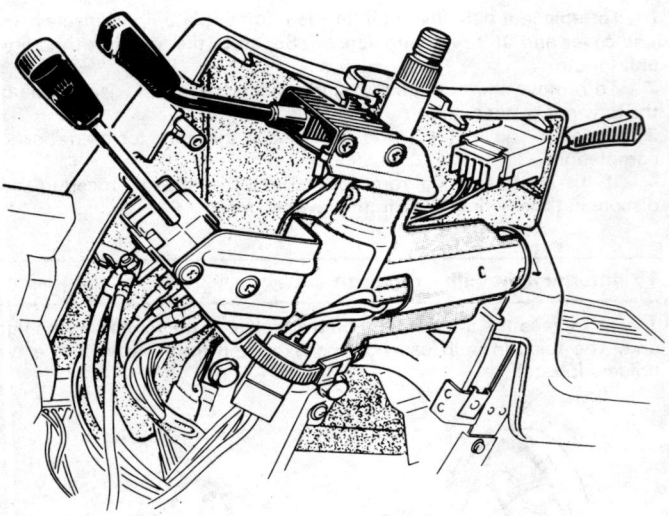

Fig. 9.8 The combination switch and indicator switch shown with lower cowling removed (LH drive illustrated)

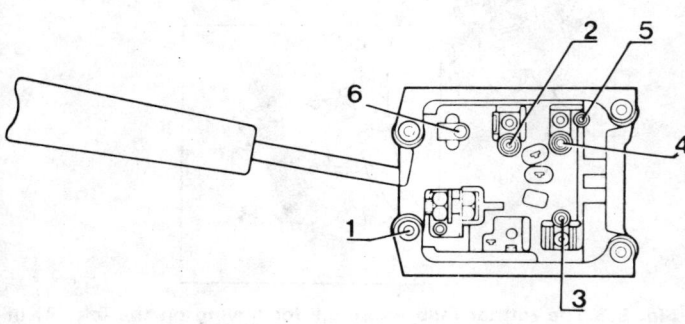

Fig. 9.9 The combination switch terminal positions

1 Positive feed
2 Side and tail lights
3 Headlights – dip
4 Headlights – main beam
5 Headlights – main beam warning light
6 Horn

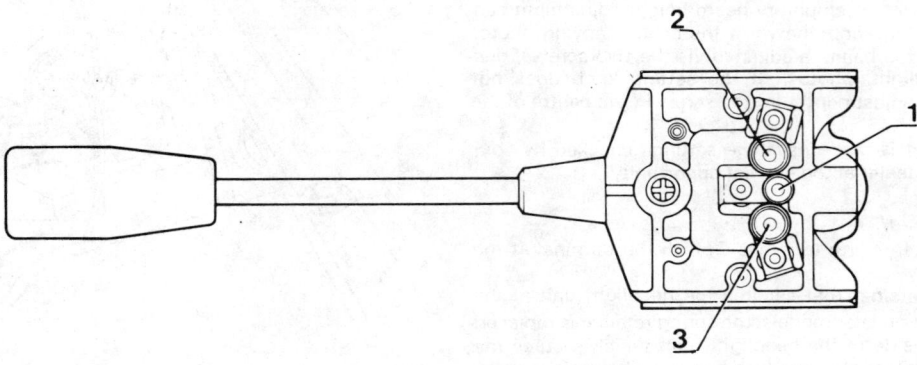

Fig. 9.10 The indicator switch terminal positions

1 Positive feed   2 Right-hand indicator   3 Left-hand indicator

19 Windscreen wipers and drive motor – fault diagnosis

1 If the wipers fail to operate first check that current is reaching the motor. This can be done by switching on and using a voltmeter or 12 volt bulb and two wires between the (+) terminal on the motor and earth.
2 If no current is reaching the motor check whether there is any at the switch. If there is then a break has occurred in the wiring between switch and motor.
3 If there is no current at the switch go back to the ignition switch and so isolate the area of the fault. The problem may simply be a blown or badly connected fuse.
4 If current is reaching the motor but the wipers do not operate, switch on and give the wiper arm a push – they or the motor could be jammed. Switch off immediately if nothing happens otherwise further damage to the motor may occur. If the wipers now run the reason for them jamming must be found. It will almost certainly be due to wear in either the linkage of the wiper mechanism or the mechanism in the motor gearbox.
5 If the wipers run too slowly it will be due to something restricting the free operation of the linkage or a fault in the motor. In such cases it is well to check the current being used by connecting an ammeter in the circuit. If it exceeds three amps something is restricting free movement. If less, then the commutator and brush gear in the motor are suspect. The shafts to which the wipers are attached run in very long bushes and often suffer from lack of lubrication. Weekly application to each shaft of a few drops of light penetrating oil, preferably, helps to prevent partial or total seizure. First pull outwards the rubber grommets, afterwards, wipe off any excess.
6 If wear is obviously causing malfunction or there is a fault in the motor it is best to remove the motor or wiper mechanism for further examination and repairs.

20 Windscreen wiper motor and mechanism – removal and refitting

1 Disconnect the battery earth terminal.
2 Remove the heater bulkhead closure plate screws. To move the plate for access to the wiper arm mechanism the heater valve must be unclipped. Slide the plate as far as possible along the heater hoses.
3 Detach the windscreen wiper arms from the pivot shafts.
4 Remove the mounting plate retaining nuts on the scuttle panel.
5 Detach the connector and remove the securing bolt (Fig. 9.11). Remove the mechanism unit.
6 To remove the motor unscrew and remove the crank arm retaining nut from the motor drive spindle. Unscrew the three motor unit retaining screws and withdraw the motor.
7 The wiper motor is normally a reliable unit. If it is known to be defective it should either be repaired by a competent auto-electrician or exchanged for a replacement unit.
8 Installation of the motor and wiper mechanism is a direct reversal of the removal procedure. Apply a smear of medium grease to the mechanism pivots and check the wiper operation before refitting the heater bulkhead closure plate.

21 Horn – general

1 The horn is normally a very reliable component and the only time that is likely to give problems is when it gets waterlogged. Being located at bumper level at the front (photo) this may be possible under adverse conditions. If it fails to operate check all the other parts in the circuit first – the push is much more likely to fail.
2 Should the horn fail to work the first thing to do is make sure that current is reaching the horn terminal. This can be done by connecting a 12 volt test lamp to the feed wire and pressing the horn button with the ignition switch on. If the bulb lights then the fault must lie in the horn or the horn mounting. The tightness and cleanliness of the horn mounting is important as the circuit is made to earth through the fixing bolt. The connections should, of course, be a clean, tight fit on the horn terminals.
3 If no current is reaching the horn check wiring connections as indicated in the wiring diagram.
4 If it is found that the fault lies in the horn unit then it will have to be replaced. Removal of all types of horn is simply the undoing of the

Fig. 9.11 The wiper motor securing bolt (A)

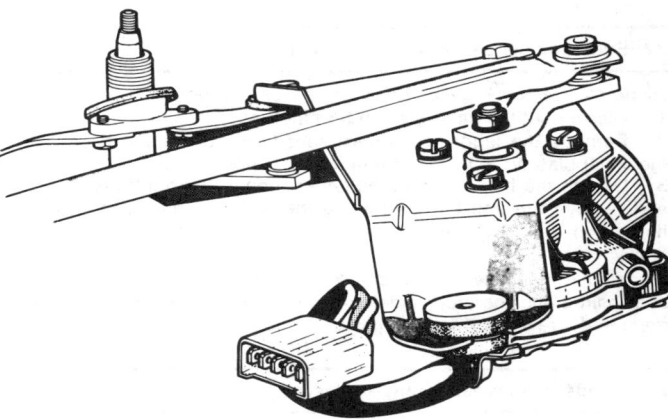

Fig. 9.12 The three wiper motor unit retaining screws and crank arm

21.1 Horn location (adjuster arrowed)

Chapter 9 Electrical system

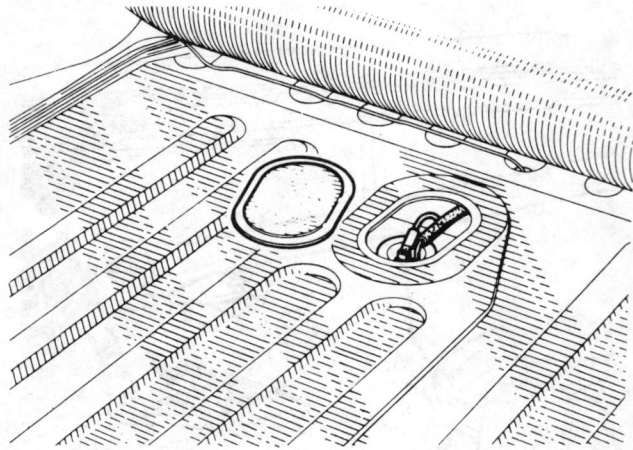

Fig. 9.13 The fuel tank sender unit aperture in floor panel

nut which secures the horn to the mounting arm, and disconnecting the wire connection from the horn terminal. The horn cannot be adjusted or repaired and must be renewed if defective.

22 Fuel tank sender unit – removal and refitting

Note: *Do not smoke when removing the sender unit!*
1 Disconnect the battery earth terminal.
2 Remove the luggage compartment floor covering and peel back the cover plate from the sender unit access hole.
3 Detach the fuel pipe and the wire connectors from the sender unit and remove the unit by unscrewing the retaining ring with a piece of flat metal and withdrawing it carefully through the aperture in the floor.
4 Your Renault dealer can check the unit for accuracy but it cannot be repaired and therefore if defective, renew it and install the replacement in the reverse order of removal.

23 Windscreen washer unit – general

1 The windscreen washer unit is operated by an electric pump (photo) which draws the cleaning fluid from the reservoir and pressurises the water nozzles.
2 Malfunction is usually due to blocked jets on the screen delivery nozzles. These can be cleared with fine wire. Other causes of failure are usually due to blocked pipes, kinked pipes or disconnected pipes. The latter is usually apparent when the pump is operated.
3 Should the pump fail to operate, first check the supply pipes to the nozzles as described above and if they are in order make an inspection of the wiring to the pump.
4 If the pump is found to be defective it must be renewed as a unit as it is not repairable.

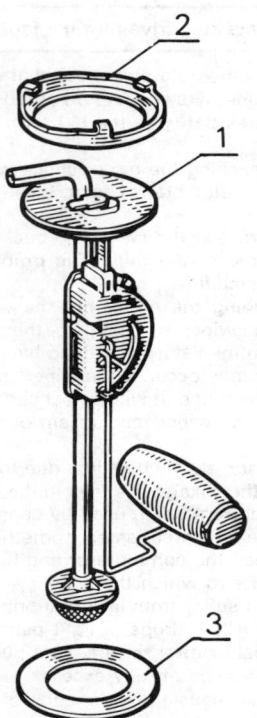

Fig. 9.13a Fuel tank sender unit

1 Sender unit   3 Sealing ring
2 Retaining ring

Note: *Do not use standard cooling system antifreeze in the washer unit. Special windscreen washer solutions to aid windscreen cleaning and prevent freezing are available from most garages and accessory shops. Washer jets are adjustable by inserting a pin into the nozzle ball assemblies.*

24 Reversing light switch – removal and refitting

1 The reversing light switch is located on the lower right-hand side of the transmission casing (photo).
2 To remove first disconnect the battery earth connection.
3 Jack the car up on the right-hand side and support with an axle stand.

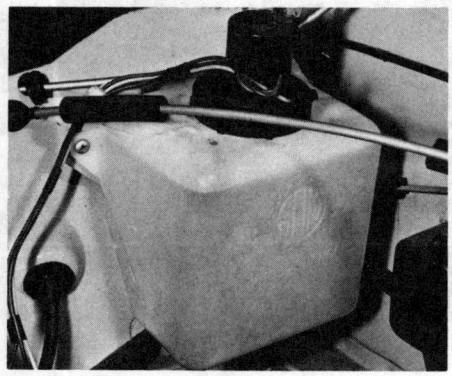

23.1 The windscreen washer unit

24.1 Reversing light switch

25.1 Oil pressure switch

Chapter 9 Electrical system

4 Drain the engine/transmission oil. If a replacement switch is at hand for a direct swap it will not be necessary to drain the oil as the oil spillage will be minimal.
5 Detach the switch wires from the terminals and unscrew the switch.
6 Install in the reverse order using a new washer.

25 Engine oil pressure warning switch – removal and refitting

1 Locate the switch next to the oil filter on the top of the engine (photo). Disconnect the lead and unscrew the switch from the crankcase.
2 Refitting is a reversal of removal.

26 Hazard warning light switch – general

1 Mounted on the dashpanel, it operates the four indicators simultaneously in the event of a roadside emergency.
2 Hopefully it will not be used very often and will therefore give little trouble. The wiring connector positions to the switch are shown in Fig. 9.14.
3 The wiring and connections can be checked in the normal way using a test light or continuity tester if available. If the switch is found to be defective it must be renewed.

27 Speedometer cable – removal and refitting

1 The speedometer drive cable is in two sections.
2 The cable may be disconnected at its coupling just above the steering rack by unscrewing the ring nut from the threaded cable sleeve.
3 The cable can be released from the transmission by unscrewing the locknut and setscrew which hold the cable sleeve in position.
4 Pull out the cable sleeve. If the drive pinion and O-ring seal must be withdrawn, use a pair of curved-nosed pliers to extract them.
5 The speedometer cable may be disconnected from the speedometer head by reaching up under the instrument panel and unscrewing the knurled ring nut or by squeezing the cable connector according to the type of connection used.

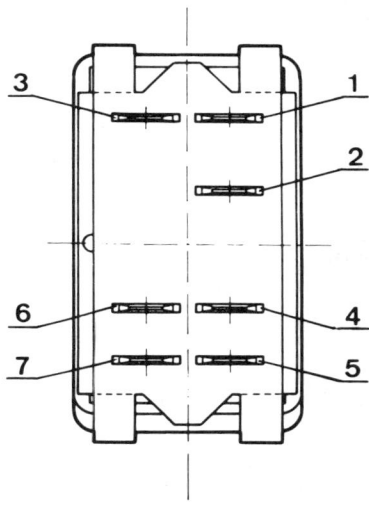

Fig. 9.14 The hazard warning light switch connections

1 Ignition switch positive (+)
2 Positive feed
3 Flasher unit (+)
4 Direction indicators
5 Right-hand hazard indicators
6 Warning light
7 Left-hand hazard indicators

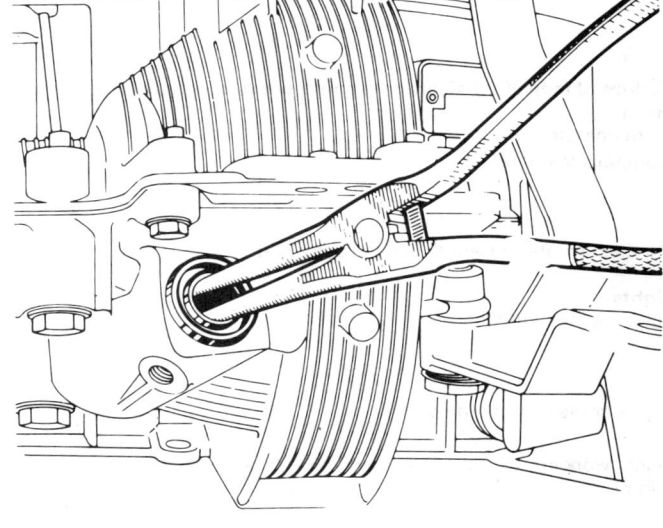

Fig. 9.15 Removing speedometer drive pinion

28 Fault diagnosis – electrical system (general)

Symptom	Reason(s)
Starter fails to turn engine	Battery discharged Battery defective internally Battery terminal leads loose or earth lead not securely attached to body Loose or broken connections in starter motor circuit Starter motor switch or solenoid faulty Starter motor pinion jammed in mesh with flywheel gear ring Starter brushes badly worn, sticking, or brush wires loose Commutator dirty, worn or burnt Starter motor armature faulty Field coils earthed
Starter turns engine very slowly	Battery in discharged condition Starter brushes badly worn, sticking or brush wires loose Loose wires in starter motor circuit

Chapter 9 Electrical system

Symptom	Reason(s)
Starter spins but does not turn engine	Starter motor pinion sticking Pinion or flywheel gear teeth broken or worn Battery discharged
Starter motor noisy or excessively rough engagement	Pinion or flywheel gear teeth broken or worn Starter motor retaining bolts loose
Battery will not hold charge for more than a few days	Battery defective internally Electrolyte level too low or electrolyte too weak due to leakage Plate separators no longer fully effective Battery plates severely sulphated Drivebelt slipping Battery terminal connections loose or corroded Alternator not charging Short in lighting circuit causing continual battery drain Regulator unit not working correctly
Ignition light fails to go out, battery runs flat in a few days	Drivebelt loose and slipping or broken Alternator brushes worn, sticking, broken or dirty Alternator brush springs weak or broken Internal fault in alternator Regulator incorrectly set Open circuit in wiring of regulator unit

Failure of individual electrical equipment to function correctly is dealt with alphabetically, item-by-item, under the headings listed below

Horn

Symptom	Reason(s)
Horn operates all the time	Horn push stuck down
Horn fails to operate	Cable or cable connection loose, broken or disconnected Horn has an internal fault Blown fuse
Horn emits intermittent or unsatisfactory noise	Cable connections loose

Lights

Symptom	Reason(s)
Lights do not come on	If engine not running, battery discharged Wire connections loose, disconnected or broken Light switch shorting or otherwise faulty
Lights come on but fade out	If engine not running battery discharged
Lights work erratically – flashing on and off, especially over bumps	Battery terminals or earth connections loose Lights not earthing properly Contacts in light switch faulty

Wipers

Symptom	Reason(s)
Wiper motor fails to work	Blown fuse Wire connections loose, disconnected or broken Brushes badly worn Armature worn or faulty Field coils faulty
Wiper motor works very slowly and takes excessive current	Commutator dirty, greasy or burnt Armature bearings dirty or unaligned Armature badly worn or faulty
Wiper motor works slowly and takes little current	Brushes badly worn Commutator dirty, greasy or burnt Armature badly worn or faulty
Wiper motor works but wiper blades remain static	Wiper motor gearbox parts badly worn

Key to wiring diagram on page 120 and 121

Component code

1. Front combination light (left-hand)
2. Headlight (left-hand)
3. Junction box
4. Horn
5. Headlight (right-hand)
6. Junction box
7. Front combination light (right-hand)
8. Earth
9. Earth
10. Engine harness to front harness junction box
11. Thermal switch
12. Starter
13. Alternator
14. Oil pressure switch
15. Battery
16. Earth
17. Gearbox
18. Reversing lights switch
19. Regulator
20. Brake fluid reservoir cap/level switch
21. Ignition coil
22. Thermal switch on radiator
23. Cooling fan motor
24. Instrument panel
25. Connector A – instrument panel
26. Connector B – instrument panel
27. Connector C – instrument panel
28. Warning light switch – choke ON
29. Heating rheostat
30. Wire junction – heating – ventilation fan motor rheostat
31. Heating – ventilation fan motor
32. Earth junction plate
33. Wiper motor junction box
34. Windscreen wiper
35. Stoplights switch
36. Cigar lighter wiring junction box
37. Cigar lighter
38. Wire junction – cigar lighter illumination
39. Junction box – direction indicators
40. Junction box – combination lighting switch harness
41. Combination lighting switch
42. Direction indicators switch
43. Hazard warning lights switch
44. Rear screen demister switch
45. Accessories plate
46. Accessories plate feed
47. Junction block – front wiring harness to accessories plate
48. Junction block – rear wiring harness to accessories plate
49. Junction block – rear wiring harness to accessories plate
50. Distributor
51. Junction box – combination windscreen wiper/washer switch
52. Junction box – front harness to rear harness
53. Junction box – front harness to rear harness
54. Junction box – anti-theft switch harness
55. Ignition/starter switch
56. Wire junction – interior light door pillar switch
57. Interior light
58. Left-hand door pillar switch
59. Right-hand door pillar switch
60. Handbrake ON warning light switch
61. Wire junction – demister via tailgate stay
62. Tailgate stay
63. Rear screen
64. Earth
65. Wire junction – fuel gauge tank unit
66. Fuel gauge unit
67. Junction box – left-hand rear light assembly
68. Left-hand rear light assembly
69. Junction box – right-hand rear light assembly
70. Right-hand rear light assembly
71. Number plate light
72. Wire junction
73. Earth
74. Junction – cooling fan motor

Colour code

- B – Blue
- Bc – White
- Be – Beige
- C – Clear
- G – Grey
- J – Yellow
- M – Maroon
- N – Black
- Or – Orange
- R – Red
- S – Pink
- V – Green
- Vi – Violet

Harness code

- A   Engine front
- B   Rear
- C   Starter positive
- D   Charging
- E   Starter negative
- F   Engine
- G   Cigar lighter
- H   Combination lighting switch
- J   Direction indicator switch
- K   Interior light
- L   Door pillar switches earth

Wire identification code

Example   162 N-5 74
- 162 – Wire identification number
- N – Wire colour
- 5 – Wire diameter (see table)
- 74 – Wire destination

Wire diameter

No	1	2	3	4	5	6	7	8	9	10
Diameter (mm)	7/10	9/10	10/10	12/10	16/10	20/10	25/10	30/10	45/10	51/10

120

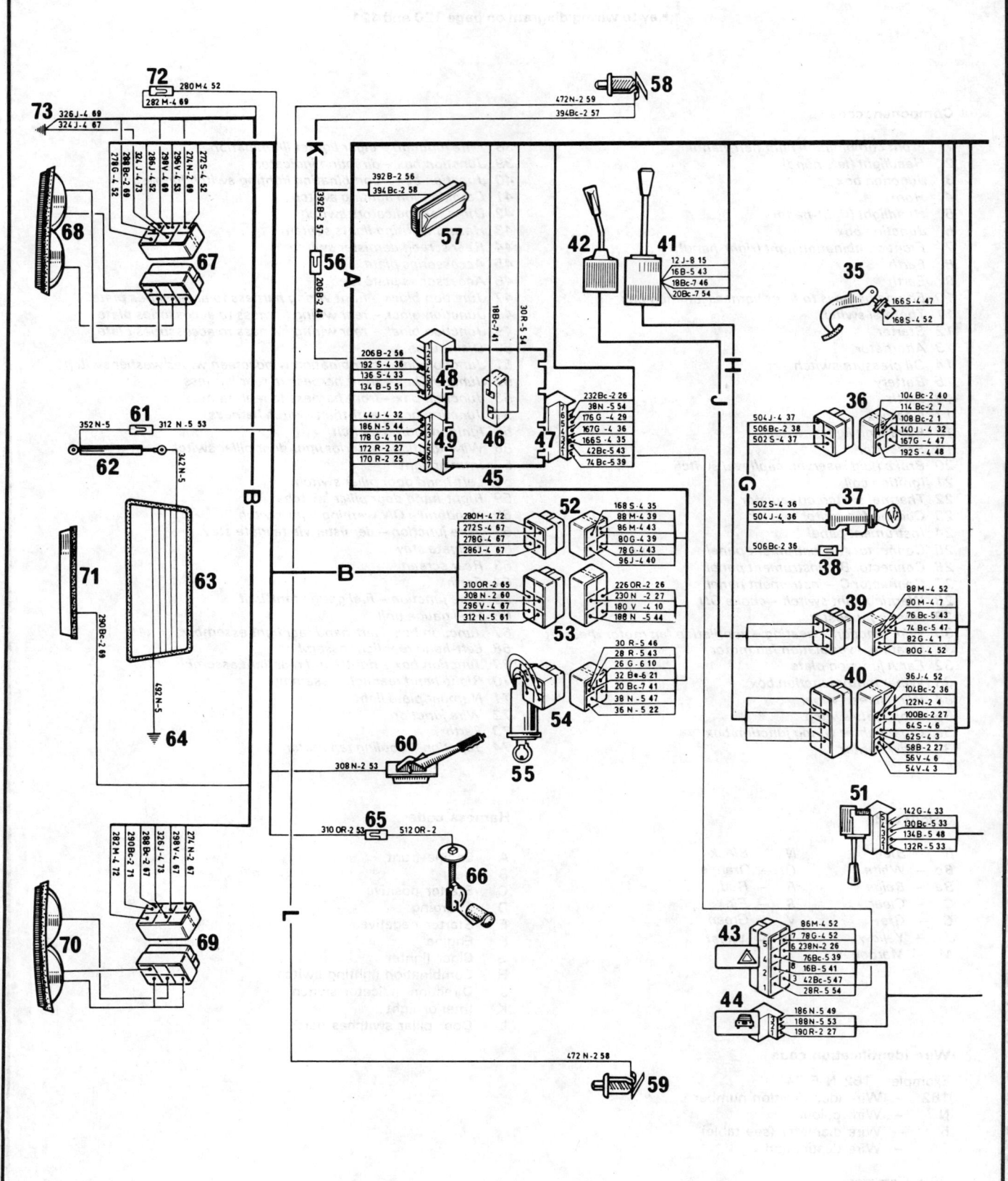

Fig. 9.15A Wiring diagram, 1977 to 1979 R.1210. Key on page 119

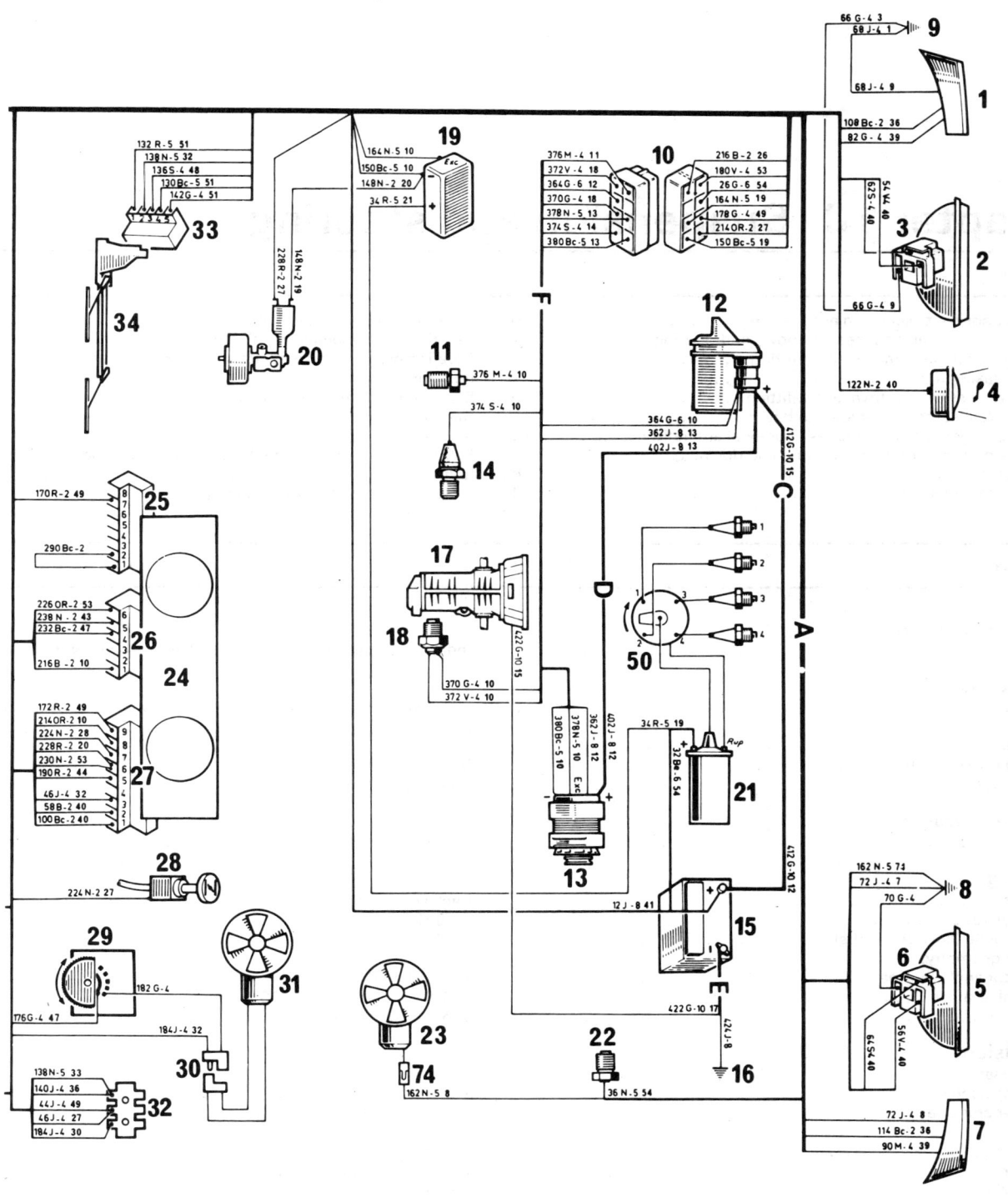

Fig. 9.15B Wiring diagram, 1977 to 1979 R.1210. Key on page 119

Chapter 10 Suspension and steering

Contents

Fault diagnosis – suspension and steering	16
Front anti-roll bar and bushes – removal and refitting	4
Front coil springs – removal and refitting	6
General description	1
MacPherson strut – removal and refitting	5
Rear shock absorbers – removal and refitting	7
Rear suspension arm and bushes – removal and refitting	9
Rear torsion bars – removal, refitting and ride height adjustment	8
Routine maintenance	2
Steering arms – removal and refitting	15
Steering and suspension – alignment and geometry	10
Steering column intermediate shaft – removal and refitting	12
Steering rack – removal and refitting	14
Steering shaft lower flexible coupling – removal and refitting	13
Steering wheel and column bushes – removal and refitting	11
Suspension and steering – testing	3

Specifications

Front suspension
Type .... Independent, coil springs and MacPherson strut, with anti-roll bar

Rear suspension
Type .... Independent single pivot trailing arms. Double transverse torsion bars

Alignment:
  Negative camber .... 0° to 1°30'
  Toe-out .... $\frac{3}{64}$ in $\pm \frac{5}{64}$ in (1 mm $\pm$ 2 mm)

Shock absorbers
Type .... Telescopic double-acting front and rear

Steering
Type .... Rack and pinion
Reduction ratio .... 21.6 :1
Turning circle (between kerbs) .... 32.8 feet
Steering geometry:
  Castor angle (normal) .... 3° $\pm$ 30' to 0°30' $\pm$ 30'
  Camber angle .... 0°30' $\pm$ 30' to 3°30' $\pm$ 30'
Toe-out .... 0 to 0.078 in (0 to 2 mm)

Dimensions
Wheelbase:
  Right-hand side .... 8 feet 2¼ inches (2.498 m)
  Left-hand side .... 8 feet 3½ inches (2.530 m)
Track:
  Front .... 4 feet 5¼ inches (1.352 m)
  Rear .... 4 feet 6¼ inches (1.378 m)
Ground clearance .... 5.875 in (0.150 m)

Torque wrench settings

	lbf ft	Nm
Anti-roll bar nuts	40.0	55
Anti-roll bar clamp bolts	37.5	50
Strut top retaining nut	30.0	40
Strut top flange nuts	7.5	10
Rear shock absorber-to-bottom arm nut	60.0	80
Steering shaft universal joint key bolt	11.25	15
Steering wheel retaining nut	34.0	45

Chapter 10 Suspension and steering

Torque wrench settings (continued)	lbf ft	Nm
Steering arm-to-rack bolt	26.25	35
Steering arm-to-strut nut	26.25	35
Steering rack-to-crossmember	26.25	35
Intermediate steering shaft flexible coupling bolts	11.25	15
Rear suspension arm mounting nuts (outer)	55.5	75
Rear suspension arm mounting bolts (inner)	30.0	40

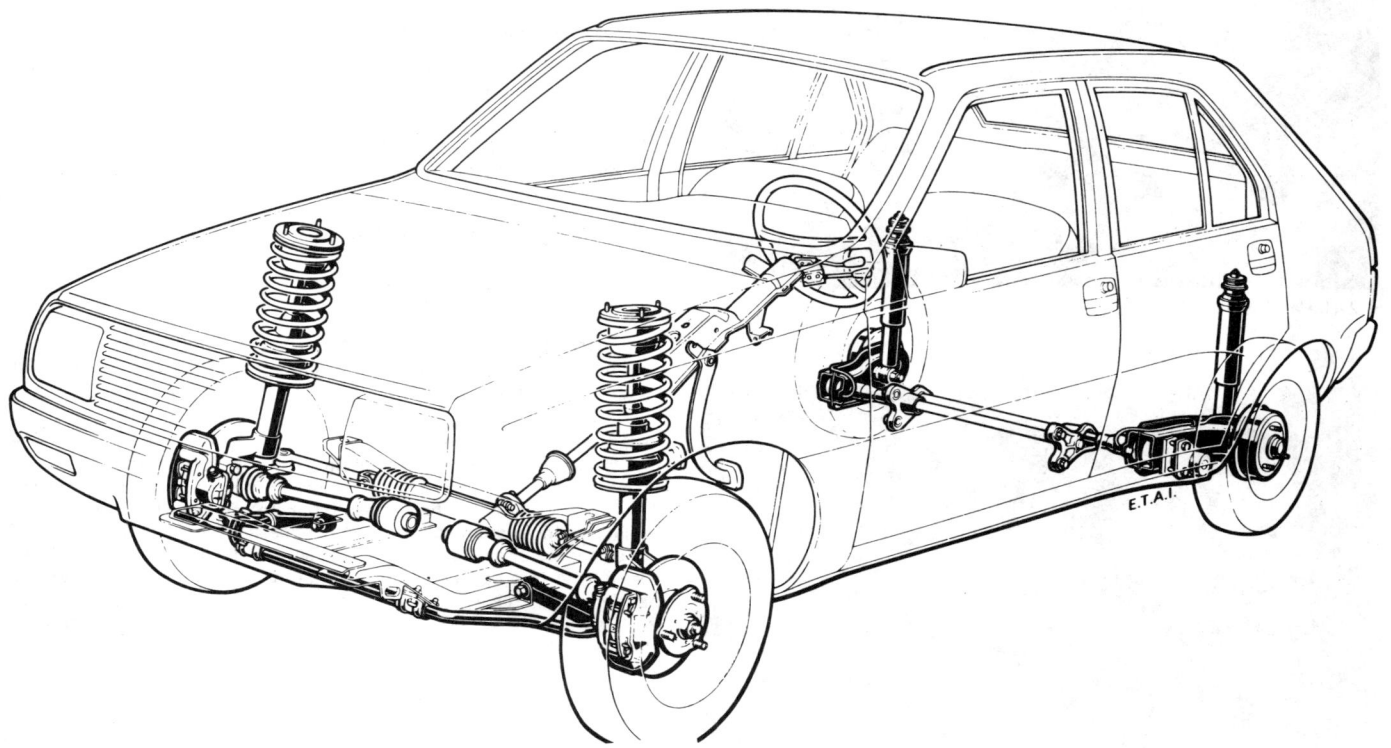

Fig. 10.1 Cutaway view showing the front suspension, steering and rear suspension layout

1 General description

The Renault 14 has independent suspension at the front and rear. At the front MacPherson struts incorporate telescopic double-acting hydraulic shock absorbers and coil springs. The struts are located at the top end within the inner wing panel. At their bottom end the struts are located by the lower suspension arms and the anti-roll bar. The front wheel bearings, steering arm and suspension arm mountings, and brake mountings are all contained in the lower portion of the MacPherson strut.

Single trailing suspension arms are used each side at the rear wheels. These pivot on the axis of the torsion bars. Double-acting telescopic shock absorbers control the wheel movement.

A rack and pinion steering unit is fitted, with self-centring action.

Some of the necessary repair and maintenance work can be undertaken on the suspension and steering by the home mechanic although some special tools, which may possibly be home made, will be necessary. Before starting any such work, read through the instructions and be absolutely sure that you have the knowledge and facilities to finish and have the correct replacement parts for your model. There have been detailed changes during production, so ensure that you have the correct replacement part before fitting it. Do not start a job you cannot finish – asking the local Renault garage to come out and replace a torsion bar which you have removed and find you cannot replace will be very expensive!

2 Routine maintenance

1 Whilst the maintenance to the steering and suspension components has been reduced to the minimum, it does not mean that it can be ignored completely! A periodic manual and visual inspection should be made.
2 Inspect the suspension joints and their fixings for security and excessive play.
3 Check the steering components and connections for signs of wear and security.
4 Inspect the shock absorbers for looseness in their mountings – look at the top and bottom rubber bushes and check for leakage of the unit itself. If it leaks replace it and the one on the opposite side also.
5 Check the tightness of the steering rack mountings. Check the steering joints for signs of excessive wear and the rack gaiters for splits and leakage. Check the rubber/canvas steering joints and the column universal joints and replace as and when necessary.

3 Suspension and steering – testing

1 Because of the construction of these vehicles the suspension cannot be viewed in isolation from the steering and vice versa. The safety of a car depends more on the steering and suspension than anything else and this is the reason why the compulsory tests made for

Chapter 10 Suspension and steering

vehicles over three years old pay attention to the condition of all these components.

2 Any parts which are weak or broken must be replaced immediately. Take great care to check the following. The first list is of those parts which will wear with use and the second list is of special check points:

Check for wear:
(a) All balljoints
(b) MacPherson strut and coil springs
(c) Shock absorbers and/or their mounting bushes
(d) Steering arm inner bushes

All these points may be tested with a tyre lever or screwdriver to see whether there is any movement between them and a fixed component.

Check:
(e) Rear torsion bars for security
(f) Steering rack mounting bolts for looseness
(g) Steering rack to column coupling
(h) Steering wheel to column
(i) Steering column bush
(j) Front and rear hub bearings (see Chapter 7)
(k) Anti-roll bar bushes

There should be no play nor failure in any single part of any of the forementioned components. It is dangerous to use a vehicle in a doubtful condition of this kind.

4 Front anti-roll bar and bushes – removal and refitting

1 When removing the anti-roll bar to replace the bushes or to replace other suspension parts, it is necessary to raise the front of the car on ramps or axle stands, or to position the car over a pit. Make sure the vehicle is properly secure if on stands or on the ground with the handbrake on. It is not always necessary to remove the front wheels.

2 Remove the retaining bolts and washers securing the anti-roll bar clamps to the front sub-frame members on each side. Take note of any shims located between the clamps and frame members (photo).

3 Before disconnecting the ends of the anti-roll bar from the lower suspension arms it is advisable to take up the tension of the roll bar across the car, just forward of the lower suspension arm (see Chapter 7, Section 2 paragraph 6 for details).

4 Remove the retaining pins (photo) and unscrew the retaining nuts.

4.2 The anti-roll bar locating clamp

4.4 Retaining pin and nut to lower suspension arm

5.4 The strut retaining nuts at the top

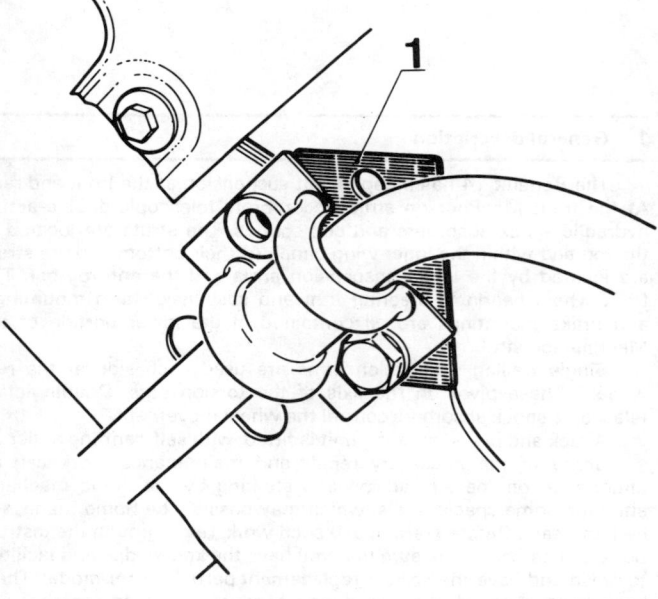
Fig. 10.2 Anti-roll bar shim/s (1)

Chapter 10 Suspension and steering

Fig. 10.3 The correct bush and washer locations

Withdraw the anti-roll bar noting the positions of the bushes and washers.

5   Carefully clean and inspect the rubber bushes and if worn or perished they must be renewed.

6   *Do not drive the car without the anti-roll bar fitted!*

7   Installation of the anti-roll bar is a direct reversal of the removal procedure. Ensure that the bushes and washers are fitted in the correct order (Fig. 10.3) and lightly smear the bearing areas with a medium grease. Lower the car onto its wheels before tightening the retaining bolts/nuts to the specified torque settings. Do not forget to refit the shims between the clamps and the sub-frame members.

5 MacPherson strut – removal and refitting

1   Raise the front end of the car and support on axle stands or blocks. Check that the handbrake is firmly applied.

2   Remove the roadwheel at the suspension unit to be dismantled.

3   Remove the brake caliper unit but do not disconnect the hydraulic brake line. Suspend the caliper unit from a body member to prevent damage to the brake line. See Chapter 8 for full details.

4   The preliminary operations required for the removal of the MacPherson strut are described in Chapter 7, Section 2 paragraphs 4 to 9. When these operations are completed it only remains to unscrew the three strut retaining bolts on the inner wing panel in the engine compartment (photo) and lower the unit.

5   The lower suspension arm can be disconnected from the strut by unscrewing the retaining nut and using a balljoint separator or wedge to free the joint.

6   To remove the coil spring from the strut refer to the next Section.

7   It is not possible to dismantle the shock absorber without specialised equipment. This work should be entrusted to your Renault dealer.

8   Installation of the MacPherson strut is a direct reversal of the removal procedure. Note the following:

  (a) *If a replacement strut is to be fitted check that the correct type is obtained. Steering arm balljoint taper angles can differ. (See Fig. 10.23 for identification markings)*
  (b) *Lightly smear pivot pins, bushes and driveshaft splines with a medium grease*
  (c) *When refitting the lower suspension arm use a suitable wrench to twist it so that the pivot pin can be inserted*
  (d) *Do not finally tighten suspension fastenings until the weight of the car is on the wheels*

6 Front coil springs – removal and refitting

1   Refer to Section 5 and remove the MacPherson strut.

2   A suitable coil spring compressor will be required to enable the spring to be removed from the strut (Fig. 10.3A).

3   Unscrew and free the bellows from between spring coils and slide up the strut (Fig. 10.4).

4   Position the spring compressor to relieve the expansion pressure of the spring between the strut lower flange and the upper mounting flange. Ensure that the spring compressor is fully and securely located on the spring bottom cup, and through the coils of the spring near the top. Compress the spring sufficiently to enable the top mounting to be removed.

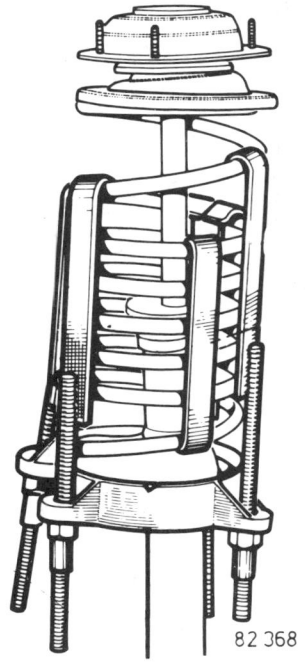

Fig. 10.3a Typical coil spring compressor

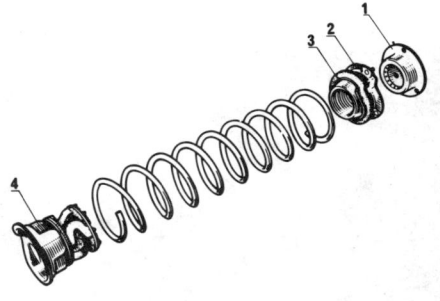

Fig. 10.5 The coil spring assembly

1 Top mounting flange   4 Bottom cup and bump stop
2 Top cup   3 Top rubber

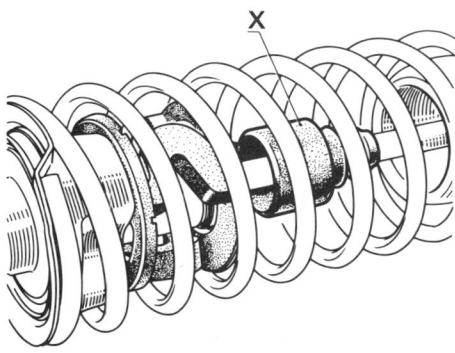

Fig. 10.4 Free the bellows (X) and slide them up the strut

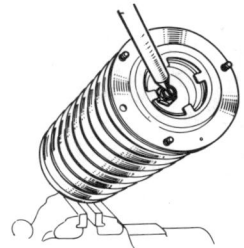

Fig. 10.6 Stake punch the retaining nut to secure

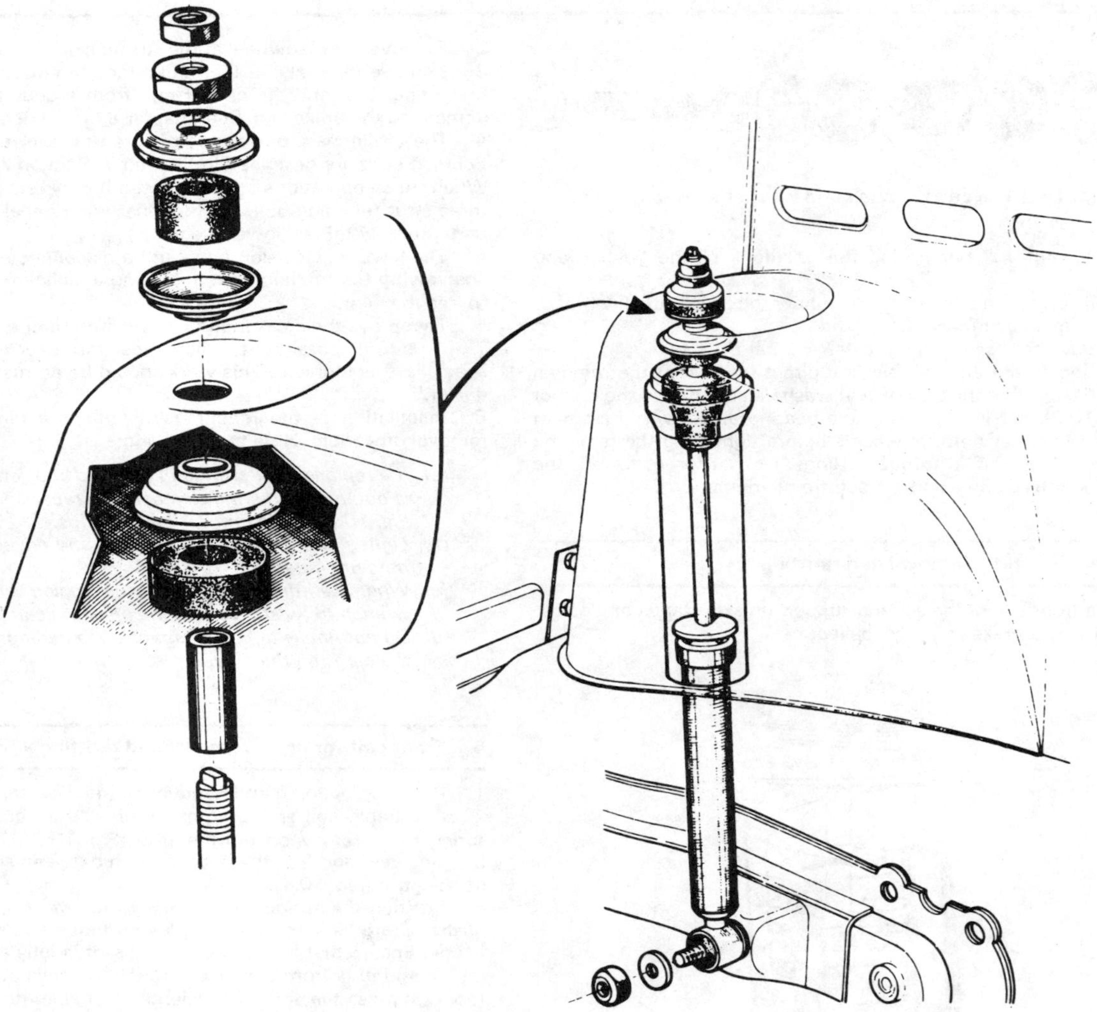

Fig. 10.7 Rear shock absorber fixing

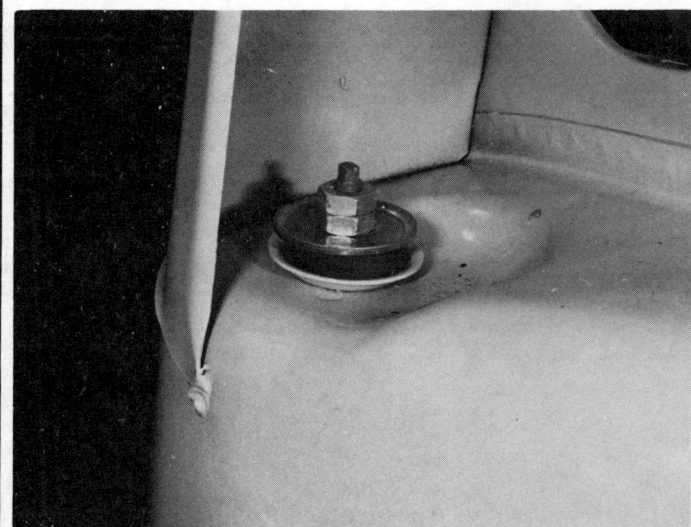

7.3 Rear shock absorber top mounting

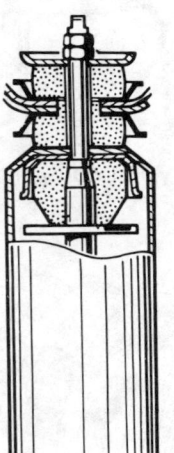

Fig. 10.8 Rear shock absorber upper mounting showing bush positions

Chapter 10 Suspension and steering

5 Use a box-wrench or socket and unscrew the strut rod nut at the top end. The nut will have been staked to lock it and the indent must be relieved if the nut is reluctant to unscrew.
6 Now remove the top mounting flange, cup and rubber cone. The spring can be withdrawn together with the bottom cup and bump stop rubber.
7 If a new spring is being fitted it will need to be compressed using the spring compressor prior to assembly.
8 The anti-friction washer must always be renewed.
9 Replace the top cup rubber and bottom cup unit if the originals are worn or defective in any way.
10 Installing the spring into position on the strut is a reversal of the removal procedure. Remember to lubricate and reposition the rubber bellows over the shock absorber. Always use a new retaining nut and stake the nut to the rod when tightened to the specified torque.

7 Rear shock absorbers – removal and refitting

1 The principle of checking the condition of the rear shock absorbers as frequently as the fronts is valid, although they do not generally take such abuse.
2 Jack-up one side of the vehicle at a time and remove the wheel. Undo the nut on the lower attachment point and remove.
3 Support the suspension arm and undo the top mounting nuts (photo). Lower the suspension arm and remove the shock absorber noting the order in which the washers and spacers are fitted.
4 Refit in the reverse sequence. Remember to smear the lower mounting pin with grease and to ensure that the washers and spacers are replaced in the correct order.
5 Tighten the lower retaining nut to the specified torque.
6 **Note**: *Before fitting a new shock absorber stand it vertically and pump it a few times to prime the mechanism.*

8 Rear torsion bars – removal, refitting and ride height adjustment

1 Place the rear of the vehicle on stands over a pit or on a hydraulic ramp with both roadwheels clear. Remove the relevant wheel (on the right-hand side it is necessary to remove the brake limiter valve protective cover).
2 Remove the shock absorber as described in Section 7.

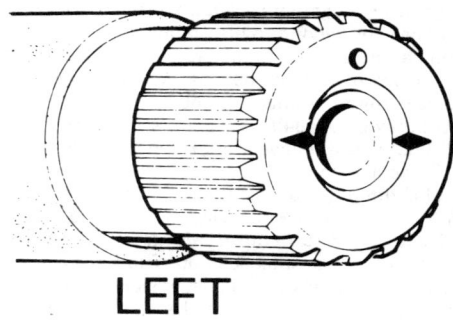

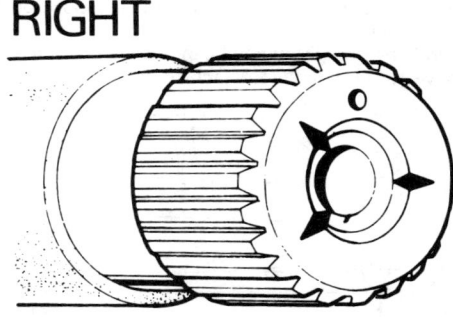

Fig. 10.9 The right and left-hand torsion bar identification marks

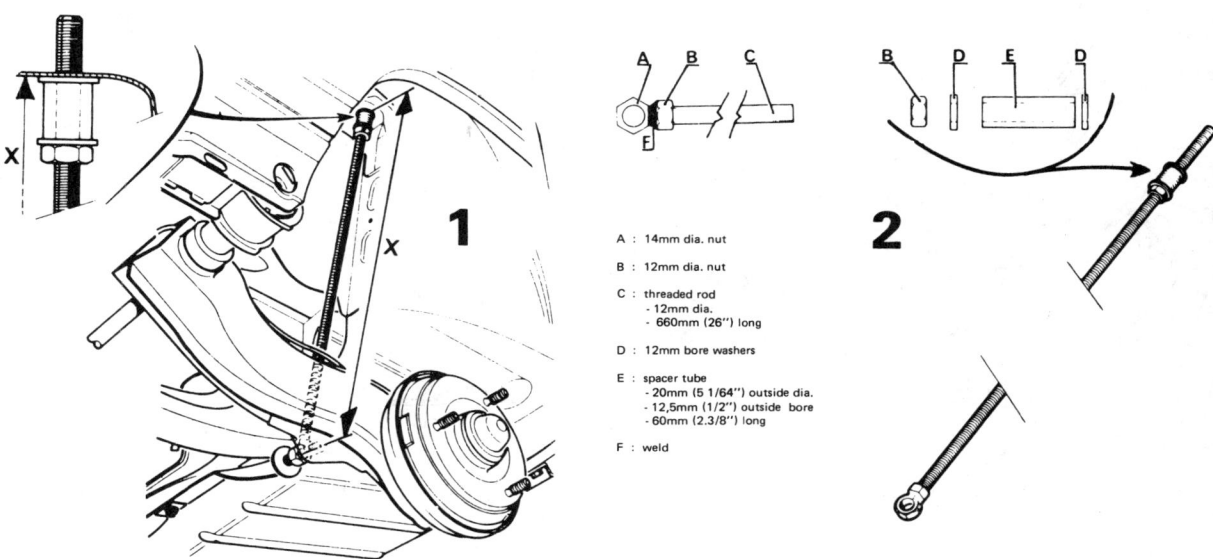

Fig. 10.10 Special tool to check and adjust vehicle height

1 Fit the tool in place of the shock absorbers
2 Special tool details

A : 14mm dia. nut

B : 12mm dia. nut

C : threaded rod
   - 12mm dia.
   - 660mm (26") long

D : 12mm bore washers

E : spacer tube
   - 20mm (5 1/64") outside dia.
   - 12,5mm (1/2") outside bore
   - 60mm (2.3/8") long

F : weld

3 The torsion bar can now be tapped out from the opposite side of the car.
4 If both torsion bars are removed, note their respective identity markings. Each are twisted in different directions so they must not be interchanged. If replacements are to be fitted ensure that the markings on the new ones correspond to the originals.
5 To install a torsion bar the rear suspension arm must be located in the correct position so that the specified ride height is achieved. A special tool is required for this, it is shown in Fig. 10.10.
6 This tool may be available for loan or hire from your local Renault dealer, but if not you will have to make one. The dimensions of the tool are shown in the illustration.
7 The tool is fitted in place of the shock absorber and is adjusted to the length X according to which side of the vehicle you are working – Right-hand side $23\frac{1}{4}$ in (590 mm), Left-hand side $23\frac{5}{8}$ in (600 mm).
8 Slide the torsion bar into position. The bar has 25 splines at the anchor end and 24 splines at the suspension arm end. The diameter at the anchor end is reduced so that the splines will pass through the suspension arm. The differing number of splines at each end enables a fine adjustment of the ride height to be made. It also means that when the suspension arm is set at the specified position the torsion bar will only slide into position when the splines are correctly aligned at each end. Keep rotating the bar until the position is discovered where the bar slides freely home.
9 If the body ride height has to be increased or decreased the special tool will have to be extended or shortened (length X) by $\frac{1}{8}$ in (3 mm) increments, which is the minimum variation which can be obtained for one differential spline. The ride heights are measured from the positions shown in Figs. 10.11 and 10.12. The H5 measurement should be $\frac{7}{16}$ in or 11 mm less than H4 ($\pm \frac{11}{32}$ or $\pm$ 9 mm). The difference between the right and left-hand settings should not exceed $\frac{3}{8}$ in (10 mm).
10 Remember when checking or resetting the ride height that the vehicle should be unladen (full fuel tank) and the tyre pressures must be correct.
11 If the vehicle height is altered then the headlight beams will need to be reset (see Chapter 9).

9 Rear suspension arm and bushes – removal and refitting

1 Whilst the removal of the rear suspension arm is a relatively simple task, the torsion bar will have to also be extracted and this requires the use of a special tool (see previous Section). If the bushes are to be renewed they will need to be pressed into position, necessitating access to a suitable press.
2 With the above in mind the suspension arm and bush overhaul is not really a job for the home mechanic, but is best entrusted to your Renault dealer. However brief details are given.
3 The rear of the vehicle must be raised and supported on axle stands or blocks.
4 Remove the rear shock absorbers (both sides), see Section 7.
5 On the appropriate side, disconnect and remove the handbrake secondary cable and the hydraulic brake pipe (see Chapter 8). On the left-hand side the brake limiter must also be disconnected.
6 Remove the torsion bar from each side, see Section 8.
7 Unscrew and remove the three inner and two outer suspension arm retaining nuts and bolts (Fig. 10.13), and then withdraw the suspension arm.
8 Note that two types of rear suspension arm have been fitted, either cast or of fabricated sheet metal construction. Although the removal and installation procedure is the same for each type, the cast type, fitted on cars up to and including vehicle number 253000 must only be fitted to the Girling brake backplate. From vehicle number 253001 they are fully interchangeable.
9 Installation is a direct reversal of the removal process, referring to

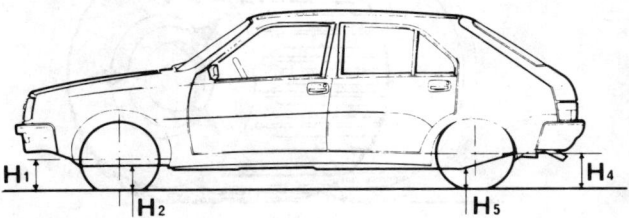

Fig. 10.11 Ride height check points

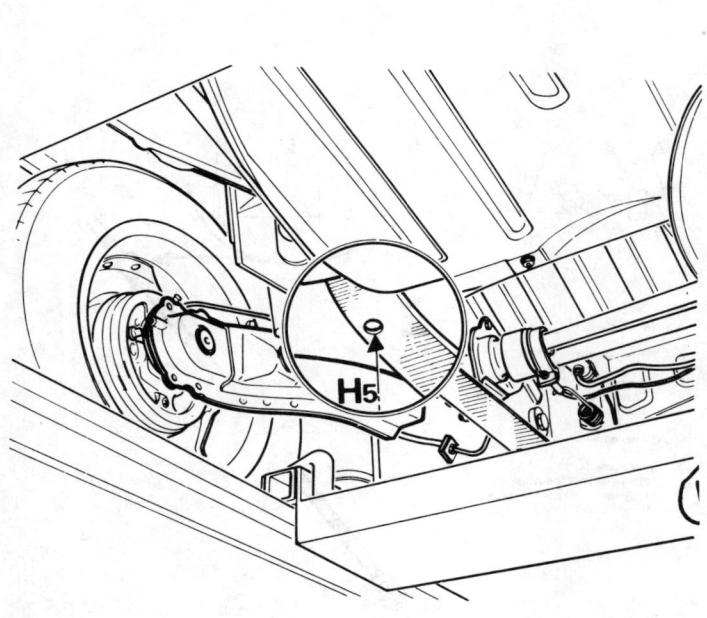

Fig. 10.12 Measure H5 from point indicated and calculate difference from H4 which should be $-\frac{7}{16}$ in $\pm \frac{11}{32}$ in (–11 mm $\pm$ 9 mm)

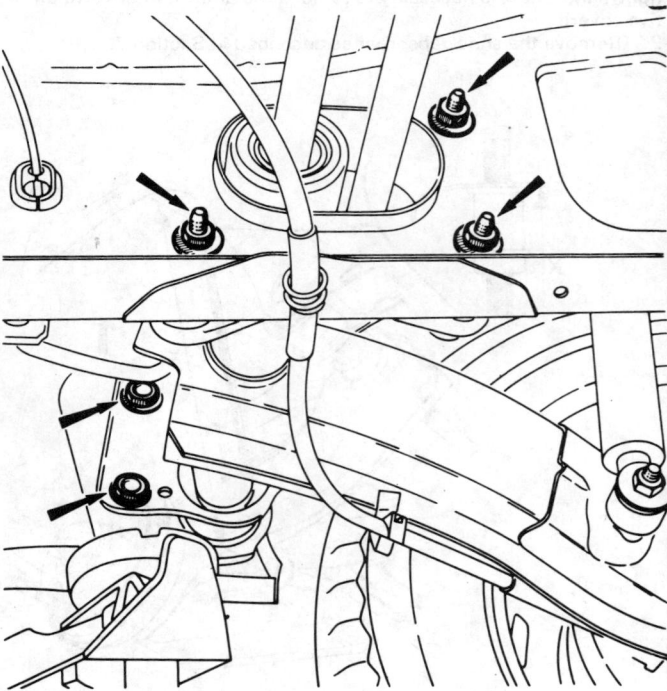

Fig. 10.13 The suspension arm inner and outer retaining bolt positions

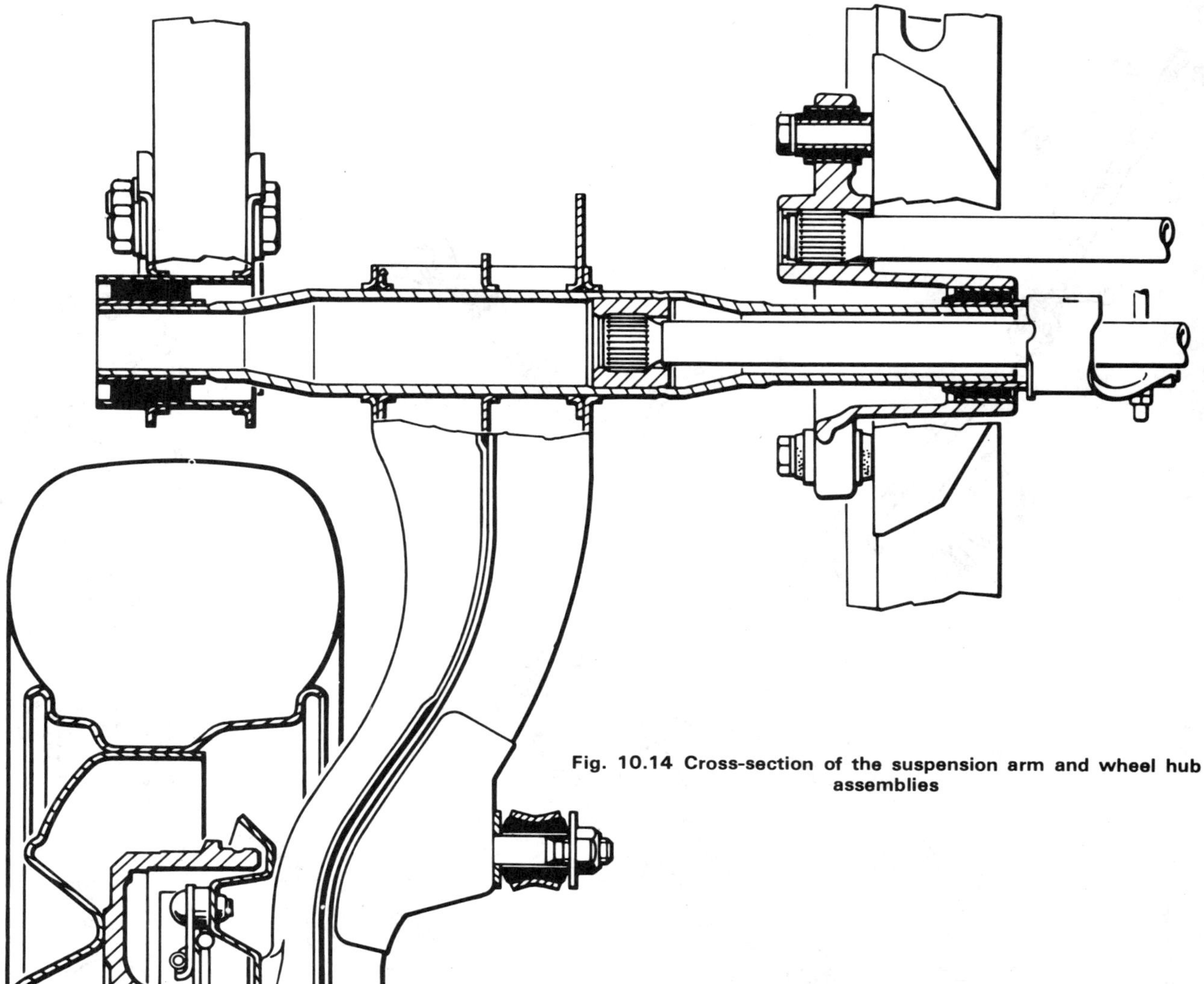

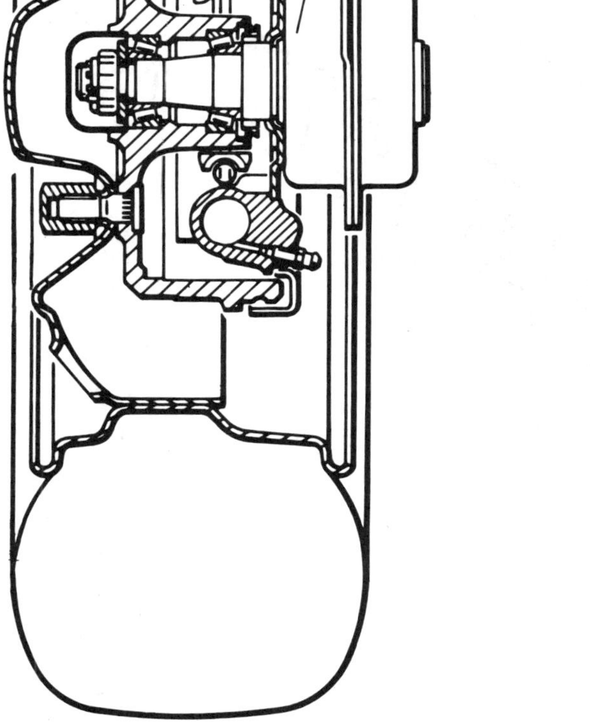

Fig. 10.14 Cross-section of the suspension arm and wheel hub assemblies

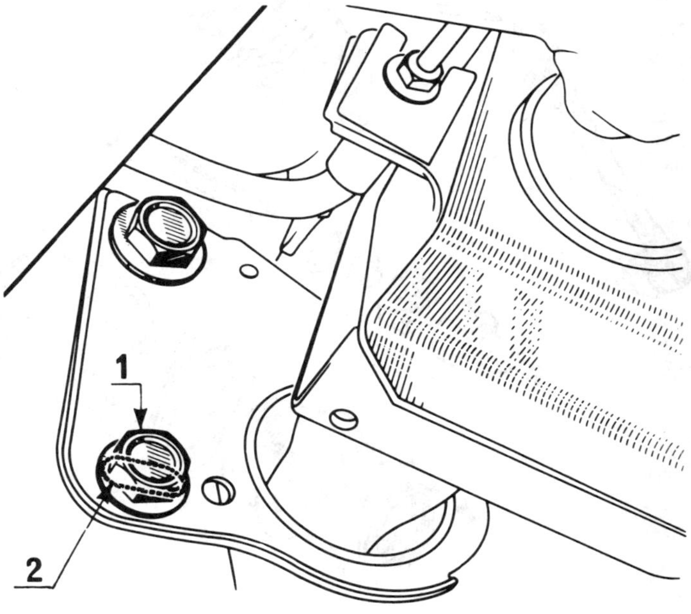

Fig. 10.15 Rear suspension arm outer bearing bolts

1 Bolts   2 Slot

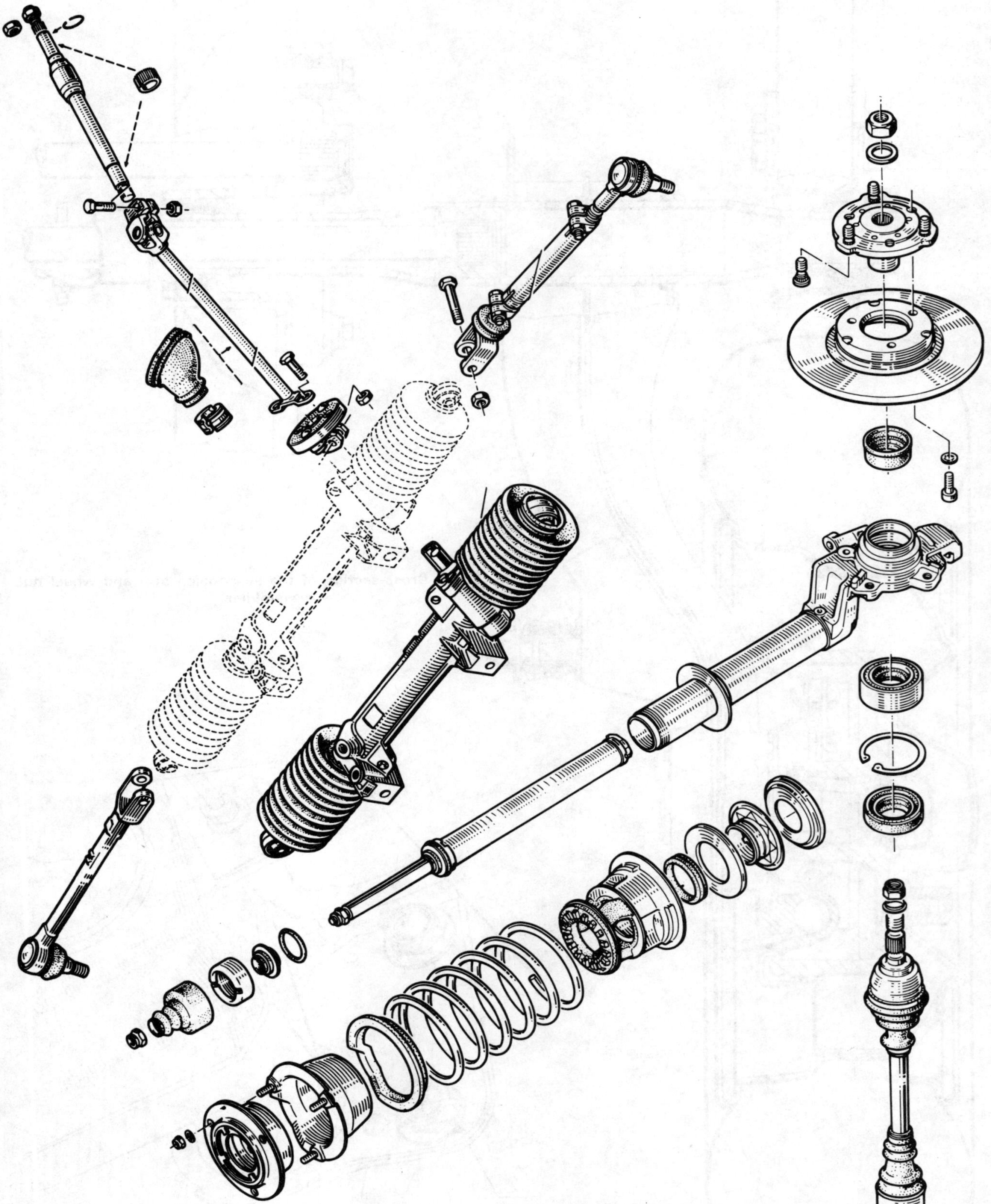

Fig. 10.15A The steering, suspension and driveshaft assemblies (left-hand drive shown)

Chapter 10 Suspension and steering

the respective Chapters/Sections concerned. If the brake limiter has been removed, have your Renault dealer check the cut-off pressure when it is reassembled. Another essential check which only your Renault dealer can do is to check the rear wheel alignment. Should this be out of adjustment he will be able to accurately reset the defective component. Without specialised equipment you can only hazard a guess and that is not good enough!

10 Steering and suspension – alignment and geometry

1 Such is the complexity of steering and suspension adjustment and its relative importance to the total correct functioning of the vehicle it is advised that any adjustment of this type be made by the local Renault garage which will be equipped with the necessary optical and measuring jigs. If any major suspension or steering part is removed and/or replaced it is absolutely necessary to have the vehicle checked. This of course also applies if the handling of the vehicle deteriorates and the inspection and maintenance checks detailed in Section 2 have not revealed an apparent reason.

2 If you have carried out any repair work to the steering or suspension components the appropriate items listed below should be checked by your Renault dealer. Tell them the work you have carried out and they will assess what checks are necessary.

3 Since they are checking the vehicle it obviously makes sense for them to make any necessary adjustments, the additional cost of which should be relatively low. The most likely items that they will check are:

(a) The castor angle
(b) The camber angle
(c) The front axle geometry
(d) The wheel positions in relation to the steering centre point

Front suspension

4 The camber, castor or kingpin inclination settings are set in production and are non-adjustable. Any deviation from specified angles discovered will be due to collision damage.

5 The front wheel alignment (tracking or toe) may be altered by releasing the clamp or locknut on the tie-rods and turning the rod tubes an equal amount, but in opposite directions. The steering should be centred for this adjustment.

Rear suspension

6 The camber angle is non-adjustable, but the tracking (toe) is adjustable by moving the suspension arm outer bearings within the limits of their mounting bolt slots, equally.

11 Steering wheel and column bushes – removal and refitting

1 Prise free the central embellisher from the steering wheel to gain access to the retaining nut (photo).

2 Use a suitable box or socket spanner and unscrew the nut. Mark the relative positions of the steering wheel and shaft.

3 Use a steering wheel puller and withdraw the wheel from the shaft.

4 Unscrew and remove the column lower trim panel. Note the length and location of the various retaining screws to ensure correct

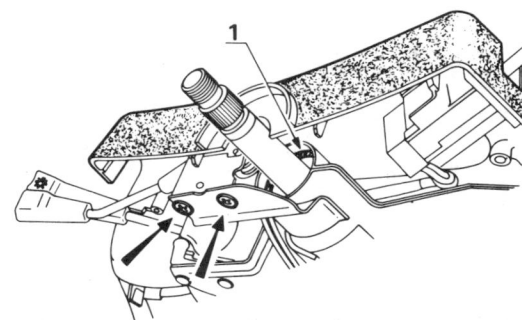

Fig. 10.16 Remove the indicator switch and snap-ring

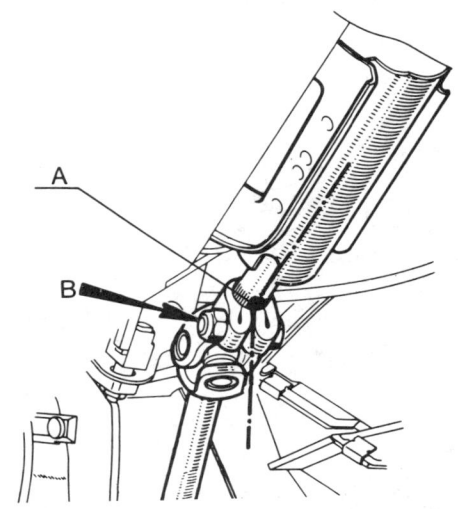

Fig. 10.17 The intermediate shaft and universal joint. Align top shaft flat with the UJ as indicated (A). The key-bolt is also shown (B)

11.1 Prise the central embellisher free to gain access to the steering wheel nut

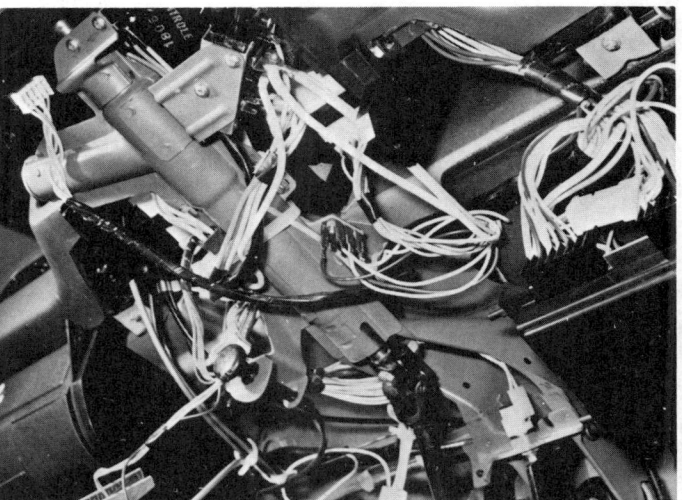

11.4 General view with lower trim panel removed

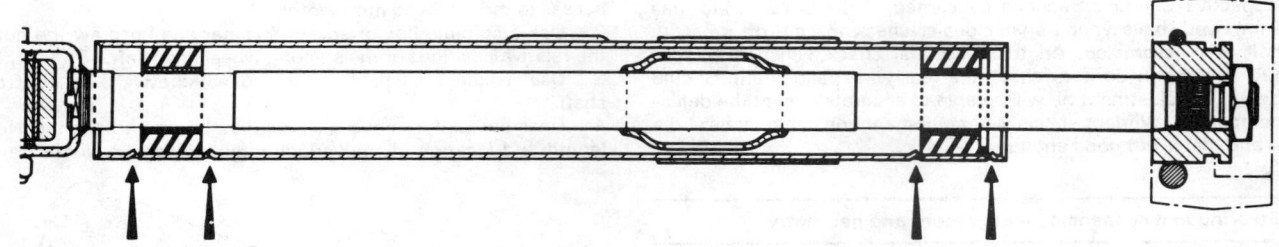

Fig. 10.18 Cross-section view of the steering shaft showing the bush locations

Fig. 10.20 Measure distance between rod centres

Note: The rods must be parallel to each other and horizontal as shown

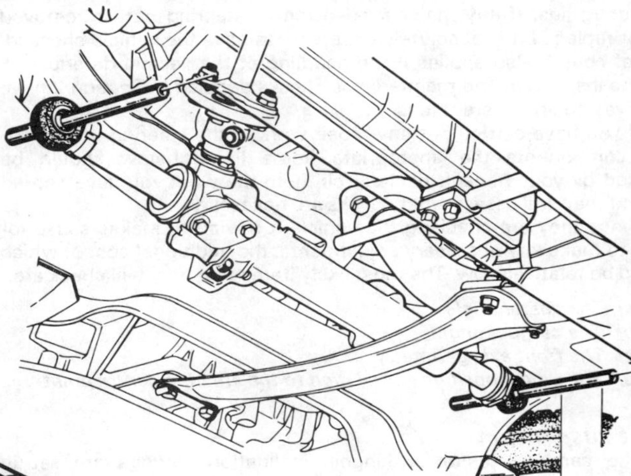

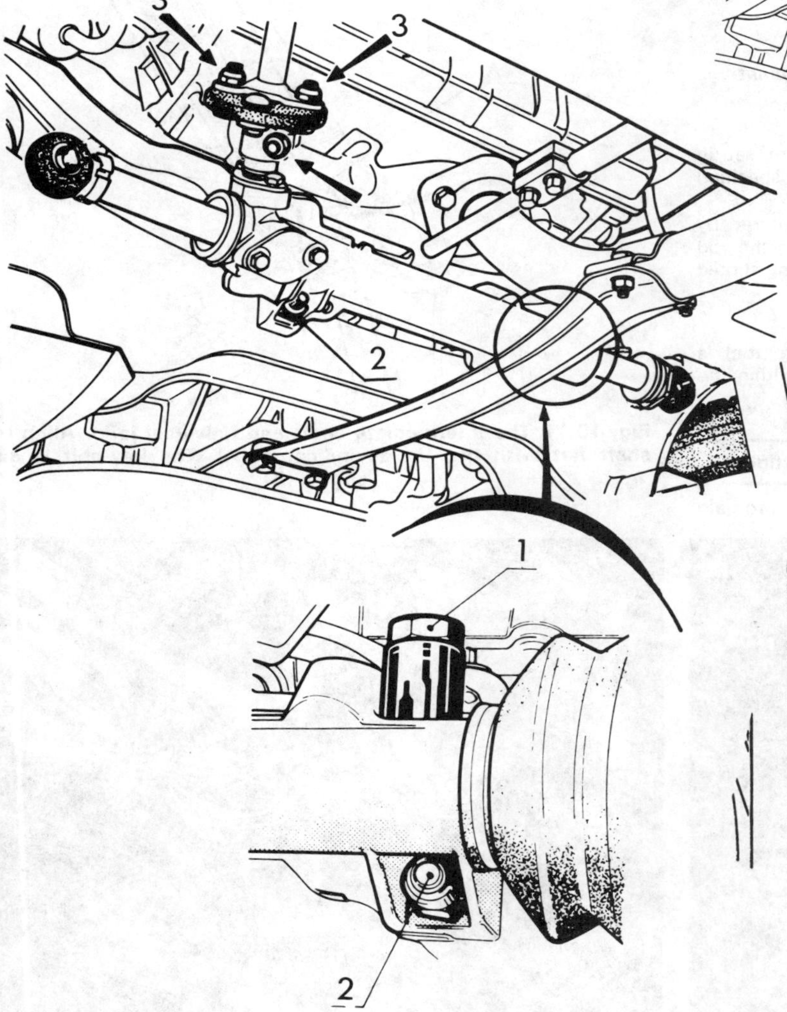

Fig. 10.19 Remove the gear selector quadrant retaining bolt (1), the rack bolts (2) and flexible coupling bolts (3) (left-hand drive shown)

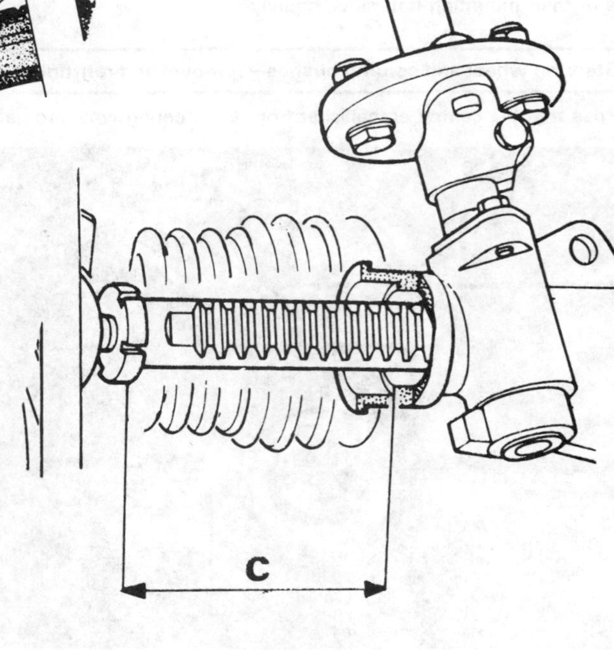

Fig. 10.21 Measure between points shown (C)

Chapter 10 Suspension and steering

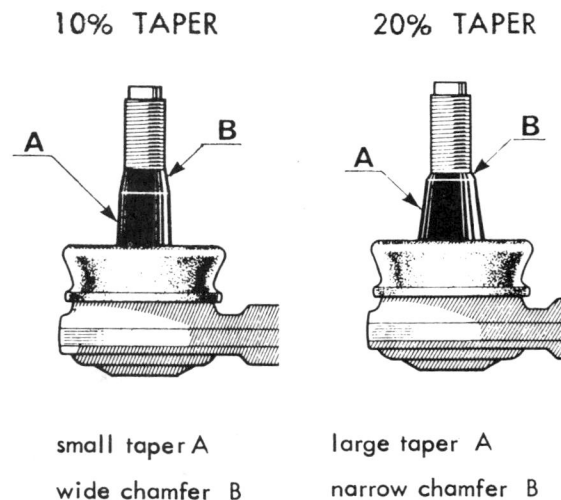

Fig. 10.22 The two types of steering arm fitted
small taper A, wide chamfer B
large taper A, narrow chamfer B

12.5 Steering shaft flexible coupling (view from above – engine removed)

reassembly (photo).
5 Detach the battery earth terminal connection, then disconnect and remove the indicator switch.
6 Extract the top bush snap-ring from the column.
7 Unscrew and remove the special key-bolt from the universal joint unit, then with the steering wheel temporarily refitted to act as a puller, withdraw the shaft sufficiently to remove the top bush. The steering lock must be in the *unloaded* position.
8 Now tilt the steering column shaft and use it to push out the bottom bush.
9 Before installing the replacement bushes they must be lubricated with a lithium-based molybdenum disulphide grease (Duckhams LBM 10). Similarly lubricate the bearing surface of the shaft.
10 Install the bottom bush using a length of suitable $1\frac{3}{8}$ in (35 mm) diameter tube. With the new upper bush in position on the shaft, reassemble into the column and through the bottom bush.
11 Carefully align the shaft splines with those of the universal joint. The flat section of the shaft must correspond to that of the universal joint as in Fig. 10.17. Reinsert the key-bolt and tighten it to the specified torque.
12 Tap the top bush into its position in the column using the tube used for inserting the lower bush. When fully in position retain with the snap-ring.
13 Realign and refit the steering wheel to the shaft splines and retain with the nut which must be tightened to the specified torque.
14 Reassemble the indicator switch and lower column trim panel. When the battery is reconnected operate the indicators to ensure correct functioning.

12 Steering column intermediate shaft – removal and refitting

1 Unscrew and remove the steering column lower trim panel. Note the length and position of the various retaining screws to ensure correct installation.
2 Refer to Fig. 10.17, unscrew and remove the steering shaft universal joint key-bolt.
3 Grip the steering wheel and carefully pull the column just sufficiently to release the steering shaft from the universal joint.
4 The car must now be either run over an inspection pit or jacked up and supported with axle stands to enable the intermediate shaft flexible coupling to be released.
5 Unscrew and remove the two bolts from the flexible coupling (photo).
6 Free the sealing bellows from the scuttle and withdraw the shaft.
7 Installation is a direct reversal of the removal procedure. Tighten the coupling retaining bolts to the specified torque. Align the flat section of the shaft with that of the universal joint (see Fig. 10.17).

13 Steering shaft lower flexible coupling – removal and refitting

1 Raise the vehicle at the front and support on axle stands, or position the vehicle over an inspection pit.
2 Unscrew and remove the gear selector quadrant retaining bolt from the steering box. Loosen the clamp-bolt holding the coupling to the pinion shaft.
3 Unscrew and remove the bolts holding the intermediate shaft to the flexible coupling.
4 Unscrew and remove the two bolts securing the steering rack to the crossmember, then lower the steering rack and remove the coupling.
5 Install in the reverse order and tighten the securing bolts to the specified torque.

14 Steering rack – removal and refitting

1 Referring to the previous Section, disconnect the intermediate shaft at the flexible coupling, and detach the steering rack from the crossmember.
2 Slide the rubber gaiter along the steering arm at the pinion and detach the steering arms from the rack.
3 Remove the steering rack.
4 The installation procedure is a reversal of the removal but note the following:

(a) Renew the rubber gaiters if torn or perished
(b) If a new steering box has been fitted then the steering 'parallel' must be checked. To do this you will need two 10 mm diameter rods. Insert one in each rack end fitting and measure the distance between the rod centres. This must be set at $21\frac{1}{32}$ in (534 ± 0.5 mm), and is adjusted on the pinion side.
(c) It is essential that the rack center point is correct and this can be checked by measuring between the points shown in Fig. 10.21. The inner face of the rack end nuts must be 2·906 in (74 mm) from the steering box abutment indicated. If adjustment is necessary, this requires the use of a special C spanner to set the nut accordingly and is a task best entrusted to your Renault dealer
(d) Tighten all bolts and nuts to the specified torque settings

15 Steering arms – removal and refitting

1 Raise and support the car at the front end.
2 To remove the steering arm remove the through bolt connecting it to the rack and then with a balljoint remover or two wedges release the balljoint from the suspension strut.

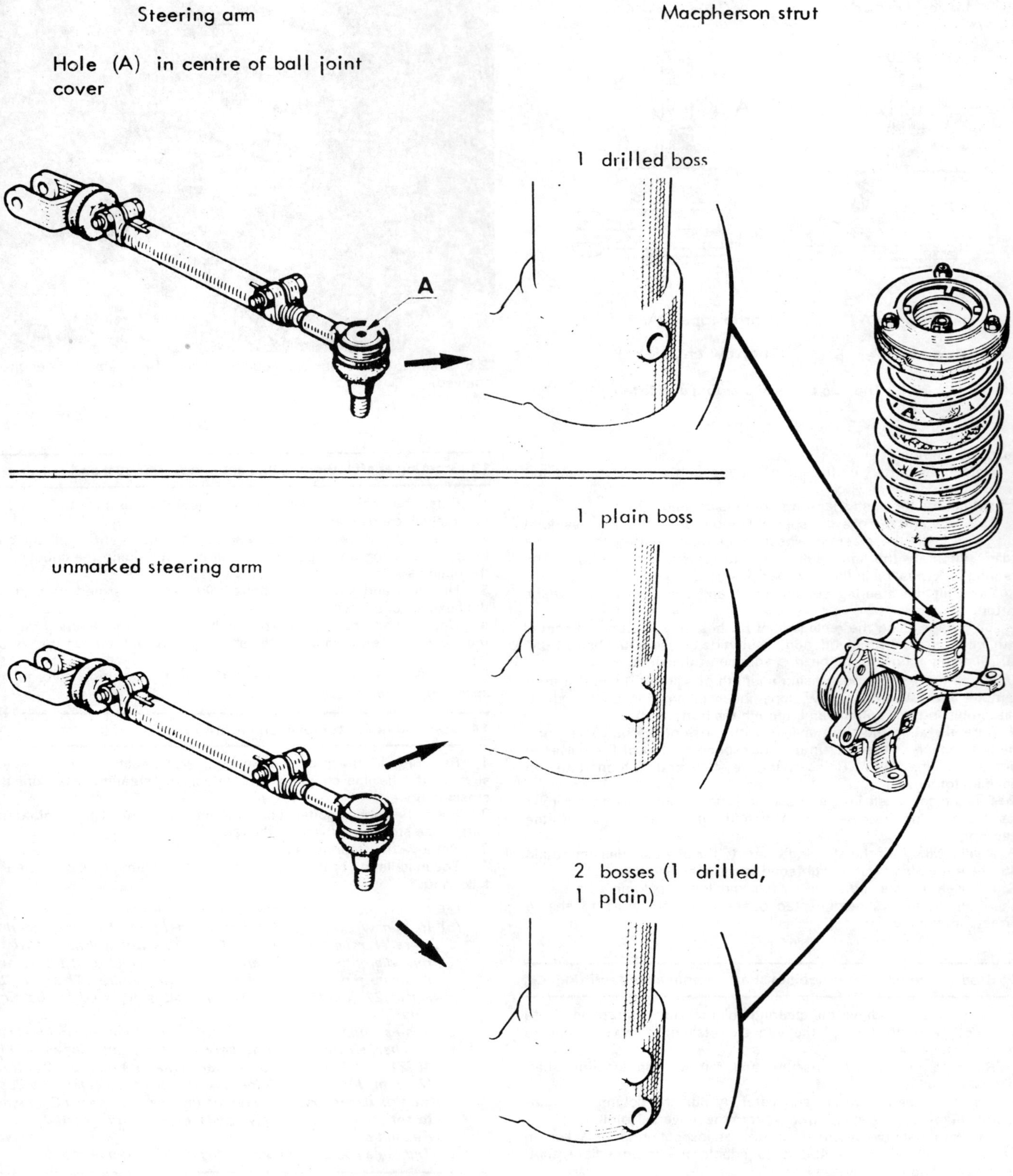

Steering arm

Hole (A) in centre of ball joint cover

Macpherson strut

1 drilled boss

unmarked steering arm

1 plain boss

2 bosses (1 drilled, 1 plain)

Fig. 10.23 The identification features of the two types of ball-joint and strut

Chapter 10 Suspension and steering

3 Check the condition of the steering arm and balljoint. No maintenance is available for this joint except the replacement of the rubber dust cover. If it is worn replace the whole steering arm. Make sure that the replacement part is of the right specification, as these arms have been modified. Two types have been produced which have different taper angles and it is of utmost importance that the correct replacement type is fitted. The two types are shown in Figs. 10.22 and 10.23.
4 Replacement is a reversal of the removal sequence. On completion have the front wheel track checked.

15 Fault diagnosis – suspension and steering

Before diagnosing faults from the following chart, check that any irregularities are not caused by:

    (a) Binding brakes
    (b) Incorrect tyres
    (c) Incorrect tyre pressures
    (d) Misalignment of the bodyframe or rear suspension

Symptom	Reason/s
Steering wheel can be moved considerably before any sign of movement of the wheels is apparent	Wear in the steering linkage, gear and column coupling
Vehicle difficult to steer in a consistent straight line – wandering	As above Wheel alignment incorrect (indicated by excessive or uneven tyre wear) Front wheel hub bearings loose or worn Worn balljoints or suspension arms
Steering stiff and heavy	Incorrect wheel alignment (indicated by excessive or uneven tyre wear) Excessive wear or seizure in one or more of the joints in the steering linkage or suspension arm balljoints Excessive wear in the steering unit Defective driveshaft joints
Wheel wobble and vibration	Roadwheels out of balance Roadwheels buckled/damaged Wheel alignment incorrect Wear in steering linkage, suspension arm balljoint or suspension arm inner bushes
Excessive pitching and rolling on corners and during braking	Defective shock absorbers Incorrect tyre inflation pressures Anti-roll bar broken away

Chapter 11 Bodywork and fittings

For modifications, and information applicable to later models, see Supplement at end of manual

Contents

Bonnet release cable – removal and refitting 9	General description 1
Bonnet – removal and refitting 8	Maintenance – body exterior 2
Bumpers – removal and refitting 14	Maintenance – interior 3
Dashboard – removal and refitting 17	Major body damage – repair 5
Door locks – removal and refitting 12	Minor body damage – repair 4
Doors – removal, refitting and adjustment 10	Roof embellisher strips – removal and refitting 18
Doors – tracing and silencing rattles 6	Tailgate – removal and refitting 13
Door trim and windows – removal and refitting 11	Windscreen and tailgate glass – preparation and
Front grille panel – removal and refitting 15	replacement 16
Front wing panels – removal and refitting 7	Windscreen wiper arms and blades – replacement 19

Specifications

Body type .... Steel monocoque construction

Dimensions
Length (overall) .... 13 ft 2¼ in (4025 mm)
Width (overall) .... 5ft 4 in (1624 mm)
Height (overall) .... 4ft 7⁵⁄₁₆ in (1405 mm)

Total kerb weight .... TL, GTL 1907 lbs (865 kg), LS, TS 1962 lbs (890 kg)

Towing weights
Trailer with brake .... 1764 lbs (800 kg)
Trailer without brake .... 937 lbs (425 kg)

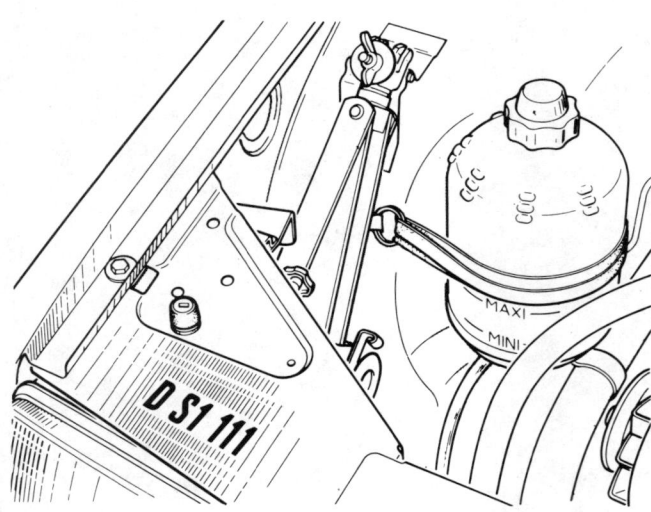

Fig. 11.1 Paint colour code position

1 General description

The Renault 14 is of monocoque construction, with separate front wings, bonnet, doors and tailgate. The engine and transmission unit are mounted in a subframe at the front which is bolted to the bodyshell.

The cavities of the lower bodyshell sections have been injected with special anti-corrosive sealants to prevent rusting in and around the respective structural joints and panels.

Where a car is to have body repairs undertaken or is to be resprayed in its original colour scheme, the paint colour code is stamped into the right-hand headlight support channel as shown in Fig. 11.1 Always quote this number to ensure getting the correct colour match when ordering the paint.

Note: *The Renault 14 is subject to an anti-corrosion warranty for a period of 5 years from delivery. The validity of this warranty, is subject to additional treatment being carried out by a Renault agent at 18 months, and after a further 24 months, from delivery.*

2 Maintenance – body exterior

1 The general condition of your car's bodywork is the one thing that significantly affects its value. Maintenance is easy but needs to be

Chapter 11 Bodywork and fittings

regular and particular. Neglect – particularly after minor damage – can quickly lead to further deterioration and costly repair bills. It is important also to keep watch on those parts of the bodywork not immediately visible, for example the underside, inside all the wheel arches and the lower part of the engine compartment.

2   The basic maintenance routine for the bodywork is washing – preferably with a lot of water from a hose. This will remove all the loose solids which may have stuck to the car. It is important to flush these off in such a way as to prevent grit from scratching the finish. The wheel arches and underbody need washing in the same way, to remove any accumulated mud which will retain moisture and tend to encourage rust.

3   Paradoxically enough, the best time to clean the underbody and wheel arches is in wet weather when the mud is thoroughly wet and soft. In very wet weather the underbody is usually cleaned of large accumulations automatically and this is a good time for inspection.

4   It is a good idea to jack up the car, remove all the roadwheels and clean their inner sides. The wheel offset keeps mud captive for a long time and could unbalance the wheel! **Do not** hose excessive quantities of water at the windows, heater vents etc.

5   Periodically have the whole of the underside steam cleaned, engine compartment as well so that a thorough inspection can be carried out to see what minor repairs and renovations are necessary. Steam cleaning is available at some garages and is necessary for removal of the accumulation of oily grime which sometimes collects thickly in areas near the engine and gearbox. If steam facilities are not available there are one or two grease solvents available which can be brush applied. The dirt can then be simply hosed off. Any signs of rust on the underside panels must be attended to immediately. Thorough wire brushing followed by treatment with an anti-rust compound, primer and underbody sealer will prevent continued deterioration. If not dealt with the car could eventually become structurally unsound and therefore unsafe.

6   After washing the paintwork wipe off with a chamois leather to give a clear unspotted finish. A coat of clear wax polish will give added protection against chemical pollutants in the air and will survive several subsequent washings. If the paintwork sheen has dulled or oxidised use a cleaner/polisher combination to restore the brilliance of the shine. This requires a little effort but is usually because regular washing has been neglected! Always check that door and drain holes and pipes are completely clear so that water can drain out. Brightwork should be treated the same way as paintwork. Windscreens and windows can be kept clear of the smeary film which often appears if a little ammonia is added to the water. If glasswork is scratched a good rub with a proprietary metal polish will often clean it. Never use any form of wax or other paint/chromium polish on glass.

3   Maintenance – interior

1   The flooring cover should be brushed or vacuum cleaned regularly to keep it free of grit. If badly stained, remove it from the car for scrubbing or sponging and make quite sure that it is dry before replacement. Seat and interior trim panels can be kept clean with a wipe over with a damp cloth. If they do become stained (which can be more apparent on light coloured upholstery especially when of the nylon 'cloth' type) use a little liquid detergent and a soft nail-brush to scour the grime out of the grain of the material. Do not forget to keep the headlining clean in the same way as the upholstery. When using liquid cleaners inside the car do not over-wet the surface being cleaned. Excessive damp could get into the upholstery seams and padded interior, causing stains, offensive odours or even rot. If the inside of the car gets wet accidentally it is worthwhile taking some trouble to dry it out properly. **Do not** leave oil or electric heaters inside the car for this purpose. If, when removing mats for cleaning, there are signs of damp underneath, all the interior of the car floor should be uncovered and the point of water entry found. It may only be a missing grommet, but it could be a rusted through floor panel and this demands immediate attention as described in the previous Section. More often than not both sides of the panel will require treatment. On cars fitted with the factory sunroof avoid touching the interior canvas. Keep it clean and rectify all tears immediately. Consult your local Renault agent as to the most suitable type of repair depending on the material used. Keep the stays and fixings very lightly but frequently oiled, particularly at the front of the roof, and periodically release and roll back the roof so that it does not become too stiff and weak.

4   Minor body damage – repair

The photo sequences on pages 142 and 143 illustrate the operations detailed in the following sub-sections.

Note: *For more detailed information about bodywork repair, the Haynes Publishing Group publish a book by Lindsay Porter called The Car Bodywork Repair Manual. This incorporates information on such aspects as rust treatment, painting and glass fibre repairs, as well as details on more ambitious repairs involving welding and panel beating.*

Repair of minor scratches in the car's bodywork

If the scratch is very superficial, and does not penetrate to the metal of the bodywork, repair is very simple. Lightly rub the area of the scratch with a paintwork renovator, or a very fine cutting paste, to remove loose paint from the scratch and to clear the surrounding bodywork of wax polish. Rinse the area with clean water.

Apply touch-up paint to the scratch using a thin paint brush; continue to apply thin layers of paint until the surface of the paint in the scratch is level with the surrounding paintwork. Allow the new paint at least two weeks to harden: then blend it into the surrounding paintwork by rubbing the paintwork, in the scratch area, with a paintwork renovator or a very fine cutting paste. Finally, apply wax polish.

Where the scratch has penetrated right through to the metal of the bodywork, causing the metal to rust, a different repair technique is required. Remove any loose rust from the bottom of the scratch with a penknife, then apply rust inhibiting paint to prevent the formation of rust in the future. Using a rubber or nylon applicator fill the scratch with bodystopper paste. If required, this paste can be mixed with cellulose thinners to provide a very thin paste which is ideal for filling narrow scratches. Before the stopper-paste in the scratch hardens, wrap a piece of smooth cotton rag around the top of a finger. Dip the finger in cellulose thinners and then quickly sweep it across the surface of the stopper-paste in the scratch; this will ensure that the surface of the stopper-paste is slightly hollowed. The scratch can now be painted over as described earlier in this Section.

Repair of dents in the car's bodywork

When deep denting of the car's bodywork has taken place, the first task is to pull the dent out, until the affected bodywork almost attains its original shape. There is little point in trying to restore the original shape completely, as the metal in the damaged area will have stretched on impact and cannot be reshaped fully to its original contour. It is better to bring the level of the dent up to a point which is about $\frac{1}{8}$ in (3 mm) below the level of the surrounding bodywork. In cases where the dent is very shallow anyway, it is not worth trying to pull it out at all.

If the underside of the dent is accessible, it can be hammered out gently from behind, using a mallet with a wooden or plastic head. Whilst doing this, hold a suitable block of wood firmly against the impact from the hammer blows and thus prevent a large area of the bodywork from being belled-out.

Should the dent be in a section of the bodywork which has double skin or some other factor making it inaccessible from behind, a different technique is called for. Drill several small holes through the metal inside the area – particularly in the deeper section. Then screw long self-tapping screws into the holes just sufficiently for them to gain a good purchase in the metal. Now the dent can be pulled out by pulling on the protruding heads of the screws with a pair of pliers.

The next stage of the repair is the removal of the paint from the damaged area, and from an inch or so of the surrounding sound bodywork. This is accomplished most easily by using a wire brush or abrasive pad on a power drill, although it can be done just as effectively by hand using sheets of abrasive paper. To complete the preparation for filling, score the surface of the bare metal with a screwdriver or the tang of a file, or alternatively, drill small holes in the affected area. This will provide a really good key for the filler paste.

To complete the repair see the Section on filling and respraying.

Repair of rust holes or gashes in the car's bodywork

Remove all paint from the affected area and from an inch or so of the surrounding sound bodywork, using an abrasive pad or a wire brush on a power drill. If these are not available a few sheets of abrasive paper will do the job just as effectively. With the paint removed you will be able to gauge the severity of the corrosion and therefore decide whether to renew the whole panel (if this is possible) or to repair the affected area. New body panels are not as expensive as most people think and it is often quicker and more satisfactory to fit a new panel than to attempt to repair large areas of corrosion.

Remove all fittings from the affected area except those which will act as a guide to the original shape of the damaged bodywork (eg headlamp shells etc). Then, using tin snips or a hacksaw blade, remove all loose metal and any other metal badly affected by corrosion. Hammer the edges of the hole inwards in order to create a slight depression for the filler paste.

Wire brush the affected area to remove the powdery rust from the surface of the remaining metal. Paint the affected area with rust inhibiting paint; if the back of the rusted area is accessible treat this also.

Before filling can take place it will be necessary to block the hole in some way. This can be achieved by the use of aluminium or plastic mesh, or aluminium tape.

Aluminium or plastic mesh is probably the best material to use for a large hole. Cut a piece to the approximate size and shape of the hole to be filled, then position it in the hole so that its edges are below the level of the surrounding body work. It can be retained in position by several blobs of filler paste around its periphery.

Aluminium tape should be used for small or very narrow holes. Pull a piece off the roll and trim it to the approximate size and shape required, then pull off the backing paper (if used) and stick the tape over the hole; it can be overlapped if the thickness of one piece is insufficient. Burnish down the edges of the tape with the handle of a screwdriver or similar, to ensure that the tape is securely attached to the metal underneath.

Bodywork repairs – filling and respraying

Before using this Section, see the Sections on dent, deep scratch, rust holes and gash repairs.

Many types of bodyfiller are available, but generally speaking those proprietary kits which contain a tin of filler paste and a tube of resin hardener are best for this type of repair. A wide, flexible plastic or nylon applicator will be found invaluable for imparting a smooth and well contoured finish to the surface of the filler.

Mix up a little filler on a clean piece of card or board – use the hardener sparingly (follow the maker's instructions on the pack) otherwise the filler will set very rapidly.

Using the applicator, apply the filler paste to the prepared area: draw the applicator across the surface of the filler to achieve the correct contour and to level the filler surface. As soon as a contour that approximates the correct one is achieved, stop working the paste – if you carry on too long the paste will become sticky and begin to pick up on the applicator. Continue to add thin layers of filler paste at twenty-minute intervals until the level of the filler is just proud of the surrounding bodywork.

Once the filler has hardened, excess can be removed using a Surform plane or Dreadnought file. From then on, progressively finer grades of abrasive paper should be used, starting with a 40 grade production paper and finishing with 400 grade wet-and-dry paper. Always wrap the abrasive paper around a flat rubber, cork, or wooden block – otherwise the surface of the filler will not be completely flat. During the smoothing of the filler surface the wet-and-dry paper should be periodically rinsed in water. This will ensure that a very smooth finish is imparted to the filler at the final stage.

At this stage the dent should be surrounded by a ring of bare metal, which in turn should be encircled by the finely 'feathered' edge of the good paintwork. Rinse the repair area with clean water, until all of the dust produced by the rubbing-down operation has gone.

Spray the whole repair area with a light coat of primer – this will show up any imperfections in the surface of the filler. Repair these imperfections with fresh filler paste or bodystopper, and once more smooth the surface with abrasive paper. If bodystopper is used, it can be mixed with cellulose thinners to form a really thin paste which is ideal for filling small holes. Repeat this spray and repair procedure until you are satisfied that the surface of the filler, and the feathered edge of the paintwork are perfect. Clean the repair area with clean water and allow to dry fully.

The repair area is now ready for final spraying. Paint spraying must be carried out in a warm, dry, windless and dust free atmosphere. This condition can be created artificially if you have access to a large indoor working area, but if you are forced to work in the open, you will have to pick your day very carefully. If you are working indoors, dousing the floor in the work area with water will lay the dust which would otherwise be in the atmosphere. If the repair area is confined to one body panel, mask off the surrounding panels; this will help to minimise the effects of a slight mis-match in paint colours. Bodywork fittings (eg chrome strips, door handles etc) will also need to be removed or masked off. Use genuine masking tape and several thicknesses of newspaper for the masking operations.

Before commencing to spray, agitate the aerosol can thoroughly, then spray a test area (an old tin, or similar) until the technique is mastered. Cover the repair area with a thick coat of primer; the thickness should be built up using several thin layers of paint rather than one thick one. Using 400 grade wet-and-dry paper, rub down the surface of the primer until it is really smooth. While doing this, the work area should be thoroughly doused with water, and the wet-and-dry paper periodically rinsed in water. Allow to dry before spraying on more paint.

Spray on the top coat, again building up the thickness by using several thin layers of paint. Start spraying in the centre of the repair area and then using a circular motion, work outwards until the whole repair area and about 2 inches of the surrounding original paintwork is covered. Remove all masking material 10 to 15 minutes after spraying on the final coat of paint.

Allow the new paint at least two weeks to harden, then, using a paintwork renovator or a very fine cutting paste, blend the edges of the paint into the existing paintwork. Finally, apply wax polish.

5 Major body damage – repair

1 Because the car is built without a separate chassis frame and the body is therefore integral with the underframe, major damage must be repaired by competent mechanics with the necessary welding and hydraulic straightening equipment.
2 If the damage has been serious it is vital that the body is checked for correct alignment as otherwise the handling of the car will suffer and many other faults such as excessive tyre wear and wear in the transmission and steering may occur.
3 There is a special body jig which most large body repair shops have, and to ensure that all is correct it is important that the jig be used for all major repair work.

6 Doors – tracing and silencing rattles

1 Having established that a rattle does come from the door(s) check first that it is not loose on its hinges and the latch is holding it firmly closed. The hinges can be checked by rocking the door up and down when open to detect any play. If the hinges are worn at the pin the hinge pin and possibly the inner hinge will need renewal. When the door is closed the panel should be flush. If not then the hinges or latch striker plate need adjustment. The door hinges are welded to the door and bolted to the pillars. The hinge pins can be driven out using an impact hammer. To adjust the setting of the door catch first slacken the screws holding the striker plate to the door pillar just enough so that it can be moved but will hold its position. Then close the door, with the latch button pressed, and then release the latch. This is so that the striker plate position is not drastically disturbed on closing the door. Then set the door position by moving it without touching the catch, so that the panel is flush with the bodywork and the other door. This will set the striker plate in the proper place. Then carefully release the catch so as not to disturb the striker plate, open the door and tighten the screws. Rattles within the door will be due to loose fixtures

Chapter 11 Bodywork and fittings

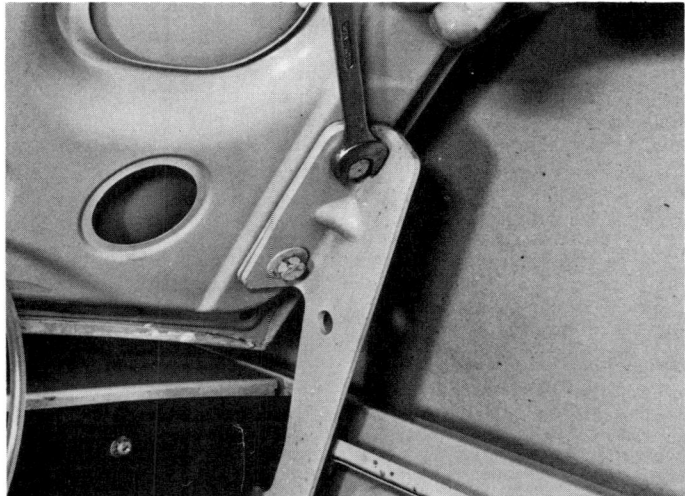

8.3 Removing the bonnet hinge bolts (note shim)

or something having been dropped inside them. Do make sure that all sealing rubbers are effective and that the prop stay is not itself loose.

7 Front wing panels – removal and refitting

1 Fortunately the front wing panels, which are probably the most exposed of the body panels, are secured in position by bolts and can therefore be easily removed for repair or replacement. Before removing a wing panel obtain a sealer which will be required for reassembly.
2 Commence by disconnecting the battery earth connection.
3 Remove the front bumper unit as described in Section 15.
4 Detach and withdraw the headlight unit from the side to be removed. This is described in Chapter 9.
5 Unscrew the retaining bolts and remove the undertray on the side concerned.
6 Drill or grind the rivet heads from the front grille trim to detach it from the wing (earlier models only).
7 Referring to Fig. 11.2, remove the outer to inner wing panel bracket, 1 at the lower front corner.
8 Unscrew the retaining nut from within the headlight aperture (Fig. 11.3). This nut retains the wing panel to the front panel crossmember.
9 Unscrew and remove the three nuts retaining the rear edge of the wing panel to the door pillar.
10 Unscrew and remove the three nuts retaining the top edge of the wing panel, within the drain channel. Remove the wing.
11 Installation is a direct reversal of the removal procedure but a suitable body sealant must be applied to the joint faces. Ensure that the wing is correctly aligned before finally tightening the securing bolts. On completion coat the underside of the panel with underseal or suitable underbody protective coating.
12 When the headlight is reassembled, check the alignment as described in Chapter 9.

8 Bonnet – removal and refitting

1 Open the bonnet and have an assistant support it.
2 Mark around the hinges to show their correct location to facilitate reassembly and then unscrew the hinge retaining screws on each side and remove the bonnet.
3 The bonnet can be removed complete with the hinges by unscrewing the retaining nuts to the inner wing panels.
4 Refit the bonnet in the reverse order and check it for alignment before fully tightening the retaining nuts. To adjust the bonnet for height in relation to the wing panels screw the rubber bump stops (snubbers) in or out to lower or raise respectively.

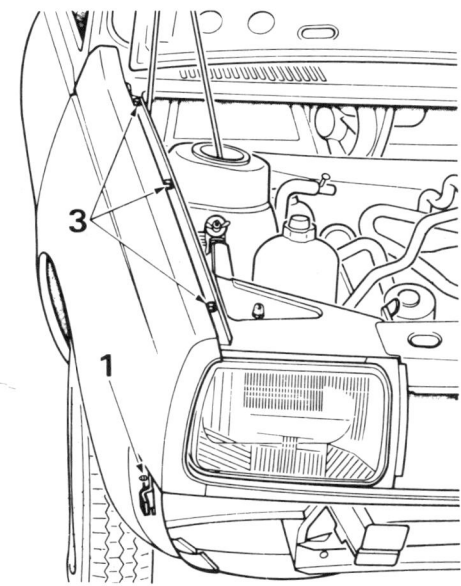

Fig. 11.2 The wing panel retaining bolt positions at the top (3) and side (1)

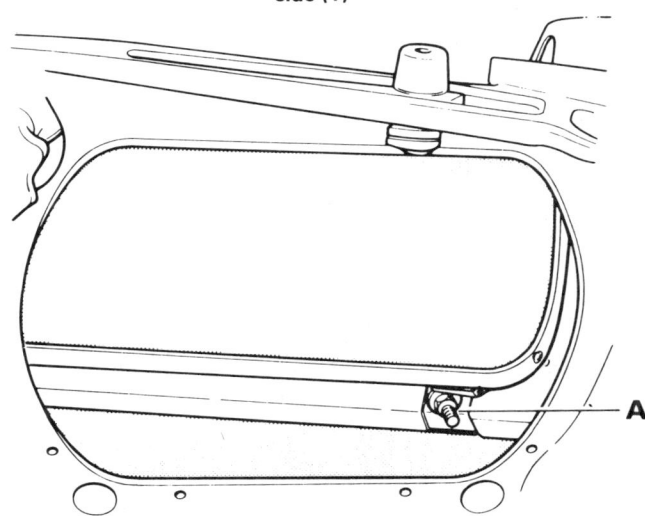

Fig. 11.3 The wing panel to crossmember retaining nut position (A)

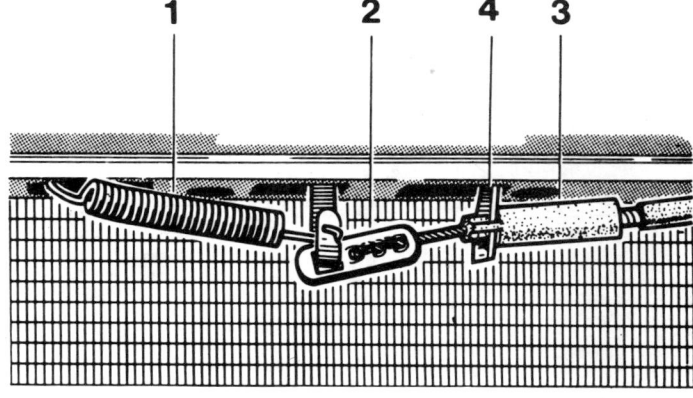

Fig. 11.4 The bonnet release cable and catch spring connection

1 Spring
2 Cable end
3 Sleeve
4 Stop

140 Chapter 11 Bodywork and fittings

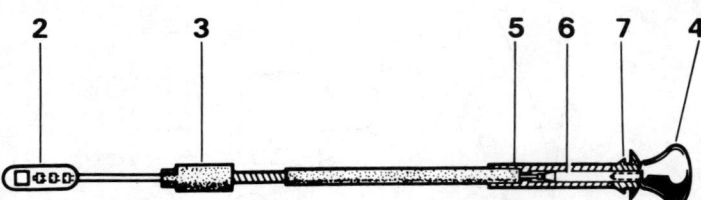

Fig. 11.5 The cable assembly

2 Endpiece   5 Guide sleeve
3 Sleeve   6 Stem
4 Knob   7 Lugs

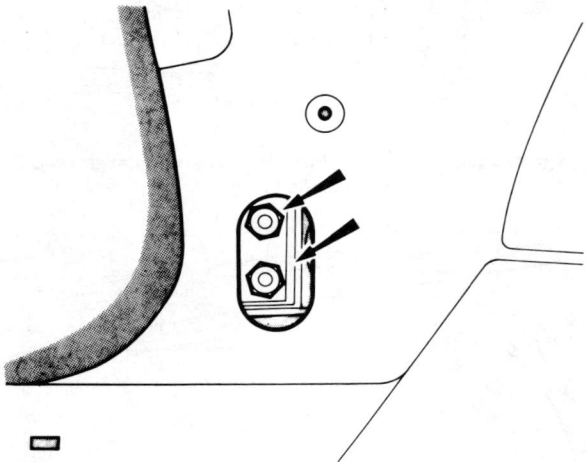

Fig. 11.6 The adjustment nuts for the front door

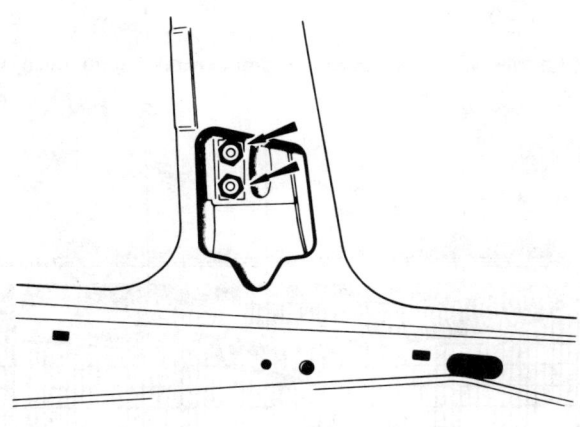

Fig. 11.7 The rear door adjustment nuts

9 Bonnet release cable – removal and refitting

1   Raise and support the bonnet. If the cable has broken, the bonnet release catch can be released by removing the front grille panel.
2   Unhook the cable return spring from the end of the cable (Fig. 11.4) and detach the sleeve from the stop.
3   Detach the cable from its retaining clips within the engine compartment.
4   From within the car, pull the cable knob so that the cable stem protrudes beyond the guide sleeve.
5   The sleeve is retained in position by two lugs. These lugs can now be pinched inwards to release the guide sleeve and cable assembly from the retaining bracket.
6   Install the replacement cable in the reverse order, but make sure that the guide sleeve is in position and secure before inserting the knob and stem. Check the bonnet release action for correct operation on completion.

10 Doors – removal, refitting and adjustment

Note: *Rear doors are fitted with childproof locks which are operative with the lever in the UP position, not in the DOWN position as stated in some Renault driver's handbooks.*

Removal and refitting

1   Both front and rear doors can be removed in the same manner. First open the door to be removed and support it underneath (without lifting it).
2   Using a suitable drift punch, tap out the upper and lower hinge pins.

Fig. 11.8 Remove winder mechanism (1) through aperture shown

Note: Mechanism retaining stud holes (2) and taped window (3)

Chapter 11 Bodywork and fittings

3 Drift out the roll-pin from the check strap using a 5 mm diameter punch and remove the door.
4 Installation is a direct reversal of the removal procedure. Use new hinge pins if the old ones are worn and lubricate before assembly. Check the door for alignment when closed.

Adjustment
5 Should the doors be in need of adjustment for height remove the trim from the central or front pillar, as appropriate, to gain access to the adjustment nuts (Figs. 11.6 and 11.7). Loosen the nuts, adjust the door accordingly and retighten the nuts. Refit the trim.
6 To adjust the door laterally, shims are inserted between the hinges and the door pillar. If the hinges are to be disturbed scribe a mark around their profiles before removing to facilitate the correct alignment on reassembly.

11 Door trim and windows – removal and refitting

Trim panel
1 To remove the inner trim panel from the door, unscrew and remove the door pull/armrest.
2 Unscrew the inner door handle control (photo) and detach catch wire.
3 Prise back the plastic cover in the window winder handle and using a suitable socket or box wrench unscrew the retaining bolt (photo).
4 Use a screwdriver with a wide blade and carefully prise the trim panel from the door which is retained in position by spring clips (photo).
5 Pull the plastic sealant from the panel to gain access to the door catch, or window winder mechanism.
6 Reassemble the trim panel in the reverse sequence.

Front door window
7 With the trim panel removed, wind the window down to approximately halfway. Wedge it or tape it to secure.
8 Refer to Fig. 11.8 and remove the four nuts that secure the winder mechanism.
9 Unscrew and remove the roller guide retaining nuts.
10 Push the winder assembly to detach it from the door and then disconnect the counterbalance rollers from the lower window channel. The mechanism can then be withdrawn through the inner panel aperture at the bottom.
11 Remove the wedge or tape from the window and, tilting it down at the front, carefully withdraw it.
12 Installation of the glass and winder mechanism is basically the reverse of removal but do not fully tighten the winder mechanism retaining bolts until the roller guide is fitted and the window is wound fully up. Lubricate the mechanism and rollers during assembly.

Rear door windows
13 Wind down the windows a distance of $5\frac{1}{2}$ inches (140 mm) measured from the top edge of the window to the underside of the top horizontal location channel (A in Fig. 11.9) Wedge or tape the window in this position.
14 Remove the door trim panel.
15 Unscrew and remove the three winder mechanism retaining screws. Push the mechanism inwards to detach.
16 Disconnect the counterbalance roller from the lower window winder channel and withdraw the mechanism unit through the aperture in the inner door panel B.
17 Unscrew and remove the screw from the top of the vertical window guide channel at the rear, and unscrew its lower retaining screw from within the door (Fig. 11.10).
18 Unclip the rubber guide from the vertical channel and also the rubber strip from the top edge of the door panel.
19 Unclip and remove the rubber strip retaining clip at the rear and remove the window by tilting it and withdrawing it upwards.
20 Installation is a direct reversal of the removal procedure. Renew the rubber strip and guide if they are perished or damaged (also their retaining clips). Lubricate the winder mechanism and roller assemblies. Check window winding operation before refitting the trim panel.

Rear door fixed window
21 Remove the main window as described above.

11.2 Inner door handle screw

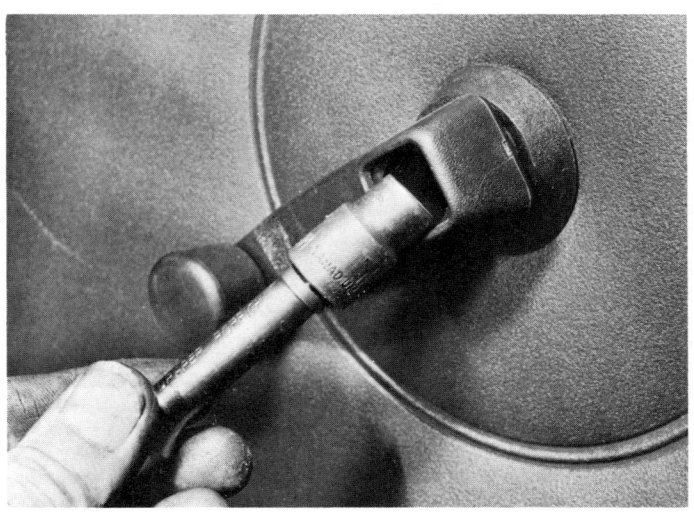
11.3 Remove window winder handle bolt

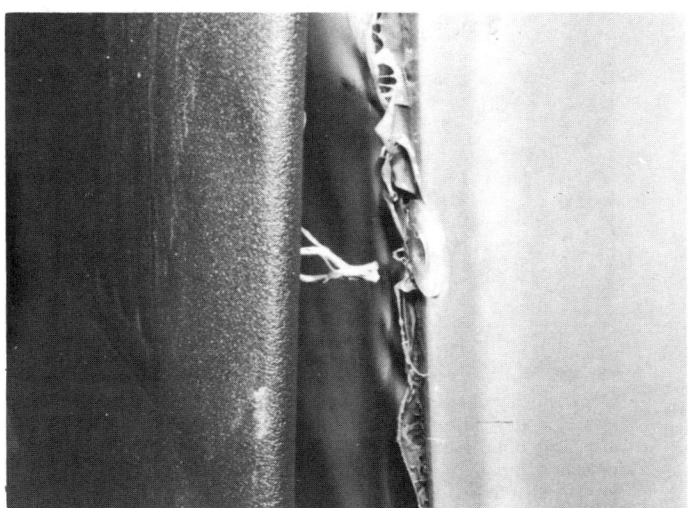
11.4 Pull the panel from the nylon clip retainers

This sequence of photographs deals with the repair of the dent and paintwork damage shown in this photo. The procedure will be similar for the repair of a hole. It should be noted that the procedures given here are simplified – more explicit instructions will be found in the text

In the case of a dent the first job – after removing surrounding trim – is to hammer out the dent where access is possible. This will minimise filling. Here, the large dent having been hammered out, the damaged area is being made slightly concave

Now all paint must be removed from the damaged area, by rubbing with coarse abrasive paper. Alternatively, a wire brush or abrasive pad can be used in a power drill. Where the repair area meets good paintwork, the edge of the paintwork should be 'feathered', using a finer grade of abrasive paper

In the case of a hole caused by rusting, all damaged sheet-metal should be cut away before proceeding to this stage. Here, the damaged area is being treated with rust remover and inhibitor before being filled

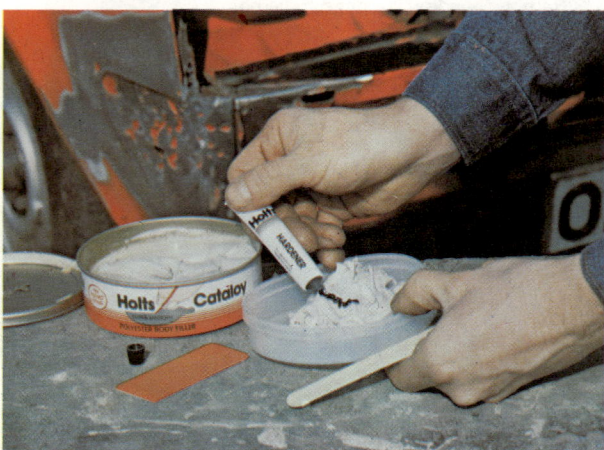

Mix the body filler according to its manufacturer's instructions. In the case of corrosion damage, it will be necessary to block off any large holes before filling – this can be done with aluminium or plastic mesh, or aluminium tape. Make sure the area is absolutely clean before ...

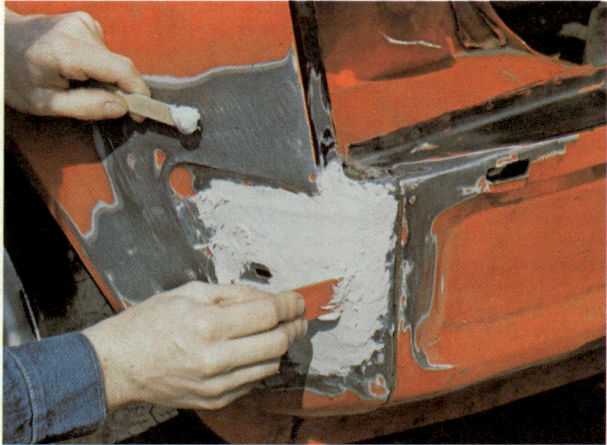

... applying the filler. Filler should be applied with a flexible applicator, as shown, for best results; the wooden spatula being used for confined areas. Apply thin layers of filler at 20-minute intervals, until the surface of the filler is slightly proud of the surrounding bodywork

Initial shaping can be done with a Surform plane or Dreadnought file. Then, using progressively finer grades of wet-and-dry paper, wrapped around a sanding block, and copious amounts of clean water, rub down the filler until really smooth and flat. Again, feather the edges of adjoining paintwork

The whole repair area can now be sprayed or brush-painted with primer. If spraying, ensure adjoining areas are protected from over-spray. Note that at least one inch of the surrounding sound paintwork should be coated with primer. Primer has a 'thick' consistency, so will find small imperfections

Again, using plenty of water, rub down the primer with a fine grade wet-and-dry paper (400 grade is probably best) until it is really smooth and well blended into the surrounding paintwork. Any remaining imperfections can now be filled by carefully applied knifing stopper paste

When the stopper has hardened, rub down the repair area again before applying the final coat of primer. Before rubbing down this last coat of primer, ensure the repair area is blemish-free – use more stopper if necessary. To ensure that the surface of the primer is really smooth use some finishing compound

The top coat can now be applied. When working out of doors, pick a dry, warm and wind-free day. Ensure surrounding areas are protected from over-spray. Agitate the aerosol thoroughly, then spray the centre of the repair area, working outwards with a circular motion. Apply the paint as several thin coats

After a period of about two weeks, which the paint needs to harden fully, the surface of the repaired area can be 'cut' with a mild cutting compound prior to wax polishing. When carrying out bodywork repairs, remember that the quality of the finished job is proportional to the time and effort expended

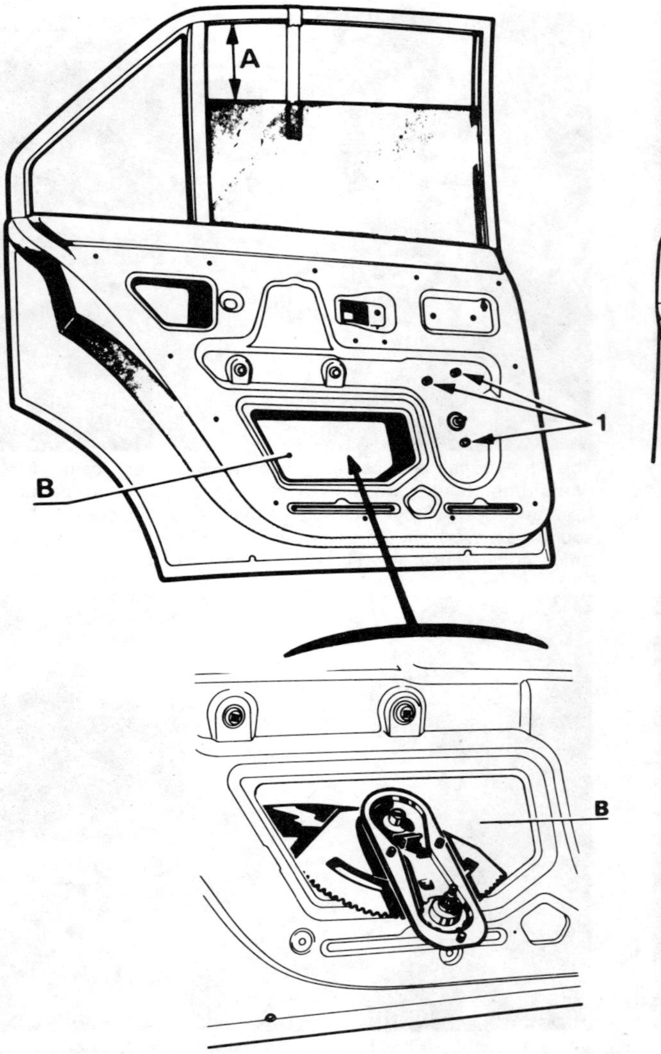

Fig. 11.9 Rear door showing window glass position (A), mechanism retaining stud holes (1) and removal through aperture (B)

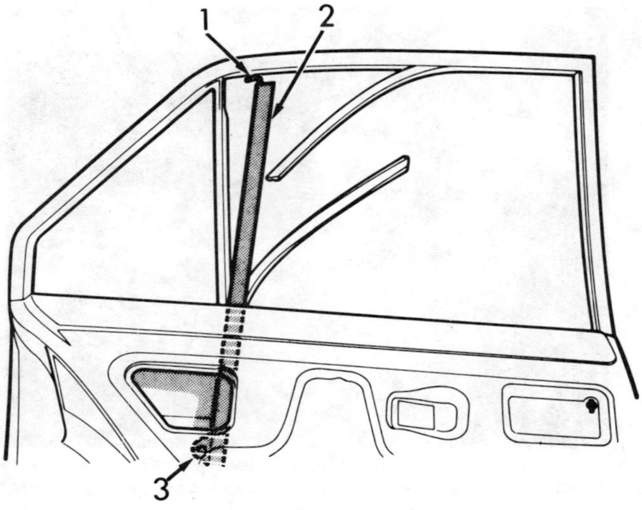

Fig. 11.10 The vertical guide

1 Top retaining screw   2 Guide frame
3 Lower retaining screw

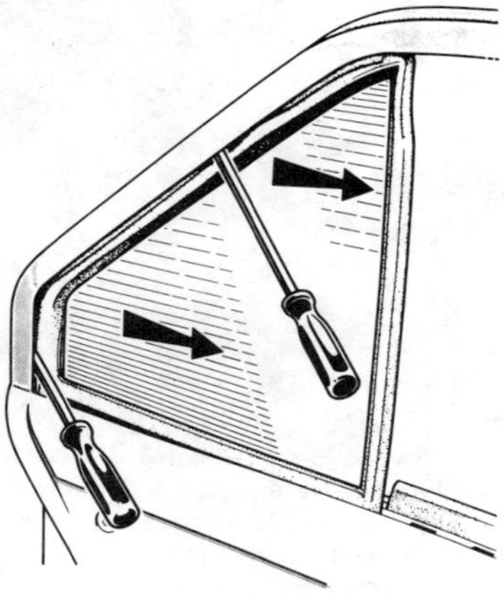

Fig. 11.11 Lever the glass as shown

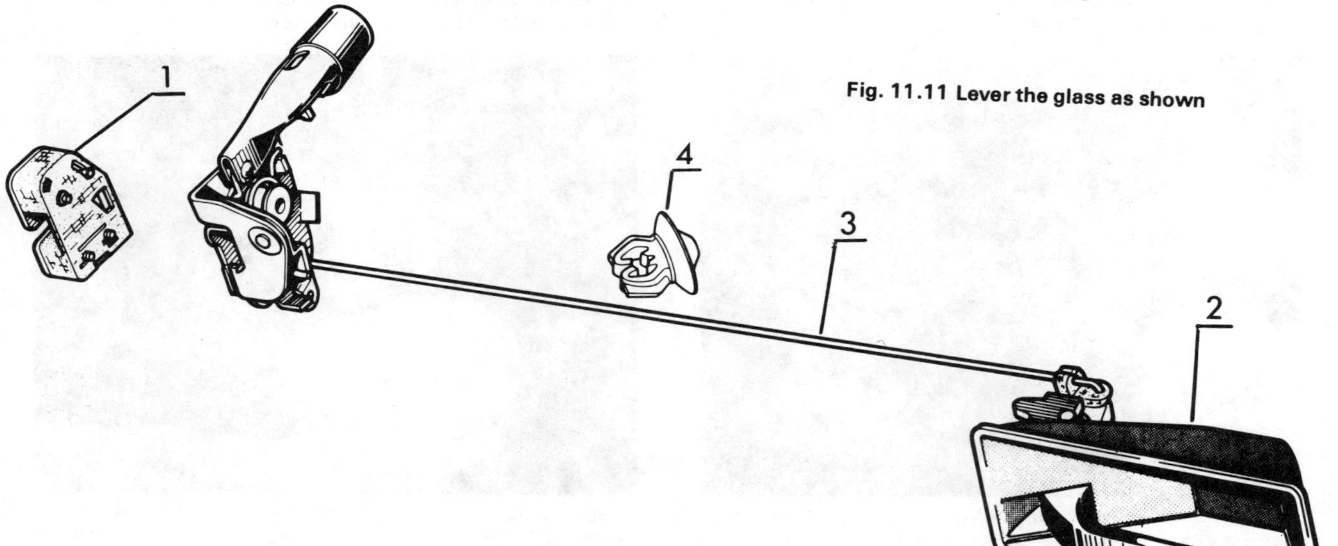

Fig. 11.12 The front door lock assembly

1 Lock   2 Control lever   3 Control rod   4 Support clips

Chapter 11 Bodywork and fittings

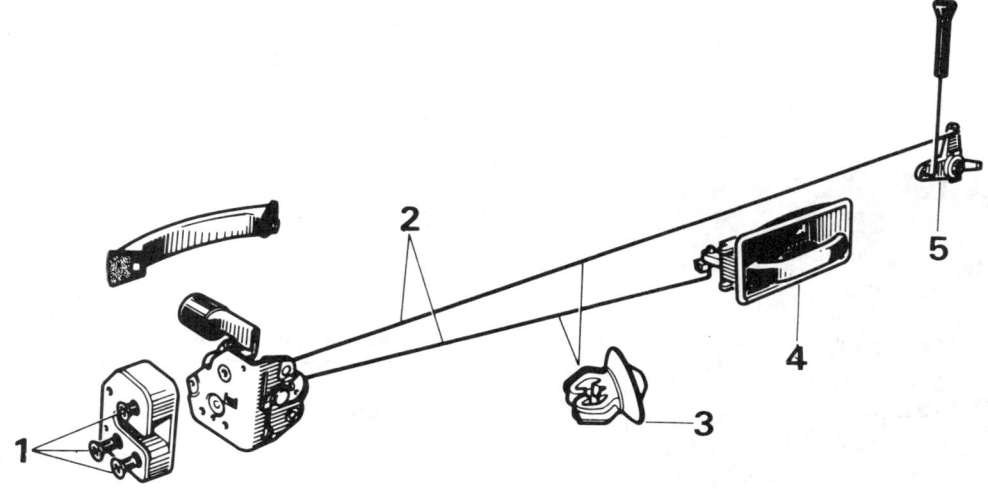

Fig. 11.13 The rear door lock assembly

1 Lock screws 2 Control rods 3 Support clip 4 Control lever 5 Lock button

22 With the vertical window guide channel removed the fixed window can be prised forwards to remove.

23 Insert two suitable screwdrivers between the frame and the glass/surround rubber as shown and lever the glass forwards to withdraw (Fig. 11.11) complete with rubber surround.

24 Install in the reverse order to removal but renew the rubber seal surround if it is perished or damaged.

12 Door locks – removal and refitting

Front doors

1 Remove the door trim as described in Section 11.
2 Unscrew the three special lock retaining screws (photo).
3 Unclip the lock control rod from the location clip. Detach the remote control.
4 Press in the lock release button and pivot it from the horizontal to the vertical plane and remove it through the aperture in the door.
5 Installation of the lock assembly is a direct reversal of the removal procedure.
6 The exterior door handles can be removed by unscrewing the retaining nut from within the door panel.

Rear doors

7 The removal procedure for the rear door locks is much the same as that for the fronts but note the following additional points.
8 The lock button is removed by unscrewing.
9 The lock button plate at the forward end of the door is released from the door by turning it to align the retaining tabs with the slotted holes in the door frame. Extract the plate unit from within the door cavity.

Tailgate lock

10 This is not easily removed without the aid of the special Renault tool (CAR 757), and considering the low relative cost, replacement is therefore best entrusted to a Renault dealer.

13 Tailgate – removal and refitting

1 Disconnect the battery earth terminal.
2 Open up the tailgate and detach the rear window demister wire at the connector (Fig. 11.15).
3 The counterbalance arms must now be disconnected at their lower end from the rear side panel. To do this simply lift the tab to release the arm from its balljoint. If the tab is stiff assist its movement with a screwdriver.

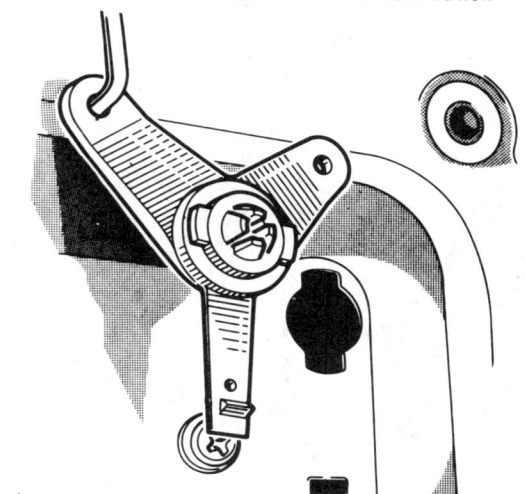

Fig. 11.14 The lock button plate and locating hole

12.2 Door lock retaining screws

Chapter 11 Bodywork and fittings

13.4 Tailgate hinge nuts aperture (covers removed)

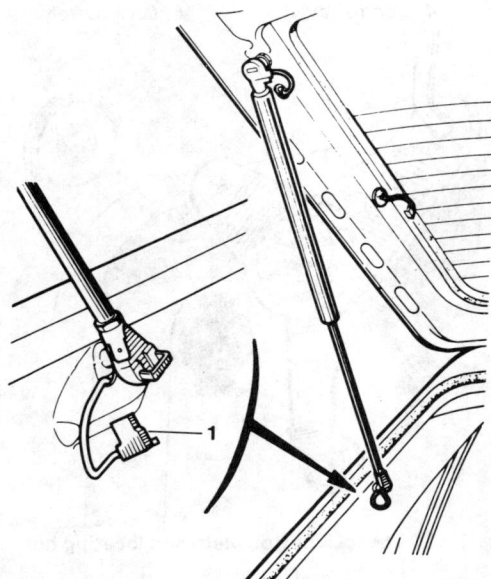

Fig. 11.15 The tailgate counterbalance arm and demister connection (1)

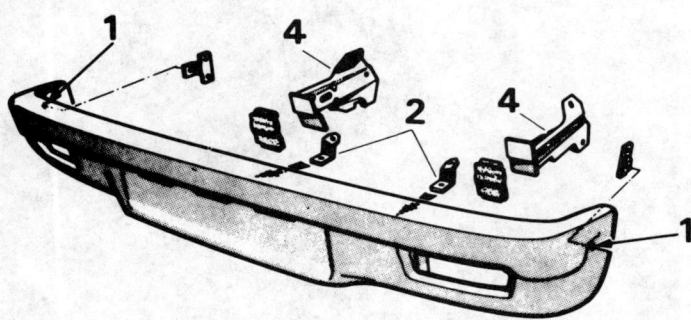

Fig. 11.16 The front bumper components

1 Side retaining brackets 4 Impact brackets
2 Centre location brackets

4 Use a screwdriver and carefully prise free the hinge covers from the top rear section of the roof panel. Using a socket or box wrench unscrew the four hinge retaining nuts (photo), and remove the tailgate. An assistant will be required to support the tailgate whilst the nuts are being removed and to help lift it clear once detached.
5 Note: *Never attempt to repair or dismantle the counterbalance arms – they contain compressed gas. Avoid scratching the piston rod as this will allow gas to escape during operation and render the arms useless. If the arms are defective they must be renewed as they are not repairable.*
6 Install in the reverse order to removal and adjust to align correctly. Shims are available to adjust the door to fit flush with the aperture whilst side play is adjustable via slotted holes for the hinge bolts.

14 Bumpers – removal and refitting

Front
1 Disconnect the battery earth terminal.
2 Detach the front combination light wires.
3 Unscrew and remove the fixing screw on each side of the bumper.
4 Unscrew and remove the two retaining bolts to the brackets on the bumper underside and remove the bumper unit complete with lights. The bumper bears on impact brackets, to remove these it will be necessary to detach the front grille panel for access to the retaining bolts.
5 Install in the reverse order of removal and on completion check the operation of the side and indicator lights. Check bumper position for correct alignment.

Rear
6 Lift open the tailgate and from inside the luggage area unscrew and remove the bumper side fixing screws (two each side).
7 Unscrew and remove the two retaining screws underneath the bumper and remove it from the car.
8 Install in the reverse order and check for correct alignment.

15 Front grille panel – removal and refitting

1 Open the bonnet and unscrew the two retaining screws from the top of the grille panel (Fig. 11.18).
2 Disconnect the central clip A and lift the grille clear of the three location dowel holes.
3 Install in the reverse order to removal.

16 Windscreen and tailgate glass – preparation and replacement

1 To replace a front or rear screen glass clean out the window surround and remove the rubber seal. Check that it is in good condition – not stretched, perished or cracked. However, do make sure that your fitting the glass is necessary – many windscreen replacement companies fit free. On windscreen fitting, if the previous glass has been broken plug the gap between the facia panel and window as well as closing the ventilator flap, to avoid splinters entering.
Special Note: Refitment of the heated rear window takes special care. If you are in doubt, don't! Leave it to one with this experience.
2 Fit the rubber seal around the glass and lay on a good flat working surface. Insert a piece of string 3 to 4 mm (0.11 to 0.15 in) in diameter into the slot of the seal. Pass it totally around the seal and exit it with about 5 inches overlap and 8 inches hanging near a bottom corner.
3 Position the glass and seal into the window from outside with the string inside hanging, with the help of an assistant.
4 Square the glass up, hold, press where the string overlaps. Inside, pull each string in turn and gradually feed into the seal. Withdraw string pressing from inside, and check for good seal.

17 Dashboard – removal and refitting

1 Disconnect the battery earth terminal.
2 Unscrew and remove the steering column lower cover. Take note of the various screw locations as they are of different lengths.
3 Disconnect the speedometer cable.
4 Remove the heater control knob retaining clips, and detach the knobs.

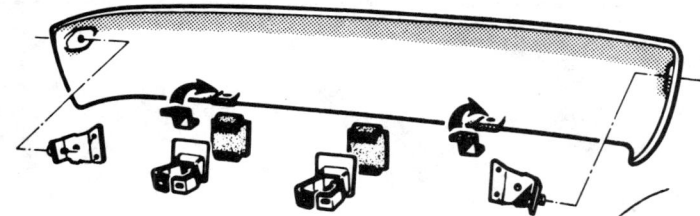

Fig. 11.17 The rear bumper and its location brackets

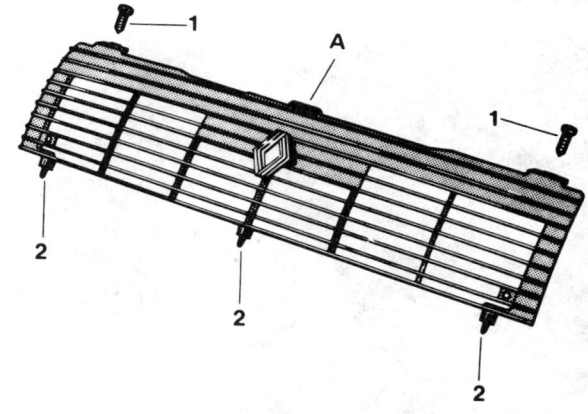

Fig. 11.18 The front grille panel showing retaining screws (1), location pegs (2) and central clip (A)

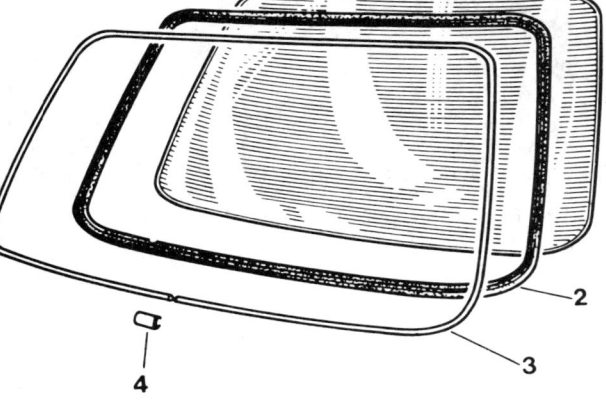

Fig. 11.19 Windscreen (1), rubber seal (2), trim (3), and joint clip (4)

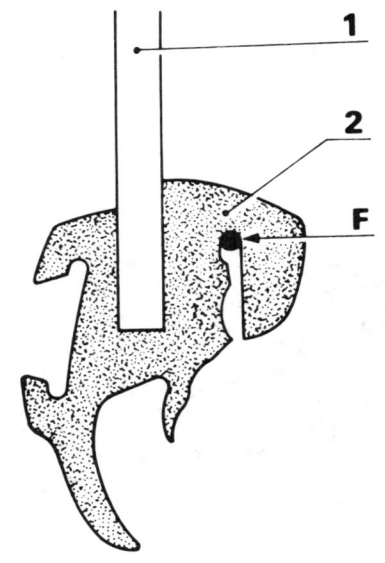

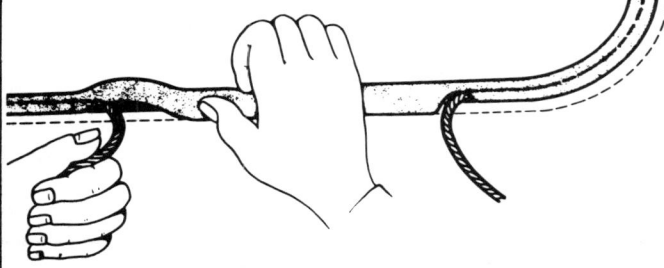

Fig. 11.21 Method of locating seal

Fig. 11.20 Cross-section of rubber seal (2), glass (1) and string position (F) ready for fitting

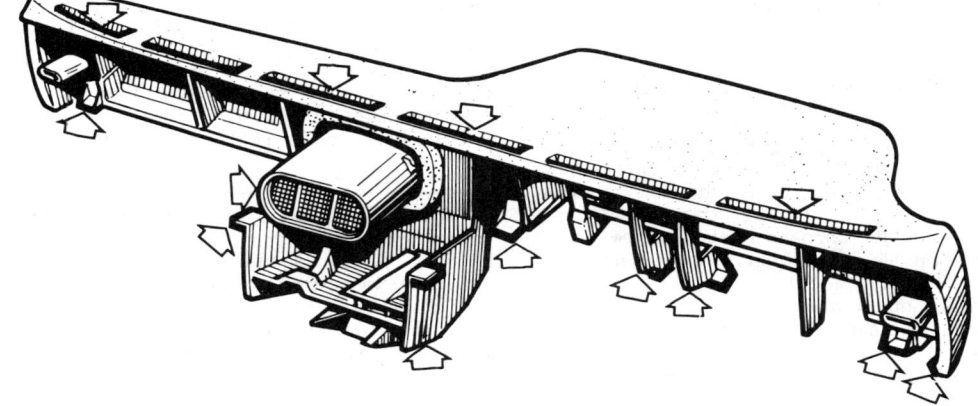

Fig. 11.22 The dashboard retaining screw positions (LH drive version shown)

Chapter 11 Bodywork and fittings

19.1 Windscreen wiper arm retaining nut

19.4 Twist arm round and pull to release

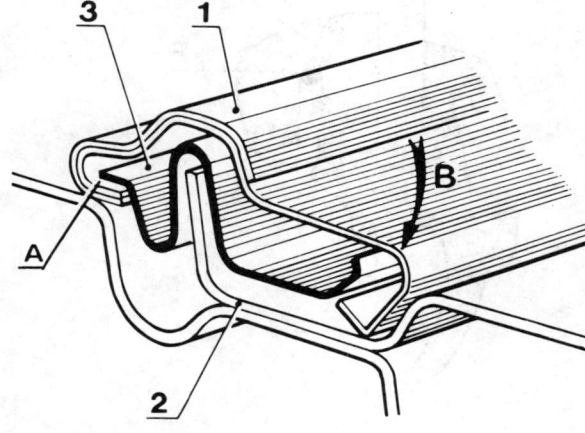

Fig. 11.23 The roof embellisher strip location

1 Moulding   2 Stiffener   3 Clip

Insert moulding at A and press into position at B

5 Detach the junction block connectors to the instrument panel.
6 Refer to Fig. 11.22 and unscrew and remove the four top panel retaining screws.
7 Unscrew and remove the four bottom panel retaining screws.
8 Unscrew and remove the retaining screw on the instrument panel side.
9 Remove the dash panel central retaining screws.
10 Remove the two screws retaining the panel to the steering column.
11 The panel can now be removed. This is made easier if the steering wheel is withdrawn – see Chapter 10.
12 Install in the reverse order of removal. When the wiring connections are in place check all the circuit operations before refitting the lower steering column cover. Don't forget to reconnect the speedometer cable.

18 Roof embellisher strips – removal and refitting

1 If for any reason the roof embellisher strips are to be removed special care must be taken not to scratch or chip the roof paint during removal and refitting.
2 Carefully prise the moulding strip from the stiffener/clip using a suitably protected screwdriver, working from one end to the other. Do not distort or bend the strip any more than is necessary when removing.
3 Install in the reverse order, fitting the retaining clips to the stiffener edge and then progressively pressing the embellisher down into position. Use new clips if the old ones are rusty or damaged.

19 Windscreen wiper arms and blades – replacement

1 Pivot the retaining nut cover up and unscrew the arm-to-pivot nut using a suitable spanner (photo).
2 Pull the arm and blade from the pivot.
3 When installing replacement, position on splines and semi tighten the retaining nut and check wiper operation. Relocate as necessary to achieve the optimum sweep and to allow correct parking, then tighten retaining nut.
4 To remove the blade only from the arm, compress the locking lever, twist the blade round and then pull it from the arm to release (photo).
5 Fit replacement in reverse sequence.

Chapter 12 Supplement:
Revisions and information on later models

Contents

Introduction	1
Specifications	2
Engine	3
Engine (1360 cc) – description	
Timing chain guide – engine Type 150	
Timing cover (5-speed gearbox) – removal and refitting	
Oil pump (Type 150 and X5J engines)	
Fuel system	4
Carburettors – general	
Carburettors (later models) – idle adjustment	
Solex 32 PBISA-11 – description	
Solex 32 PBISA-11 – adjustments	
Weber 32 IBR carburettor – description	
Weber 32 IBR carburettor – adjustments	
Solex 32-35 CICSA – description	
Solex 32-35 CICSA – adjustments	
Exhaust pipe (later models)	5
General	
Transmission (five-speed)	6
Description	
Gearchange selector mechanism – removal and refitting	
Primary shaft 5th speed gear – removal and refitting	
Secondary shaft 5th speed gear – removal and refitting	
Rectification of reverse gear fault	
Braking system	7
Braking system components – general	
Master cylinder (Bendix) – removal, overhaul and refitting	
Front disc pads (Bendix) – inspection and renewal	
Rear brake shoes (Bendix) – inspection and renewal	
Disc caliper (Bendix) – removal, overhaul and refitting	
Brake disc (Bendix) – inspection, removal and refitting	
Brake drum (Bendix) – inspection, removal and refitting	
Rear wheel cylinder (Bendix) – removal, overhaul and refitting	
Brake hydraulic system – bleeding	
Electrical system	8
Fuses	
Alternator – later models with integral regulator	
Integral regulator – renewal (Paris-Rhone)	
Brush holder – renewal (Paris-Rhone)	
Changing alternator with remotely-mounted regulator for alternator with integral regulator	
Central door locking system – description	
Electric door lock – removal and refitting	
Electrically-operated window gear – removal and refitting	
Heated rear window	
Rear screen washer/wiper	
Rear fog lamp – installation	
Auxiliary fog and driving lamps	
Radio – installation	
Aerial – installation	
Loudspeakers – installation	
Radio wiring connection	
Radio – interference suppression	
Instruments and warning lights	
Bodywork	9
Towing bracket – installation	

1 Introduction

This Supplement covers the modifications to the Renault 14 range up to and including 1982.

The main items include the Type 150 and X5J engines, the five-speed gearbox also the Bendix braking systems and supplementary electrical accessories.

1982 vehicles equipped with the X5J engine are designated R1213.

2 Specifications

Engine Type 150 and Type X5J

Capacity	1360 cc	
Bore	75.0 mm	
Stroke	77.0 mm	
Compression ratio	9.3 : 1	
Valve timing:	Type 150	Type X5J
Inlet valve opens	12° BTDC	8° BTDC
Inlet valve closes	52° ABDC	40° ABDC
Exhaust valve opens	52° BBDC	40° BBDC
Exhaust valve closes	12° ATDC	8° ATDC

Ignition timing

	Type 150	Type X5J
Initial advance setting	3° ± 1 BTDC	10° ± 1 BTDC
Cam angle	57° ± 3	57° ± 3
Dwell percentage	63% ± 3	63% ± 3

Carburettors

Solex 32 PBISA-11

	Mark 714, 770	Mark 715
Choke tube	25	25
Main jet	127.5	122.5
Air compensating jet	170	160
Idle jet	43	45
Fuel inlet valve	1.5	1.5
Throttle valve plate initial opening Mark 714	0.75	0.70
Mark 770	0.70	–
Float level adjustment	35.5 to 37.5 mm	35.5 to 37.5 mm
Auxiliary jet	30	30
Enrichment device	65	–
Accelerator pump jet	40	40
Accelerator pump stroke	3.0 mm	4.0 mm
Dashpot part-open setting	2.3 mm	–
Diaphragm take-up clearance	1.5 mm	–

Weber 32 IBR 100

Choke tube	24.5
Main jet	130
Air compensating jet	190
Idle jet	47
Diffuser	4.5
Emulsion tube	F 87
Accelerator pump	60
Accelerator pump stroke	4.5 mm
Fuel inlet needle valve	1.5
Throttle valve plate initial opening	0.75 mm
Float level adjustment	6.5 mm
Float arm stroke	7.0 mm
Choke dashpot part-open setting	4.5 mm

Solex 32 – 35 CICSA Mark 703

	Primary	Secondary
Choke tube	24	24
Main jet	117.5	120
Idle jet	39	50
Auxiliary jet	30	–
Air compensating jet	185	120
Accelerator pump jet	45	35
Fuel inlet needle valve	1.5	–
Throttle valve plate initial opening	0.80 mm	–
Float level	41.0 mm	–
Choke dash pot part-open setting	3.0 mm	–

5-Speed gearbox

Type	BH5 (415) – Five forward speeds (all synchromesh) and reverse
Application	R1212, R1213
Ratios	
1st	3.083 : 1
2nd	1.823 : 1
3rd	1.192 : 1
4th	0.892 : 1
5th	0.718 : 1
Reverse	2.833 : 1
Final drive	
R1212	3.353 : 1
R1213	3.866 : 1

3 Engine

Engine (1360 cc) – description

1 The Types 150 and X5J engines are basically identical with the Type 129 featured in Chapter 1. The difference in engine capacity is achieved by varying the stroke.

2 Reference should be made to the Specifications Section of this Supplement for other engine differences and also to the following Sections in respect of Type X5J engine overhaul procedures.

Timing chain guide – engine Type 150

3 A guide shoe is bolted to the engine cylinder block on this type of engine.

Timing cover (5-speed gearbox) – removal and refitting

4 Where a Type 150 or X5J engine is fitted in conjunction with a five-speed gearbox, before the timing cover can be removed (engine in

Chapter 12 Supplement

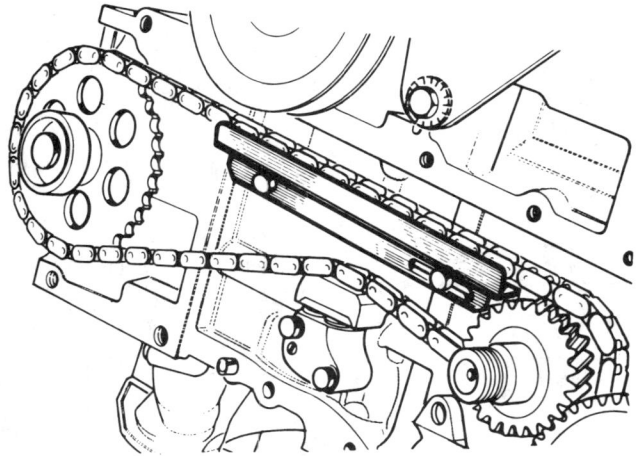

Fig. 12.1 Timing chain guide shoe on Type 150 engine

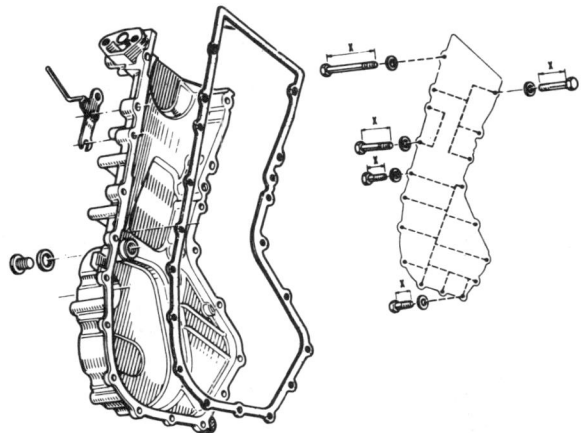

Fig. 12.3 Timing cover fixing bolt locations

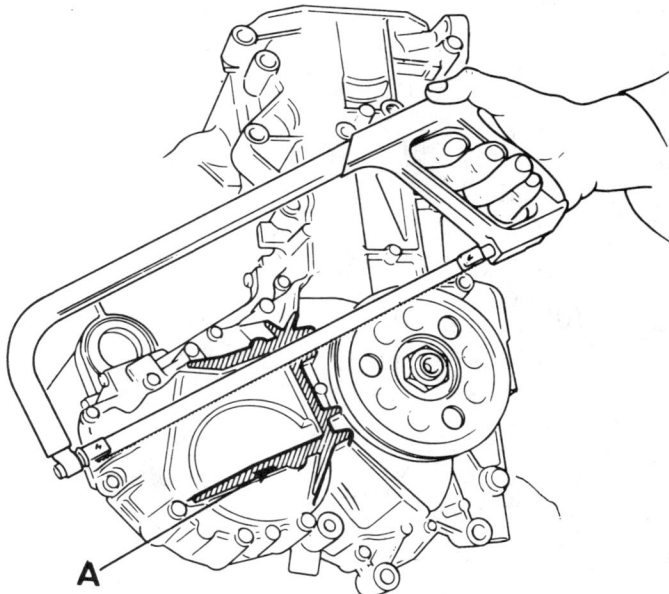

Fig. 12.2 Sawing off timing cover casting projection (A) from Type 150 engine

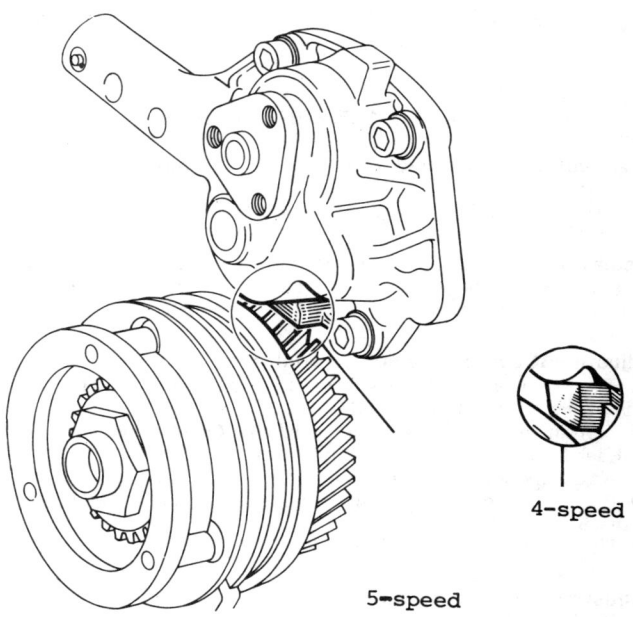

Fig. 12.4 Oil pump fitted in conjunction with 4 and 5 speed gearboxes

car) the casting projection (A) on the cover must be sawn off with a hacksaw (Fig. 12.2).

5 Now raise the engine about 20.0 mm ($\frac{3}{4}$ in) and unscrew the bottom bolt from the timing cover.

6 Unscrew and remove the two engine mounting nuts from the right-hand side.

7 Remove the retaining timing cover bolts and pull the cover forward.

8 When refitting the cover, note carefully the varying lengths of the fixing bolts which must be screwed into their original holes.

Oil pump (Type 150 and X5J engines)

9 It should be noted that where a five-speed gearbox is fitted, the boss on the oil pump is omitted to avoid fouling 5th speed gear.

10 Make sure that if a new oil pump is installed, it is of correct type.

4 Fuel system

Carburettors – general

1 Later models, particularly if equipped with a larger capacity engine may be fitted with one of three additional types of carburettor. These include the Solex 32-35 CICSA, Solex 32 PBISA, Weber 32 IBR.

Carburettors (later models) – idle adjustment

2 Before attempting to adjust the idle speed, have the engine at normal operating temperature, choke control fully off, and all electrical accessories switched off.

3 The air cleaner should be in position and the ignition system correctly adjusted.

4 Use the figures in the table as a guide to adjustment.

Carburettor	Idle speed	CO content
Solex 32-35 CICSA	850 to 900 rpm	2.0 to 2.5%
Solex 32 PBISA	775 to 825 rpm	2.0 to 2.5%
Weber 32 IBR	775 to 825 rpm	2.0 to 2.5%

5 With the engine idling, turn the idle speed (volume) screw (A) – Fig. 12.5, 12.6 or 12.7 – until the speed matches that specified in the table. If the car is not equipped with a tachometer, connect an independent instrument in accordance with the manufacturer's instructions.

6 The fuel mixture screw (B) is pre-set during production and should not normally require alteration unless the carburettor has been

Chapter 12 Supplement

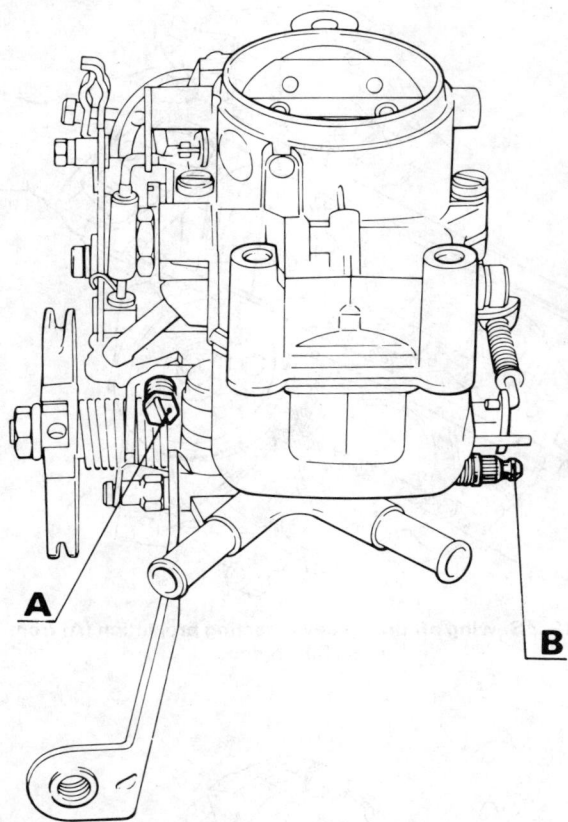

Fig. 12.5 Weber IBR carburettor

A   Idle speed (volume) screw   B   Idle mixture (fuel) screw

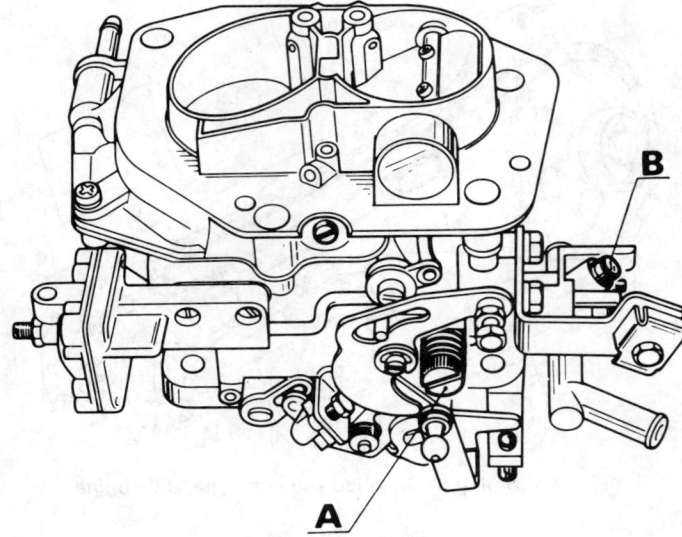

Fig. 12.6 Solex CICSA carburettor

A   Idle speed (volume) screw   B   Idle mixture (fuel) screw

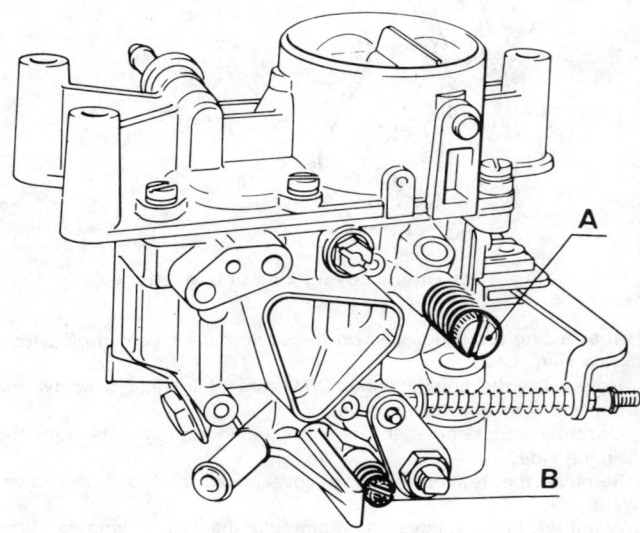

Fig. 12.7 Solex 32 PBISA II carburettor

A   Idle speed (volume) screw   B   Idle mixture (fuel) screw

overhauled or engine characteristics have changed owing to wear or carbon build-up. In this case, the tamperproof cap will have to be broken from the screws.

Adjustment using exhaust gas analyser
7 Turn the idle speed screw in or out as necessary until the engine runs at the specified idle speed.
8 Turn the mixture screw until the exhaust CO level indicated on the analyser is within the permitted tolerance.
9 Adjust the idle speed as necessary.
10 Repeat the adjustment until both CO level and idle speed conform to the Specifications.
11 Fit a new tamperproof cap.

Adjustment without exhaust gas analyser
12 Break off the tamperproof cap.
13 Turn the idle speed screw until the idle speed reaches its maximum level within its specified range (see paragraph 4).
14 Now turn the fuel mixture screw until the idle speed reaches its maximum level.
15 Repeat the two operations until the engine is idling at specified level.
16 Now screw in the mixture screw to slightly weaken the mixture and reduce the idle speed by 25 rpm but without causing the engine to run roughly.
17 Fit a new tamperproof cap.

Solex 32 PBISA-11 – description
18 This carburettor is of vertical downdraught type with manually-operated choke (cold start).
19 The main features include a main jet system, a constant richness idle circuit, a coolant warmed throttle valve block, an accelerator pump and a full throttle enrichment device.

Solex 32 PBISA-11 – adjustments
20 These operations will normally only be required after overhaul or the fitting of new parts to the carburettor.

Accelerator pump stroke
21 Insert a gauge rod or twist drill, of equivalent diameter to the accelerator pump stroke dimension given in Specifications, between the edge of the throttle valve plate and the wall of the throttle block bore.
22 Check that the accelerator pump rod is at the end of its stroke. If it is not, adjust the position of the nut on the rod.

Dashpot part-open setting
23 Close the choke flap (3) and measure the diaphragm take-up

Chapter 12 Supplement

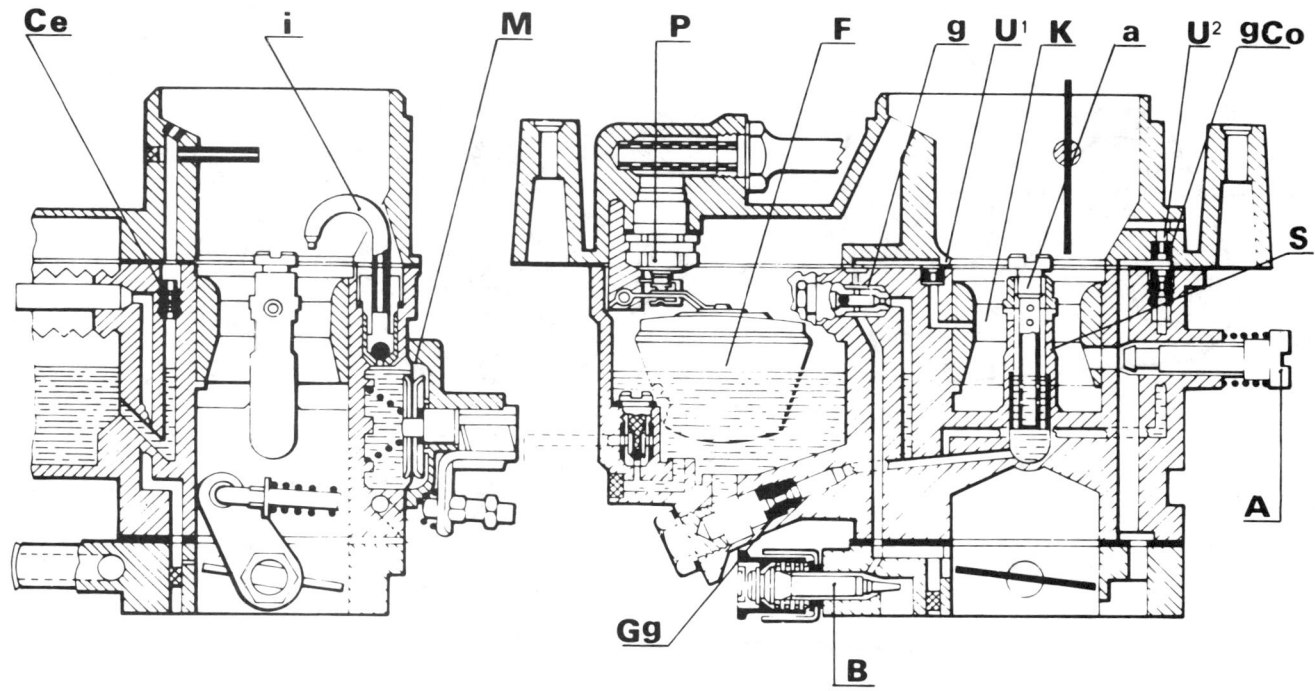

Fig. 12.8 Sectional view of Solex 32 PBISA II carburettor

K	Choke tube	U^1	Idle air calibrated orifice	Ce	Calibrated econostart jet	F	Float
Gg	Main jet	P	Needle valve	M	Accelerator pump diaphragm	A	Idle speed (volume) screw
g	Idle jet	gCO	Auxiliary jet	i	Accelerator pump jet	B	Idle mixture (fuel) screw
a	Air compensating jet	U^2	Air calibrated orifice				
S	Emulsion tube						

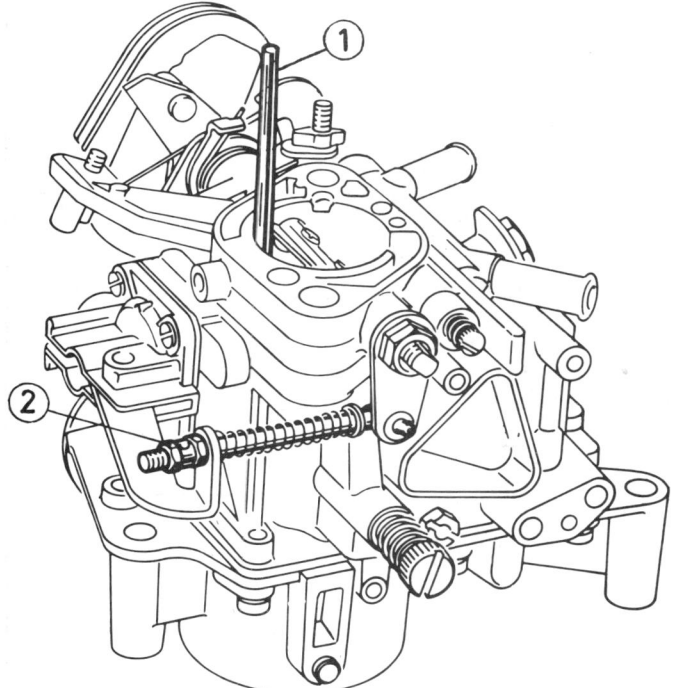

Fig. 12.9 Accelerator pump stroke adjustment (Solex 32 PBISA II)

1 Gauge rod
2 Rod adjuster nut

clearance (Y). If the dimension is outside the specified tolerance, bend the diaphragm bracket (5) (Fig. 12.10).

24 Now move the choke operating cam (3) so that the operating rod (6) is up against the adjusting screw (7) (Fig. 12.11).
25 Measure the gap between the edge of the smaller offset of the choke valve plate and the wall of the top cover bore. Use a twist drill or gauge rod to do this. If the gap is different from that specified, turn the adjuster screw.

Float level
26 Remove the top cover and hold it upside down so that the float rests under its own weight on the fuel inlet needle valve.
27 Check dimension (A) which should be as given in the Specifications. If it is not, carefully bend the float arm (1) (Fig. 12.12).

Weber 32 IBR carburettor – description
28 This carburettor is similar to the Solex 32 PBISA 11 described earlier in this Supplement and incorporates all the features of the Solex unit.

Weber 32 IBR carburettor – adjustments
Initial throttle opening
29 Using a gauge rod or twist drill, check the gap between the edge of the throttle valve plate and the wall of the throttle block whilst the choke valve plate is held closed by means of the operating lever (3).
30 Where adjustment is required, remove the tamperproof cap and turn the screw (4) (Fig. 12.14).

Dashpot part-open setting
31 Close the choke valve plate fully.
32 Move the lever (L) to set the diaphragm in the open position.
33 Measure the gap at the edge of the larger offset of the choke valve plate. If this is outside the specified limits, bend the rod (6) (Fig. 12.15).

Accelerator pump stroke
34 Adjustment is carried out in a similar way to that described for the Solex 32 PBISA 11 carburettor, but note the difference in pump strokes between the units.

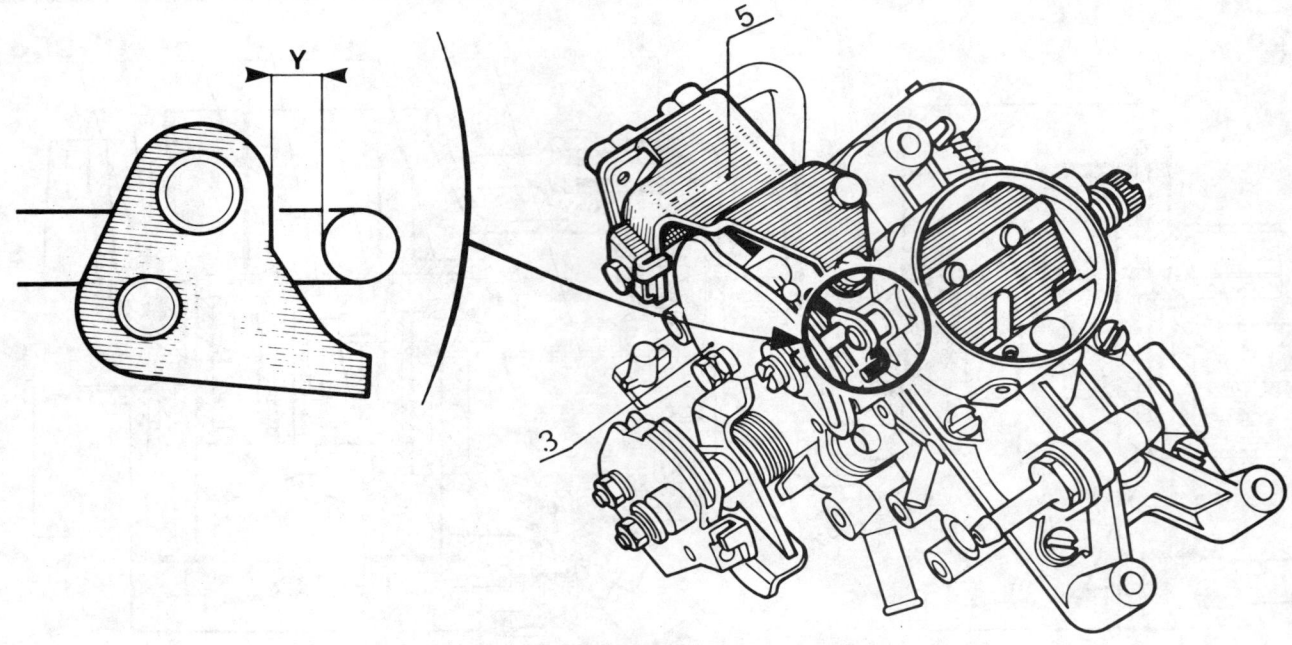

Fig. 12.10 Dashpot part-open setting (Solex 32 PBISA II)

3 Choke vacuum plate 5 Bracket Y Take-up clearance 1.5 mm

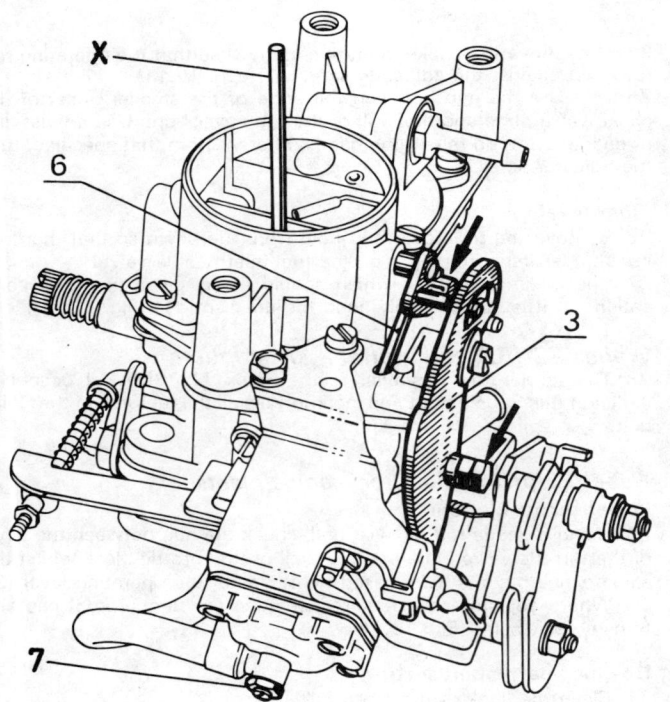

Fig. 12.11 Dashpot part-open setting (Solex 32 PBISA II)

3 Choke valve plate
6 Push-rod
7 Adjuster screw

X Gauge rod for part-open setting 2.3 mm

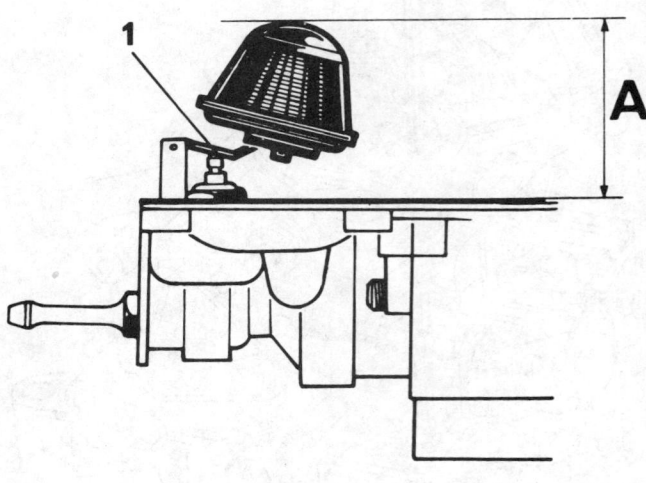

Fig. 12.12 Float setting (Solex 32 PBISA II)

1 Arm A 35.5 to 37.5 mm

Chapter 12 Supplement 155

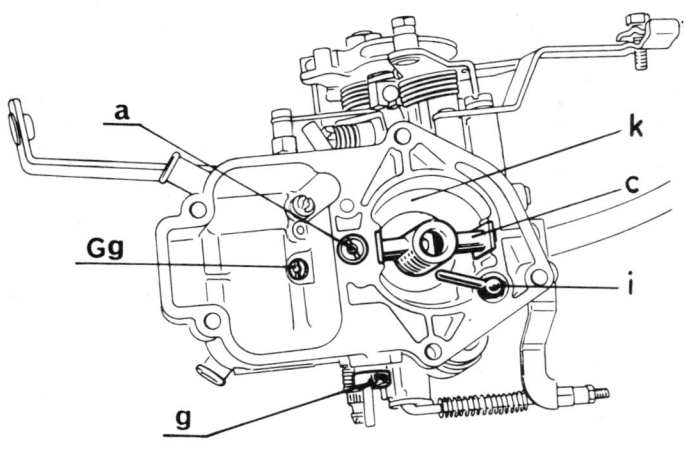

Fig. 12.13 Weber 32 IBR carburettor

k	Choke tube	g	Idle jet
Gg	Main jet	C	Diffuser
a	Air compensating jet	i	Accelerator pump

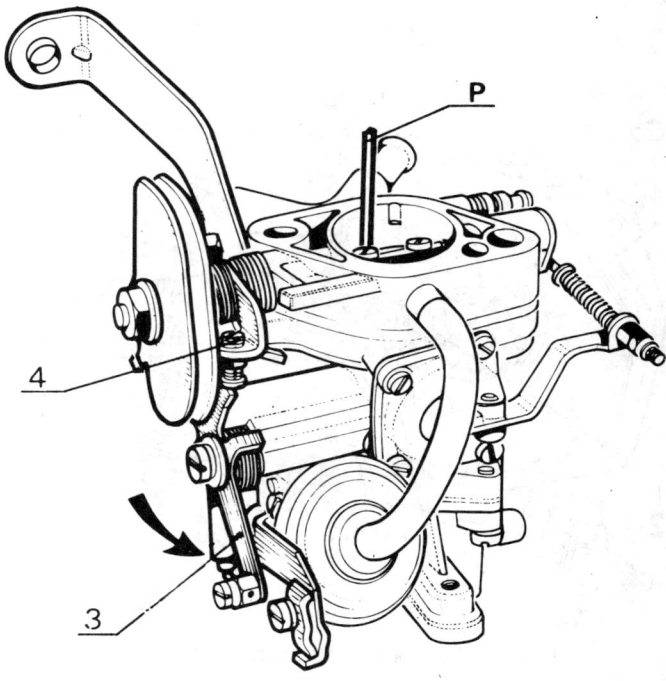

Fig. 12.14 Initial throttle opening (Weber 32 IBR)

3	Operating lever	P	Gauge rod 0.75 mm
4	Adjuster screw		(throttle plate opening)

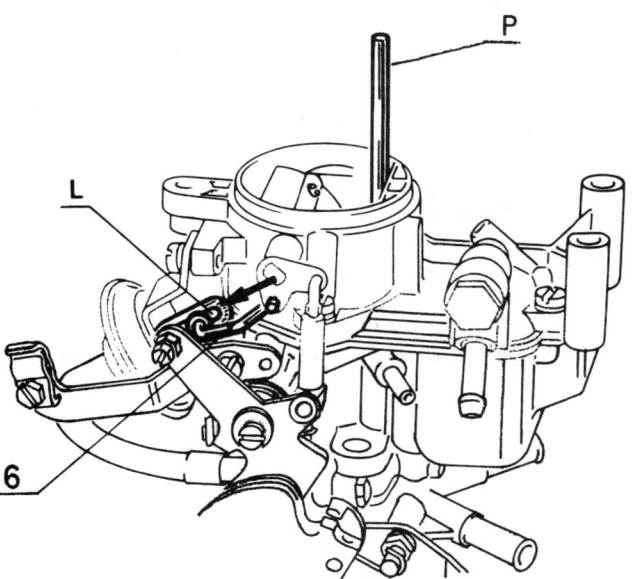

Fig. 12.15 Dash pot part-open setting (Weber 32 IBR)

6	Link rod	P	Gauge rod 4.5 mm (choke
L	Lever		plate opening)

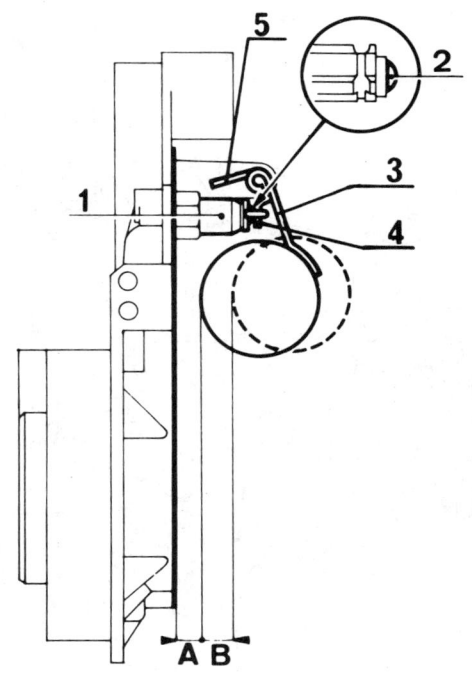

Fig. 12.16 Float setting diagram (Weber 32 IBR)

1	Fuel inlet needle valve	A	Float setting, inlet
2	Ball		valve closed 6.25 to
3	Float arm		6.75 mm
4	Tongue	B	Float stroke 7.0 mm
5	Tab		

Float level

35 Remove the carburettor top cover with gasket and hold them vertically so that the float hangs under its own weight.
36 Check dimension (A) – gasket surface to float. If it is not as specified, carefully bend the float arm (3) (Fig. 12.16).
37 Now check the float travel (B) which should be as specified. If not, bend the tab (5).

Chapter 12 Supplement

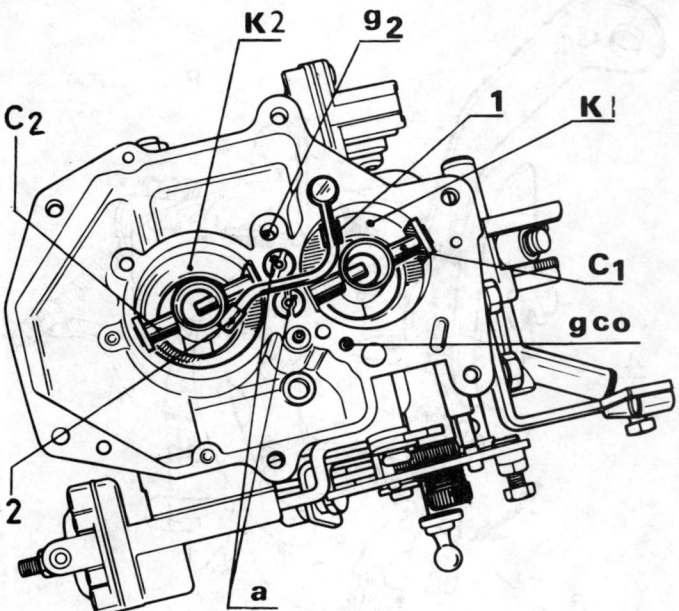

Fig. 12.17 Solex 32 – 35 CICSA carburettor

a	Air compensating jets	K1	Primary choke tube
Gg	Main jets	K2	Secondary choke tube
C1	Primary choke tube diffuser	1	Accelerator pump jet (primary)
C2	Secondary choke tube diffuser	2	Accelerator pump jet (secondary)
gco	Auxiliary idling jet		

Solex 32-35 CICSA – description

38 This is a dual barrel carburettor with a mechanically delayed opening of the secondary throttle butterfly.

39 The carburettor incorporates the following features. A manually operated choke on the primary barrel and a choke interlock mechanism to prevent secondary throttle opening whilst the choke is operative. In addition is a dash-pot part-open setting for the choke valve plate, a cam-operated accelerator pump, and a fuel return line.

Solex 32-35 CICSA – adjustments

Initial throttle opening

40 This is carried out in a similar way to that described for the Weber 321BR carburettor.

Choke butterfly part-open setting

41 Push the lever (3) as far as it will go to close the choke valve plate fully (Fig. 12.19).

42 Push the diaphragm capsule rod (T) fully in the direction of the arrow and measure the gap (P) between the edge of the choke valve plate and the wall of the bore, using a gauge rod or twist drill.

43 If the gap is outside that specified, turn the nut (J).

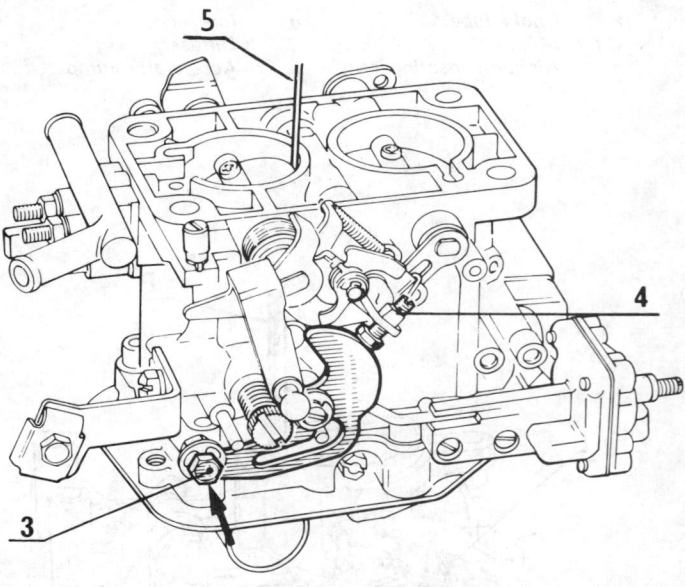

Fig. 12.18 Initial throttle opening (Solex 32 – 35 CICSA)

3 Lever
4 Adjuster screw
5 Gauge rod 0.80 mm (throttle valve plate)

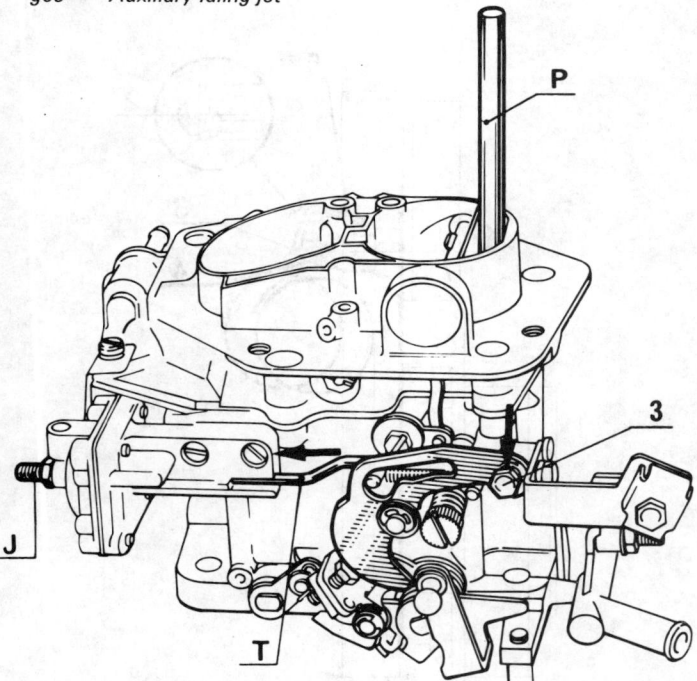

Fig. 12.19 Choke valve plate part-open setting (Solex 32 – 35 CICSA)

3 Lever
J Adjuster nut
P Gauge rod 3.0 mm (choke valve plate)
T Dashpot rod

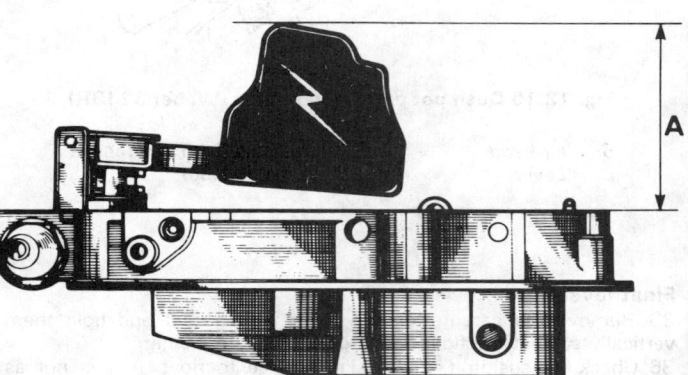

Fig. 12.20 Float setting diagram (Solex 32 – 35 CICSA)

A 41.0 mm

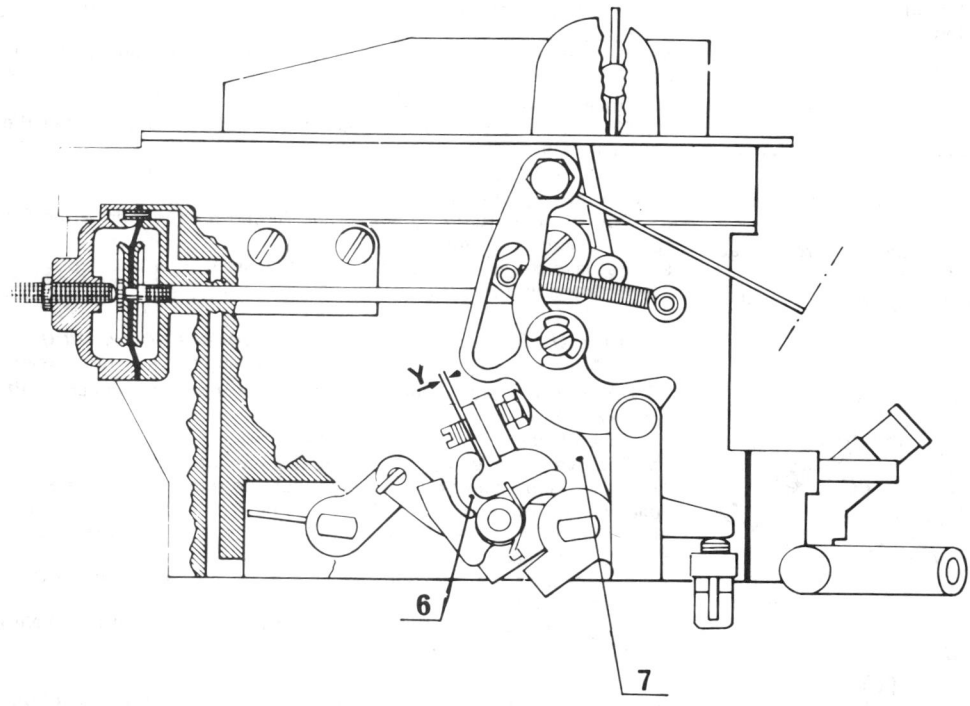

Fig. 12.21 Secondary barrel choke interlock (Solex 32 – 35 CICSA)

6 Arm
7 Primary barrel lever
Y Clearance 0.2 mm (0.008 in)

Float level

44 Remove the top cover and hold it upside down. Measure the gap (A) – joint face to float base. This should be as specified. If not, carefully bend the float arm (Fig. 12.20).

Secondary barrel choke interlock

45 All other adjustments must be completed before the secondary barrel interlock adjustment is carried out.
46 Check that the choke valve plate is fully open and both throttle plates are closed.
47 Now check that when arm (6) is in contact with the primary barrel level (7) a clearance (Y) exists of 0.2 mm (0.008 in). If not, bend the arm (6) (Fig. 12.21).
48 Move the throttle control lever and check that both valve plates open smoothly when the choke valve plate is wide open. Also check that only the primary throttle valve plate opens when the choke valve plate is even partially closed.

5 Exhaust pipe (later models)

General

1 The exhaust front pipe connection on later models is of balljoint type. Two versions of this coupling may be encountered and it is important to follow the fitting instructions precisely if a gas-tight joint is to be obtained.

Clamp bolts with locknuts

2 Make sure that the coupling mating surfaces are quite clean and apply a smear of non-hardening type jointing compound.
3 Offer up the coupling and fit the clamps.
4 Tighten the nuts evenly until the coil springs just become coil bound and then unscrew each nut 1½ turns.
5 Tighten the locknuts without disturbing the setting of the clamp nuts.

Clamp bolts with self-locking nuts

6 The first difference to note with this type of assembly is that the bolts are reversed when compared with the previous type.

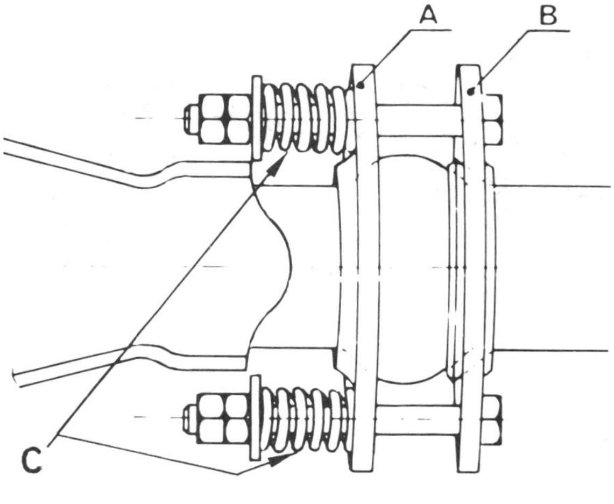

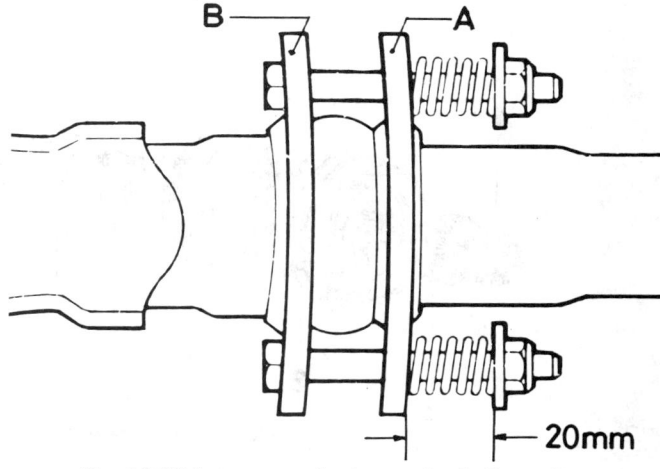

Fig. 12.22 Earlier type of exhaust pipe ball coupling

A Clamp
B Clamp
C Coil springs

Fig. 12.23 Later type of exhaust pipe ball coupling

A Clamp   B Clamp

7 Tighten the clamp nuts evenly until the length of the compressed springs is 20.0 mm (0.79 in). No further attention is required as the nuts are self-locking.

6 Transmission (five-speed)

Description
1 A five-speed gearbox is available on 1982 versions of the R1212 and R1213.
2 Although the ratios differ from the four-speed gearbox (see Specifications), overhaul and adjustment operations are similar to those described in Chapter 6 with the exception of those described in the following sub-sections.

Gearchange selector mechanism – removal and refitting
3 Select neutral.
4 Unscrew the detent plug and remove the spring and ball (10) (Fig.12.24).
5 Drive out the roll-pin (1) and remove the dog (2).
6 Drive out the roll-pin (3), unscrew the bolt (4) and remove the interlock plate (5).
7 Push reverse selector fork (6) towards 5th speed gear, but without moving the selector shaft (7).
8 Retrieve the interlock balls (8 and 9) and then withdraw the 5th speed selector shaft/fork assembly.
9 Retrieve the interlock disc from the gearbox interior.
10 Refitting is a reversal of removal.

Primary shaft 5th speed gear – removal and refitting
11 Lock up two gears simultaneously by moving two synchro-sleeves.
12 Unscrew and remove the nut (1) from the end of the primary shaft. (Fig. 12.25).
13 Take off the synchro-sleeve and hub.
14 Remove 5th speed gear (4) bush (5) and washer (6).
15 If the double taper roller bearing must be removed, use a bearing puller or press the shaft out of the bearing.
16 When refitting the bearing note that the snap-ring is towards the primary shaft gears.
17 The bearing may be refitted using a piece of 25.0 mm (1.0 in) tubing as a drift.
18 Screw on a new nut, tightening to a torque of 37 lbf ft (50 Nm) and stake the collar of the nut onto the flat of the shaft.

Secondary shaft 5th speed gear – removal and refitting
19 Unscrew the nut from the end of the shaft.
20 Using a suitable puller, remove 5th speed gear.
21 Again using the puller, draw off the bearing.
22 The 5th speed gear and the bearing may be fitted to the shaft using a piece of 25.0 mm (1.0 in) diameter tube as a drift.
23 Screw on a new nut to a torque of 37 lbf ft (50 Nm) and stake the collar of the nut onto the flat of the shaft.

Rectification of reverse gear fault
24 Where difficulty is experienced in selecting reverse gear, this condition may be due to wear in the ring (A) or the recess in the timing cover (Fig. 12.28).
25 When reverse gear is selected, the ring (A) acting as a brake, slides into a tapered recess in the timing cover and retards the rotation of the primary shaft to make gear meshing easier.
26 Check that the dowel (B) – Fig. 12.29 – is in place in the timing cover joint face of the gearbox as this is essential for centralising the ring (A).

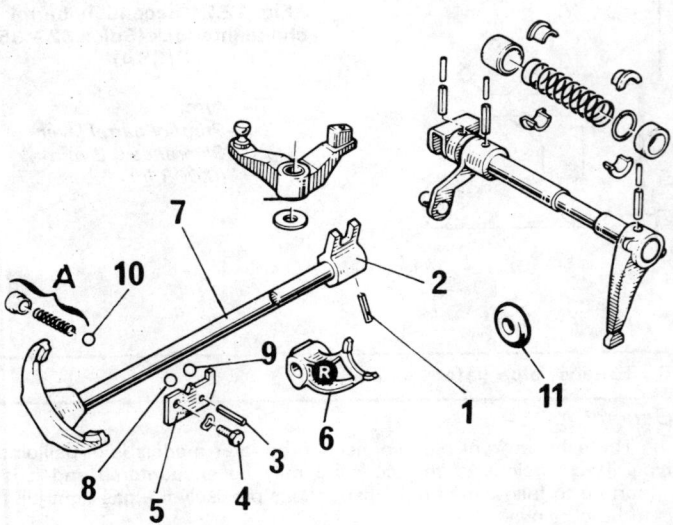

Fig. 12.24 Gear selector mechanism (5-speed transmission)

1	Roll pin	7	Reverse/5th selector shaft
2	Dog	8	Interlock ball
3	Roll pin	9	Interlock ball
4	Bolt	10	Detent ball
5	Interlock plate	11	Interlock disc
6	Reverse selector fork	A	Detent spring and plug

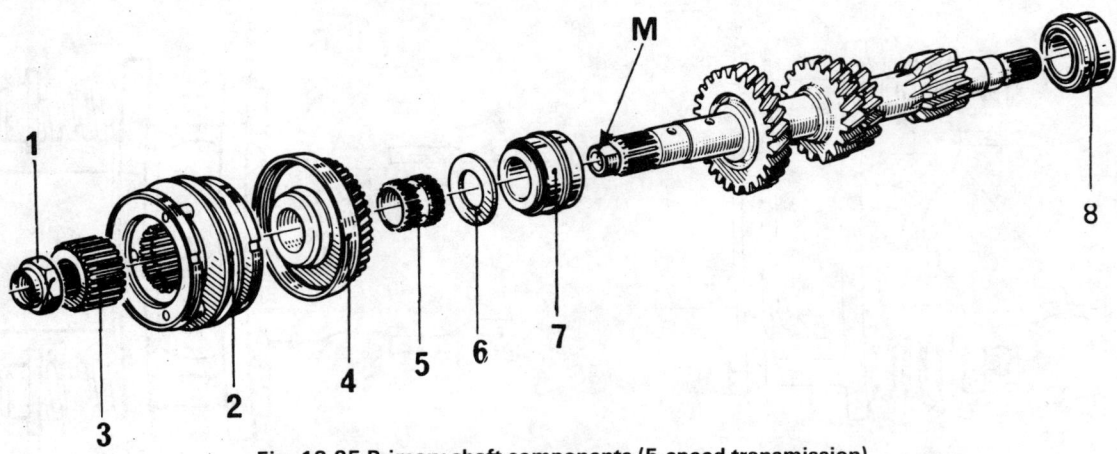

Fig. 12.25 Primary shaft components (5-speed transmission)

1	Nut	4	5th speed gear	6	Washer	8	Bearing
2	5th speed synchro sleeve	5	Bush	7	Bearing	M	Flat on shaft
3	5th speed synchro hub						

159

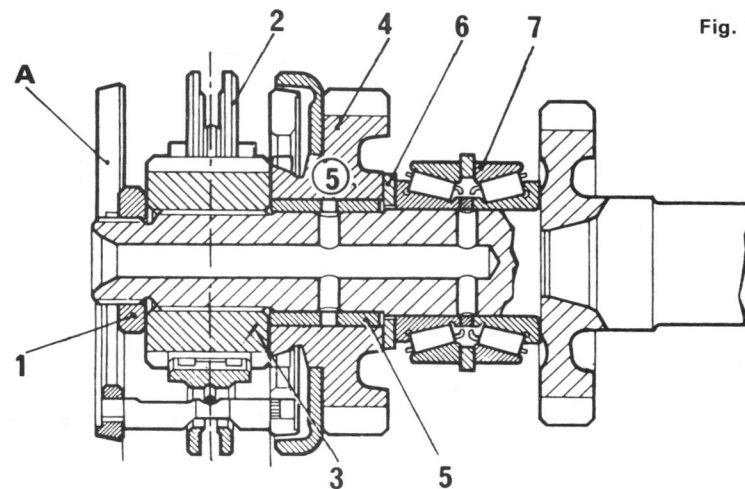

Fig. 12.26 Sectional view of 5th speed synchro

1 Nut
2 Synchro sleeve
3 Synchro hub
4 5th speed gear
5 Bush
6 Washer
7 Double roller bearing
A Primary shaft braking ring

Fig. 12.27 Secondary shaft components (5-speed transmission)

1 1st speed gear
2 1st/2nd synchro
3 Distance washer
4 2nd speed gear
5 Distance washer
6 3rd speed gear
7 Distance washer
8 3rd/4th synchro
9 4th speed gear
10 Distance washer
11 Bearing
12 Circlip
13 5th speed gear
14 Shaft nut

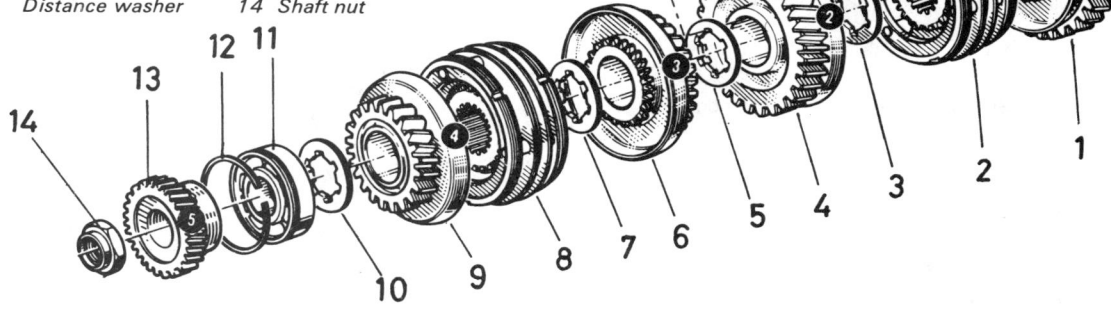

Fig. 12.28 Primary shaft reverse gear braking ring (A)

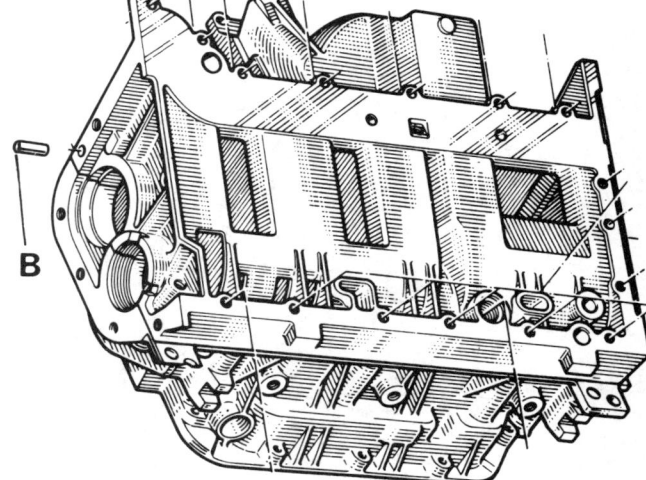

Fig. 12.29 Gearbox/timing cover centralising dowel (B)

Chapter 12 Supplement

7 Braking system

Braking system components – general

1 Components of the braking system may be of Bendix type on some vehicles instead of the ATE or Girling assemblies described in Chapter 8.
2 The following Sections cover all operations which may be required to the Bendix assemblies.

Master cylinder (Bendix) – removal, overhaul and refitting

3 Syphon the fluid from the master cylinder reservoir (an old clean battery hydrometer is useful for this).
4 Disconnect the leads from the fluid low-level switch.
5 Disconnect the pipelines from the master cylinder.
6 Unbolt the master cylinder from the front face of the servo unit and remove it, taking care not to spill any fluid onto the car paintwork. Ease the fluid reservoir out of its sealing grommets.

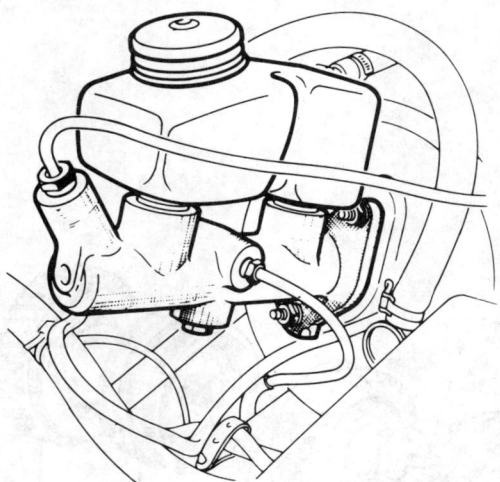

Fig. 12.30 Master cylinder mounted on servo

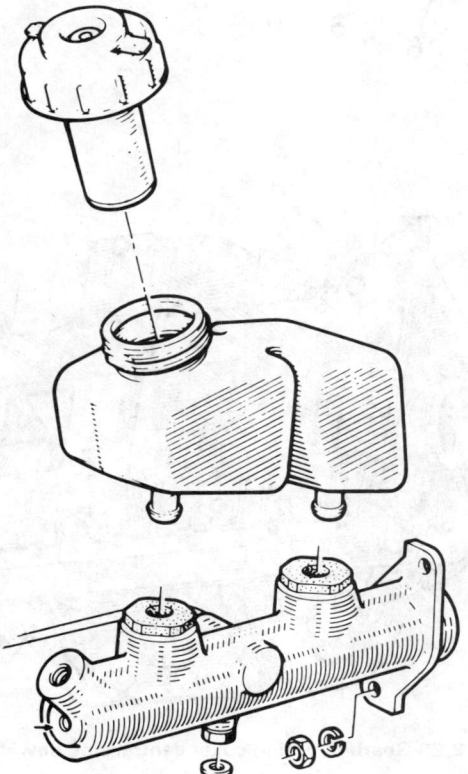

Fig. 12.31 Master cylinder and detachable fluid reservoir

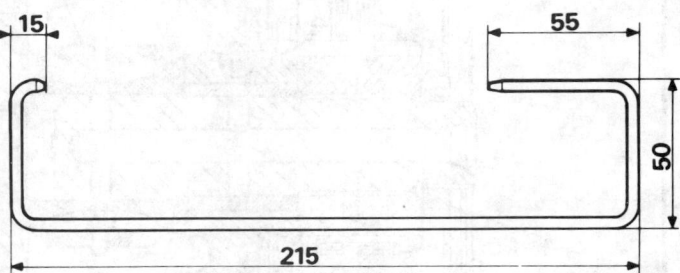

Fig. 12.32 Master cylinder piston compressing tool (dimensions in mm)

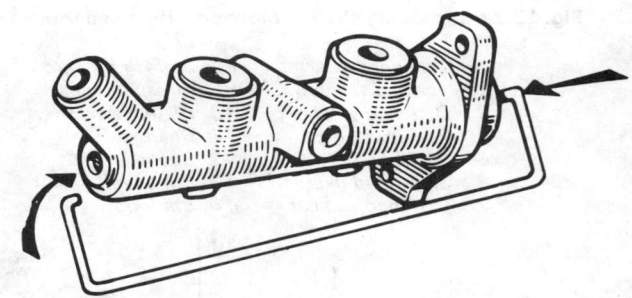

12.33 Using piston compressing tool

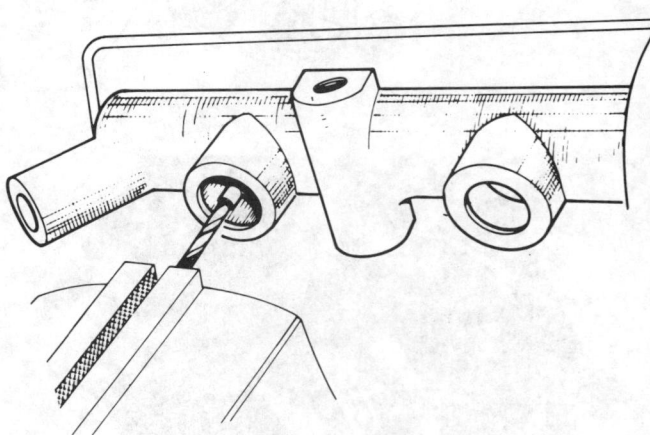

Fig. 12.34 Extracting master cylinder piston stop roll pin

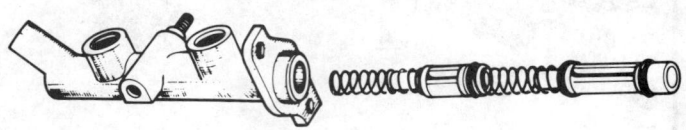

Fig. 12.35 Master cylinder primary/secondary pistons

Chapter 12 Supplement

7 Clean away all external dirt.
8 Make up a tool, similar to the one shown in the diagram, from a piece of steel rod (Fig. 12.32).
9 Use the tool to compress the primary and secondary pistons within the cylinder.
10 Now extract the secondary piston stop roll pin. Do this by gripping a 3.5 mm (0.138 in) diameter twist drill in a vice and twisting the master cylinder onto the drill. This action will extract the roll pin.
11 Remove the primary piston roll pin using the same method.
12 Remove the piston compressing tool and catch the pistons as they are ejected.
13 Examine the piston and cylinder bore surfaces. If there are any signs of scoring or metal to metal rubbing areas, renew the master cylinder complete. Where the components are in good condition, discard the seals and obtain a repair kit which will contain the new seals and other renewable items.
14 Fit the new seals using the fingers only to manipulate them into position.
15 Dip the assembled piston in clean brake fluid and insert it into the cylinder. Use a twisting motion to avoid trapping the seal lips.
16 Fit the piston compressing tool.
17 Insert the roll pins so that their slots are towards the push-rod end of the cylinder.
18 Remove the piston compressing tool.
19 Refitting is a reversal of removal, but check the adjustments described in Chapter 8, Sections 10, 11 and 12 before locating the master cylinder to the face of the servo.
20 Fill the fluid reservoir with clean fluid and then bleed the system as described later in this Supplement.

Front disc pads (Bendix) – inspection and renewal
21 Raise the front of the car and remove the roadwheels.
22 Although a brake warning lamp will normally illuminate when the disc pads have worn down to its limit, the wear in the pads can be inspected by viewing through the oval hole in the caliper.
23 If the pads are to be renewed, extract the two key retaining clips.
24 Drive out the keys with a punch.
25 Tilt the caliper and withdraw it.
26 Remove the disc pads and the spring blades.
27 Brush away all dust and dirt, taking care not to inhale, as asbestos dust can be injurious to health.
28 Using a piece of flat steel, press the piston fully back into its cylinder. This is to provide the additional clearance required for the new, thicker disc pads.
29 Insert the pads, checking that the following conditions apply:

The friction lining side is against the disc
The pads slide freely
The stop key (B) on both pads is uppermost (Fig. 12.37)

30 Engage one end of the caliper between the spring and the key face of the bracket then engage the opposite end by depressing the caliper against both springs.
31 Slide in the first key, then use a screwdriver as a lever to apply pressure to the caliper so that the second key can be engaged. Tap the keys fully home with a punch.
32 Apply the brake pedal hard several times to bring the new pads into contact with the disc.
34 Renew the pads on the opposite front brake in a similar way.

Rear brake shoes (Bendix) – inspection and renewal
35 Release the handbrake, remove the rear roadwheels.
36 Prise out the plug from the brake backplate and insert a screwdriver to move the handbrake lever and so free the peg (E) from the shoe (Fig. 12.39). Once the peg is released, push the lever rearwards to back it off.
37 Tap off the grease cap, extract the split pin, remove the nut lock and unscrew the nut and take off the thrust washer.
38 Remove the brake drum taking care not to allow the outer bearing to fall out.
39 Brush away all dust and dirt from the shoes and from the interior of the drum. Take care not to inhale, as asbestos dust is injurious to health.
40 Inspect the friction material on the shoes. If it has worn down to, or nearly down to the heads of the rivets, the shoes must be replaced

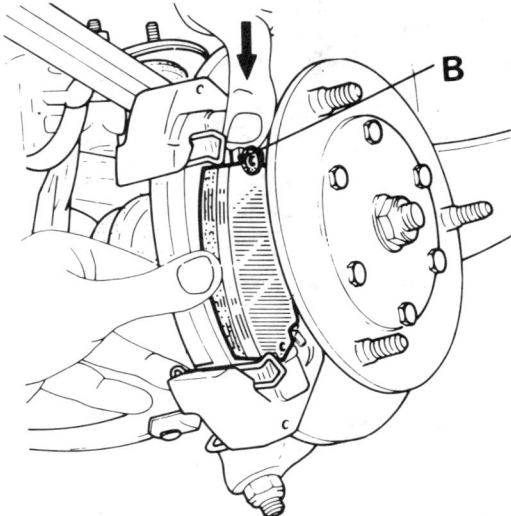

Fig. 12.37 Disc pad stop lug (B)

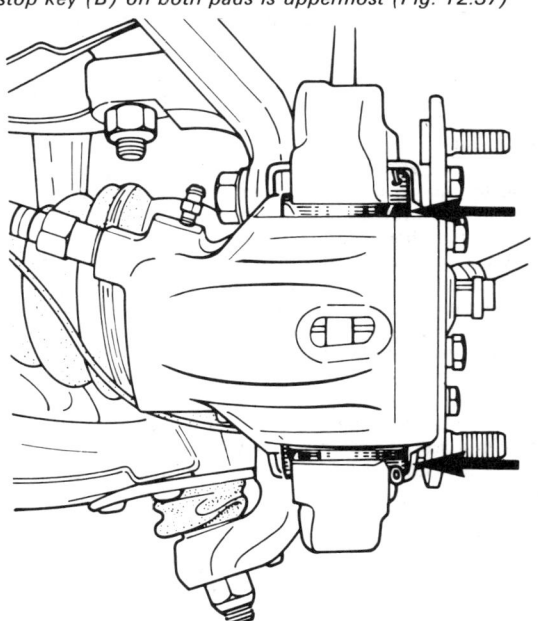

Fig. 12.36 Bendix brake caliper retaining keys (arrowed)

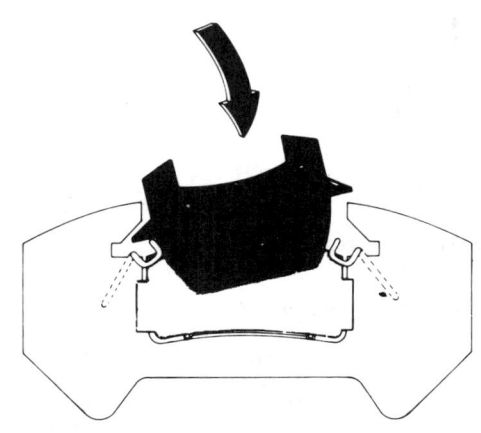

Fig. 12.38 Installing caliper to bracket

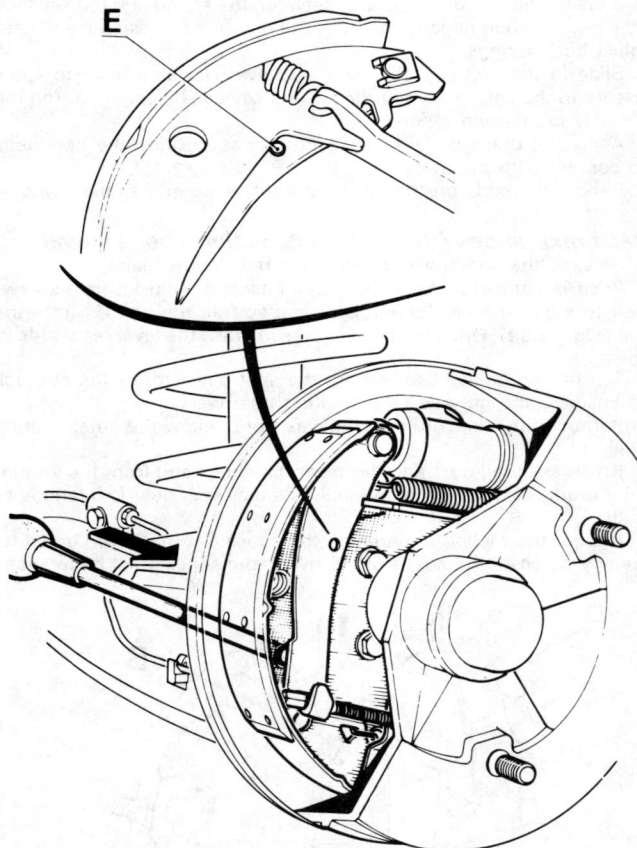

Fig. 12.39 Bendix rear brake shoe lever and peg (E)

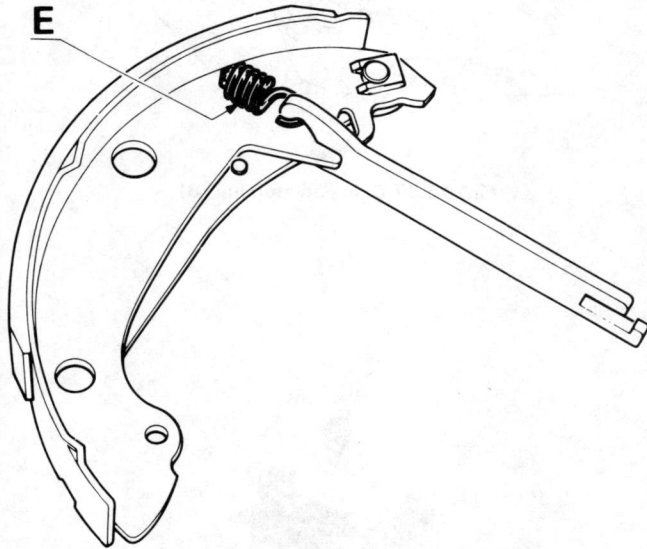

Fig. 12.41 Bendix rear brake shoe strut tension spring (E)

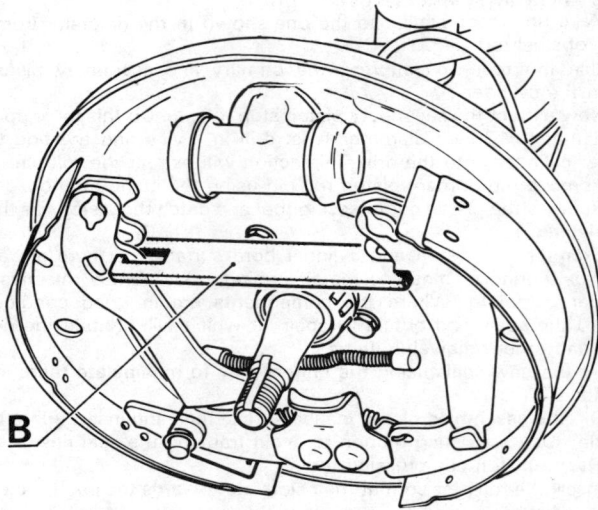

Fig. 12.40 Bendix rear brake shoe strut (B)

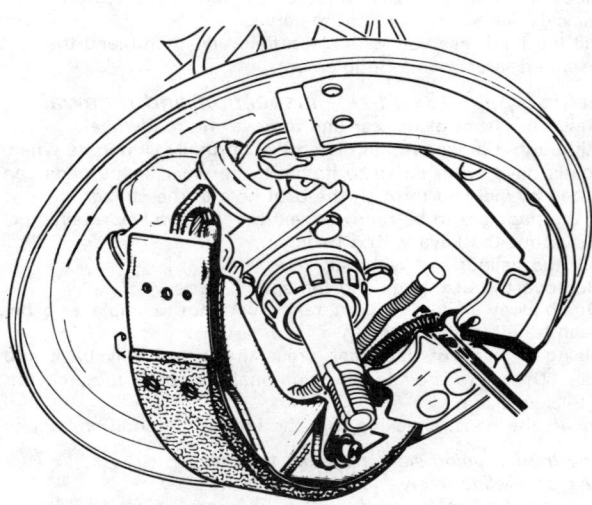

Fig. 12.42 Removing Bendix brake shoes

by factory re-lined ones. Renew all four shoes at the same time as an axle set.
41 Remove the brake shoe steady springs. Do this by pushing a suitably sized piece of tubing down inside the coil spring, compressing the spring and turning it. Slacken the handbrake cable.
42 Release the shoe upper return spring, noting carefully into which shoe holes it is engaged.
43 Prise the upper ends of the shoes apart and remove the strut (B) and tension spring (E) from the shoe (Figs. 12.40 and 12.41).
44 At this stage note, and sketch if necessary, the location of the shoes with regard to leading and trailing ends (portion of shoe not covered by friction material).
45 Release the handbrake cable from the shoe lever.
46 Move the shoes upward, at the same time slide the shoe lower return spring out from behind the shoe anchor plate.
47 Remove the shoe/lever assembly.
48 Do not touch the footbrake whilst the shoes are removed or the wheel cylinder pistons will be ejected.
49 If the wheel cylinders show signs of fluid leakage, renew their seals as described in a later Section.
50 Assemble the new shoes and levers in the same original location as the worn components and engage the lower return spring between the shoes.
51 Refit the shoes to the backplate on which the high points have been smeared with a little high melting point grease.
52 Engage the strut and connect its small tension spring.
53 Fit the shoe upper return spring.
54 Engage the handbrake cable with the shoe lever.
55 Fit the shoe steady springs.
56 Now check that there is a clearance (H) as shown in the diagram (Fig. 12.43) when the handbrake lever is in contact with the shoe. Should the clearance not be within the specified tolerance, renew the shoe return springs and the strut tension spring.
57 Engage the teeth of the sector with those on the shoe lever, but so

Chapter 12 Supplement

that the shoes are in their most retracted state.

58 Fit the drum and adjust the bearings as described in Chapter 7, Section 5.

59 Apply the footbrake several times to bring the shoes into the closest contact permissible with the drum.

60 Adjust the handbrake and fit the plug into the hole in the backplate.

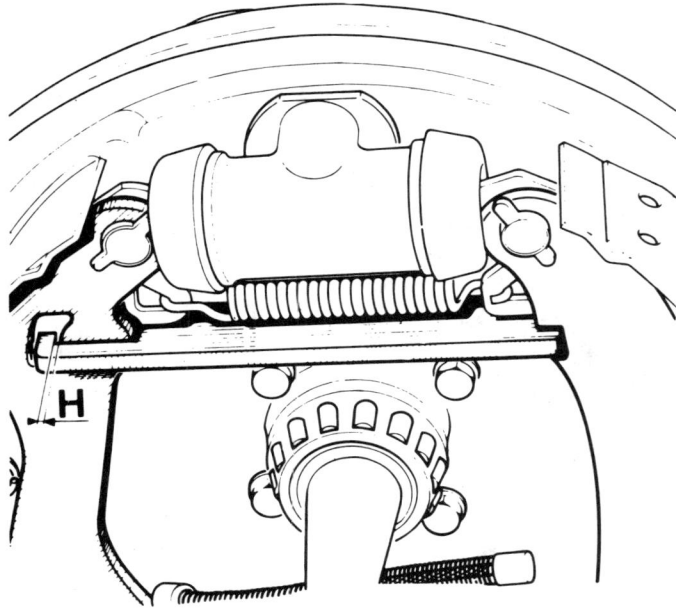

Fig. 12.43 Brake shoe strut clearance (H)
H = 1 mm

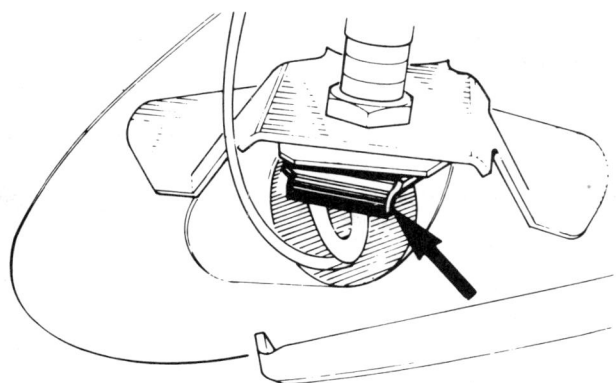

Fig. 12.44 Brake hose bracket clip

Disc caliper (Bendix) – removal, overhaul and refitting

61 Raise the front roadwheel and remove it.

62 Remove the caliper key retaining clips and drive out the keys with a punch.

63 Disconnect the rigid hydraulic pipe from the flexible hose and quickly cap the end of the pipe.

64 Remove the flexible hose clip at the support bracket.

65 Detach the caliper from the caliper bracket.

66 Hold the hose end fitting in an open-ended spanner and unscrew the caliper from it.

67 Clean away external dirt.

68 Take off the rubber dust excluder.

69 Apply low air pressure from a tyre pump to the fluid entry hole of the caliper and eject the piston. Cushion the piston as it emerges with a piece of soft wood to prevent impact damage.

70 Examine the piston and cylinder bore surfaces for scoring or metal to metal rubbed areas. If evident, renew the caliper cylinder assembly.

71 Where the components are in good condition, extract the piston seal from its cylinder groove using a sharp pointed instrument and discard the seal.

72 Obtain a repair kit and having cleaned the piston and cylinder in methylated spirit or clean hydraulic fluid, fit the new seal using the fingers only to manipulate it into position.

73 Dip the piston in clean hydraulic fluid and insert it squarely into the cylinder, but do not push it fully home until the dust excluder has been engaged to both piston and cylinder.

Brake disc (Bendix) – inspection, removal and refitting

74 The brake disc should be inspected at the same time that the pads are checked for wear.

75 If the disc is badly grooved or scored (light scoring is normal) the disc must be renewed, refinishing is not acceptable.

76 Using vernier calipers measure the thickness of the disc. Where the disc thickness is worn below 9.0 mm (0.354 in) then the disc must be renewed.

77 If uneven braking or judder has occurred during brake application, check the disc for excessive run-out. This is best checked with a dial gauge, but feeler blades can be used between the disc and a fixed point as the disc is slowly turned. The measurement should be taken near the outer edge of the disc.

78 If the run-out exceeds 0.1 mm (0.004 in) renew the disc.

79 To remove a disc, take off the caliper assembly without disconnecting the hydraulic pipeline and tie the caliper up out of the way.

80 Unscrew and remove the stub axle nut. A bar will have to be placed between the wheel hub studs to prevent the hub from turning.

81 Remove the hub/disc using either a slide hammer or a three-legged puller.

82 Separate the disc from the hub by unscrewing the connecting bolts.

83 Fitting the new disc is a reversal of removal but make sure that protective grease has been cleaned off.

84 Tighten the disc to hub connecting bolts to 61 Nm (45 lbf ft), the caliper mounting bolts to 82 Nm (60 lbf ft) and the stub axle nut to 250 Nm (184 lbf ft).

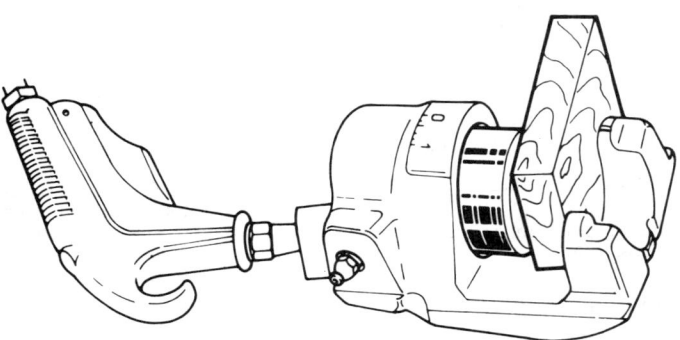

Fig. 12.45 Ejecting caliper piston

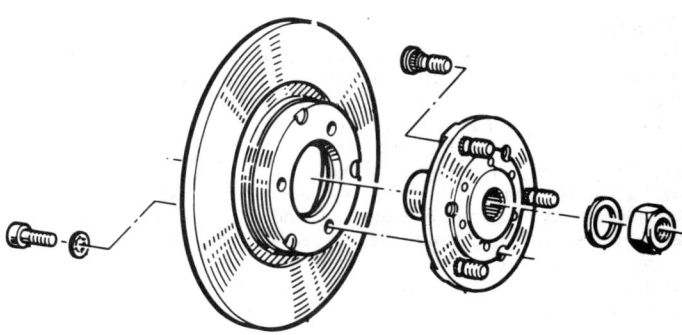

Fig. 12.46 Disc/hub components

Brake drum (Bendix) – inspection, removal and refitting

85 Whenever the rear shoe linings are being checked for wear, inspect the interior of the brake drums for scoring or grooving.
86 If uneven braking has been evident or judder has occurred during brake application, measure the interior of the drums for ovality using a pair of circlips or similar.
87 The interior of a drum can be refinished to overcome these conditions provided the specified (new) internal diameter is not increased by more than 1.0 mm (0.039 in), and both drums are refinished at the same time in order to maintain even braking.
88 Full brake drum removal, refitting and bearing adjustment procedures are described in Chapter 7 Section 5 and earlier in this Supplement for Bendix type rear brakes.

Rear wheel cylinder (Bendix) – removal, overhaul and refitting

89 The operations are very similar to those described for the Girling cylinder in Chapter 8, Section 8 but the differences in components should be noted from the illustration (Fig. 12.47).

Brake hydraulic system – bleeding

90 In addition to the two-man method of bleeding described in Chapter 8, Section 3, the following additional bleeding methods will prove to be useful alternatives if an assistant is not available.
91 The front hydraulic circuit is independent of the rear and only if a component affecting both systems (master cylinder) has been removed and refitted, need the complete system be bled.
92 If a component of only one circuit has been disturbed then only that particular circuit need be bled.

Bleeding – using one-way valve kit

93 There is a number of one-man, do-it-yourself brake bleeding kits currently available from motor accessory shops. These kits greatly simplify the bleeding operation and also reduce the risk of expelled air or fluid being drawn back into the system.
94 To use the kit, connect the tube which contains the one-way valve to the bleed screw and open the screw one half turn.
95 Depress the brake pedal fully and then slowly release it.
96 Repeat the operation several times to eject air bubbles from the system. Tighten the bleed screw and remove the tube.
97 If the brake pedal still feels spongy when depressed, air must still be trapped in the system. Repeat the bleeding operation.

Bleeding – using a pressure bleeding kit

98 These too are available from motor accessory shops and are usually operated by air pressure from the spare tyre.
99 By connecting a pressurised container to the master cylinder reservoir, bleeding is then carried out by simply opening each bleed screw in turn and allowing the fluid to run out, rather like turning on a tap until no air is visible in the ejected fluid.
100 Using this method, the large reserve of hydraulic fluid provides a safeguard against air being drawn into the master cylinder, during the bleeding operation, which very often occurs if the reservoir is not regularly topped-up.
101 Pressure bleeding is particularly effective when bleeding complex systems and when bleeding the entire system at time of routine fluid renewal. Do not depress the brake foot pedal during pressure bleeding on these model vehicles.

Bleeding – all systems

102 If the entire system is being bled, the bleeding must be carried out at each bleed screw in turn, commencing at the one furthest from the master cylinder and finishing at the one nearest the master cylinder.
103 Unless the pressure bleeding method is being employed, do not forget to keep the master cylinder fluid reservoir regularly topped-up to prevent air being drawn into the system.
104 When bleeding is completed, recheck the fluid level in the master cylinder reservoir, top-up if necessary and refit the cap. Depress the brake pedal and check that it is free from sponginess, as this would indicate air still present in the system and require further bleeding.
105 Discard any hydraulic fluid which has been expelled as it is almost certainly contaminated with moisture, air or dirt and will be unsuitable for further use. Clean hydraulic fluid should always be stored in an airtight container as the fluid absorbs moisture (ie hygroscopic) which lowers its boiling point and could affect braking performance under severe conditions.

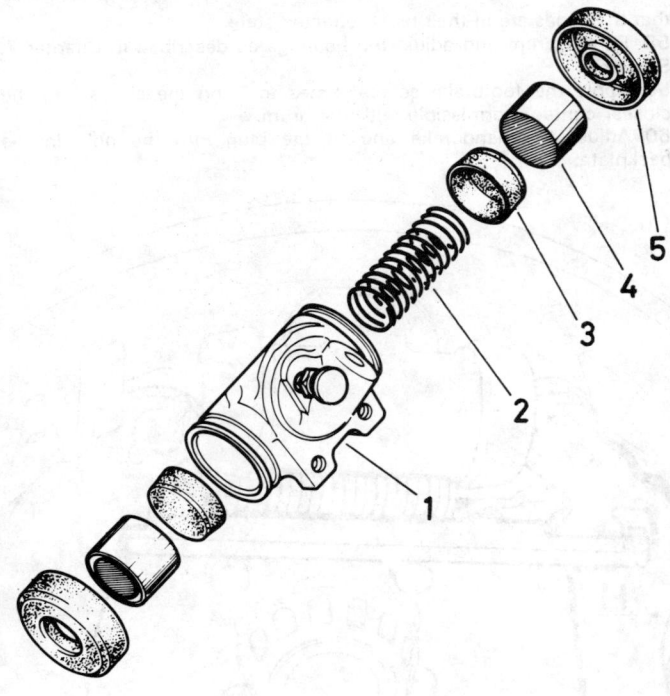

Fig. 12.47 Exploded view of Bendix rear wheel cylinder

1 Cylinder
2 Spring
3 Seal
4 Piston
5 Dust excluder

8 Electrical system

Fuses

Later models (1981–82) have a more comprehensive circuit protection fuse arrangement.

Number	Circuit protected	Fuse capacity (A)
1	LH window winder	10
2	RH window winder	10
3	Instrument panel	5
4	Spare	
5	Reverse lamps, heated rear window	16
6	Windscreen wiper	16
7	Interior lamp, cigar lighter	8
8	Spare	
9	Spare	
10	Flasher unit	5
11	Heater blower, stop lamps, radio	8
12	Spare	
13	Spare	
14	Rear fog lamp	5
15	Spare	
16	LH front side, LH tail	5
17	RH front side, RH tail	5
18	Windscreen wiper PARK	5

Alternator – later models with integral regulator

1 Later model alternators have an integral voltage regulator. The alternator is similar in all other respects to the alternators described in Chapter 9.

Integral regulator – renewal (Paris-Rhone)

2 Disconnect the battery.
3 Disconnect the junction block at the rear of the alternator.
4 Remove the plastic cover.
5 Disconnect the wire from the diode bridge.

Chapter 12 Supplement

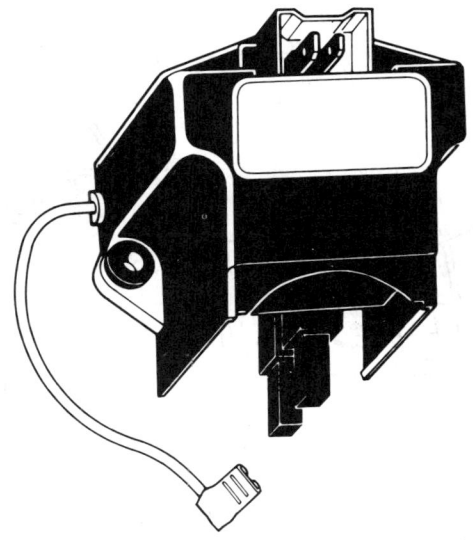

Fig. 12.48 Alternator integral regulator and brush holder

6 Unscrew the two hexagonal head regulator mounting screws and withdraw the regulator.
7 Refitting is a reversal of removal.

Brush holder — renewal (Paris-Rhone)
8 The following work can be carried out without having to remove the alternator from the car.
9 Unsolder the three connecting pins and remove the brush holder, after unscrewing the connecting bolts.

10 Refitting is a reversal of removal, but do not overtighten the fixing bolts or allow heat to transfer to the regulator when re-soldering the connecting pins.

Changing alternator with remotely-mounted regulator for alternator with integral regulator
11 Disconnect battery.
12 Remove the alternator and substitute the new integral regulator type.
13 Remove the remotely sited original regulator from its wing mounting.
14 Remove the EXC wire (3) or if this is difficult, cut and insulate each end (Fig. 12.49).
15 Connect the original + wire (2) to the 5.0 mm terminal of the built-in regulator junction block. The 6.3 mm connector on the integral regulator will not be required as the vehicle is equipped with a voltage indicator, not an ignition warning lamp.
16 The battery positive terminal is still connected to the alternator (+) BAT terminal.

Central door locking system — description
17 Certain models are equipped as standard with a central door locking system. The system provides a locking or unlocking facility for all four doors from inside or outside the car.
18 The system has in-built safety features to overcome the problem of slamming the front door when set to lock, but with the ignition key still in the switch.
19 Another safety feature is an inertia switch to unlock all doors in the event of an impact at speeds above 15 km/h (9 mph).
20 Mechanical unlocking capability is retained for use should this electrical circuit fail.
21 Components of the system comprise:

A lock barrel
Linkage and operating levers
Changeover switch
Solenoid and plunger

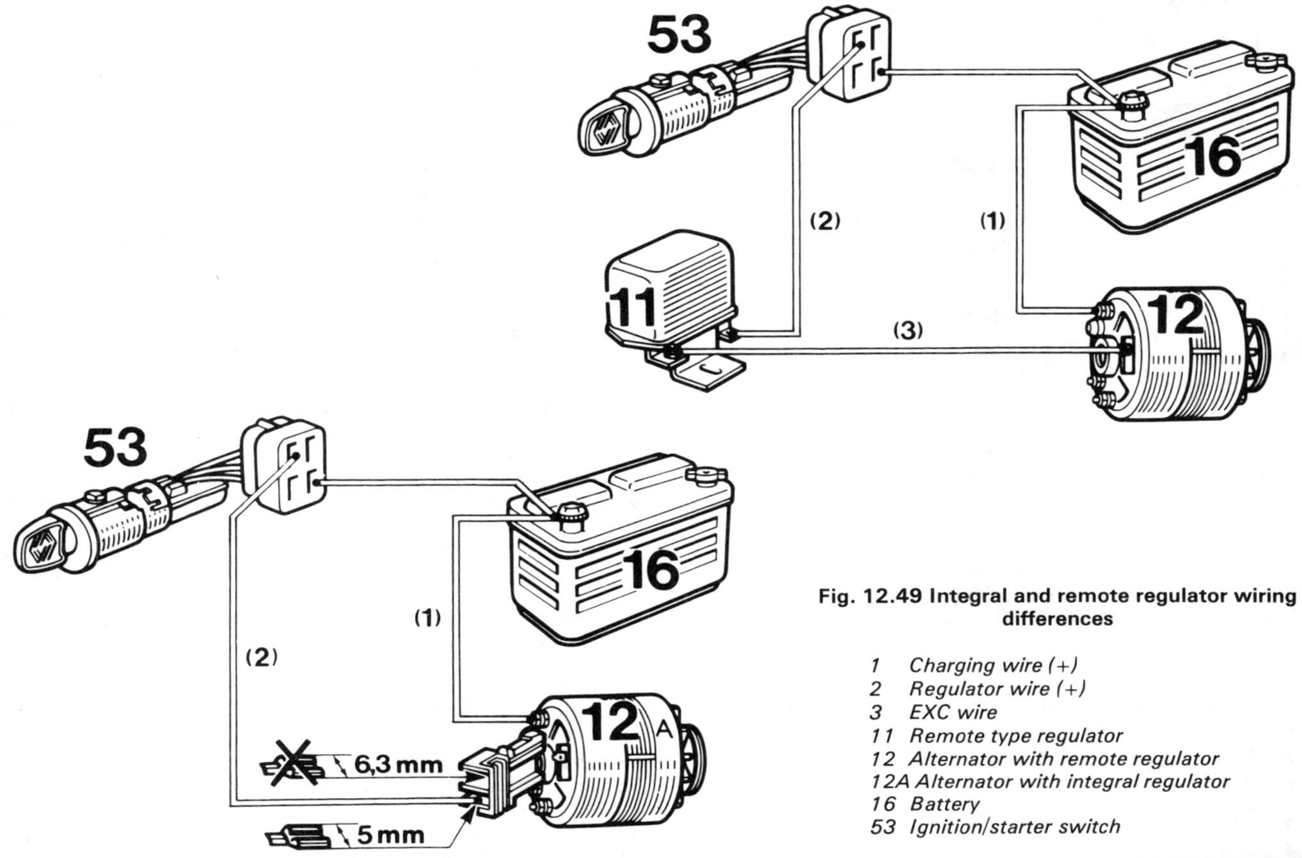

Fig. 12.49 Integral and remote regulator wiring differences

1 Charging wire (+)
2 Regulator wire (+)
3 EXC wire
11 Remote type regulator
12 Alternator with remote regulator
12A Alternator with integral regulator
16 Battery
53 Ignition/starter switch

Chapter 12 Supplement

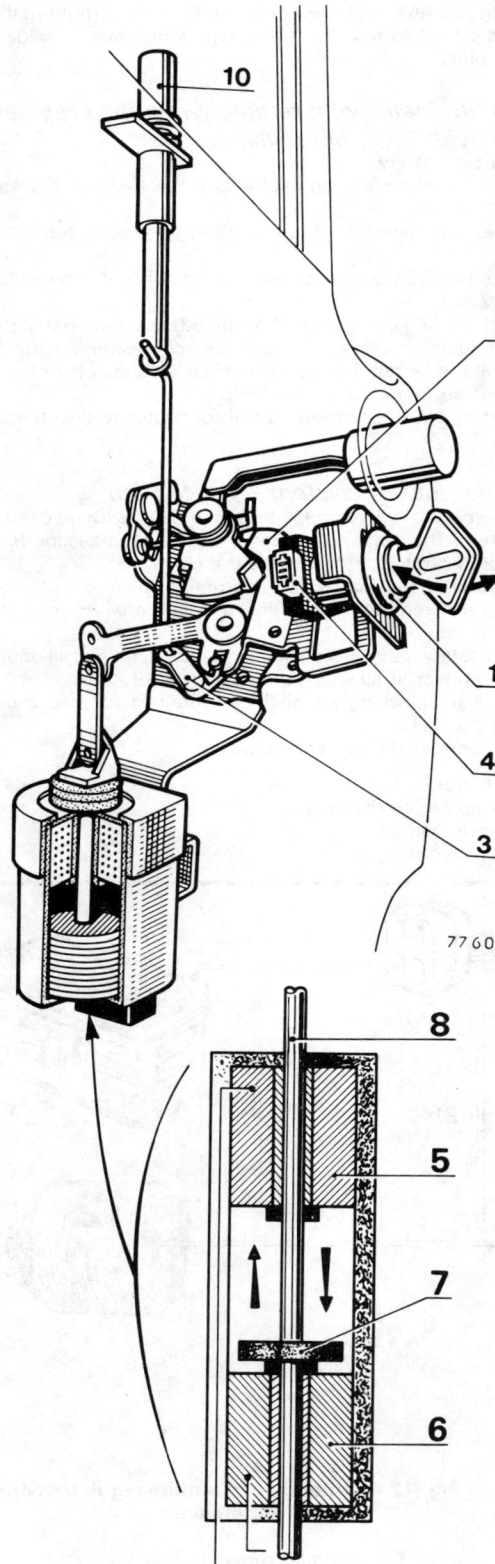

Fig. 12.50 Electrically-operated door lock

1 Lock cylinder
2 Interior spring opening control lever
3 Swing lever
4 Changeover switch
5 Solenoid coil
6 Solenoid coil
7 Ferrite disc
8 Spindle
10 Tell-tale button

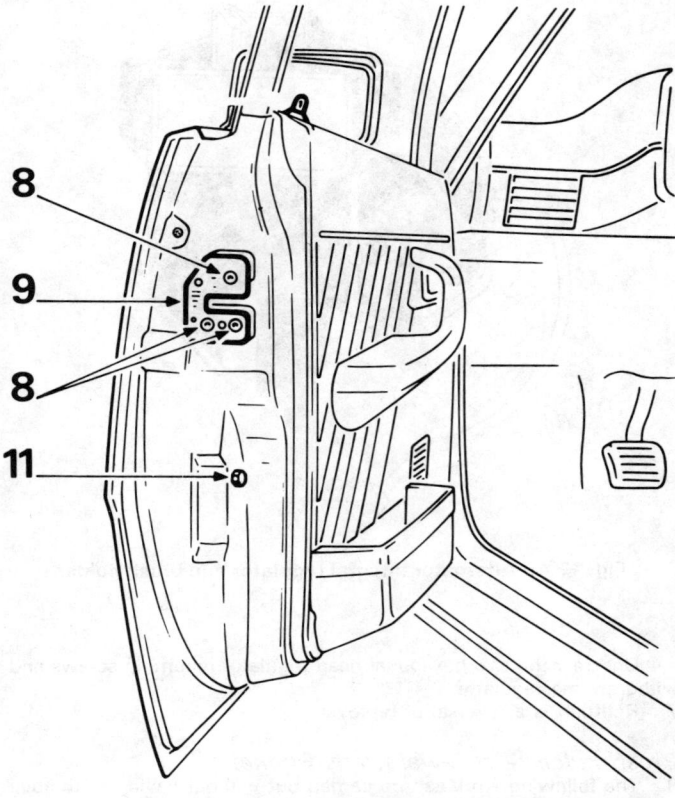

Fig. 12.51 Electrically-operated lock details

8 Fixing screws
9 Latch plate
11 Solenoid mounting bolt

Electric door lock – removal and refitting

22 Raise the window fully.
23 Remove the door interior trim panel and peel off the waterproof sheet.
24 Disconnect the wiring plug.
25 Disconnect the changeover block (4) from the lock barrel (Fig. 12.50). This is simply a press fit on the barrel.
26 Disconnect the remote control assembly by freeing the rod from the door frame clip and unscrewing the latch plate fixing screws.
27 Unbolt the solenoid.
28 Withdraw the assembly through the lower aperture in the door after pivoting it around the window guide channel.
29 Refitting is a reversal of removal.

Electrically-operated window gear – removal and refitting

30 Electrically-operated windows are fitted as standard equipment on certain models.
31 This type of window is limited to the front doors only.
32 Lower the window to the half-way position and remove the door interior trim panel.
33 Disconnect the junction block (C) (Fig. 12.52).
34 Unscrew the two nuts (2) and remove the lower glass guide rail (1).
35 Unscrew the motor mounting plate nuts, push the plate into the door cavity to clear the studs and to release the rollers from the bottom glass channel.
36 Withdraw the motor plate through the aperture in the door.
37 Refitting is a reversal of removal.

Heated rear window

38 The heated rear window element has its grid wires fixed to the inside of the glass.
39 Avoid scratching the element with rings on the fingers or luggage, and do not stick labels over the element.

Chapter 12 Supplement

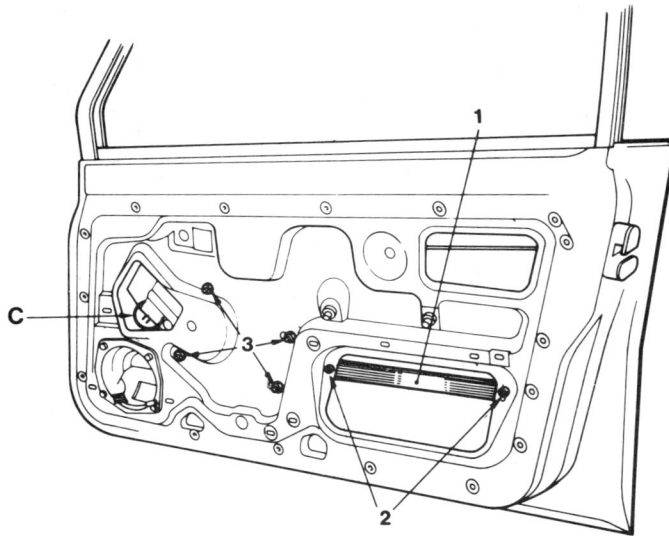

Fig. 12.52 Electrically-operated window details

1 Bottom guide rail
2 Securing nut
3 Window winder mounting plate nuts
C Wiring junction block

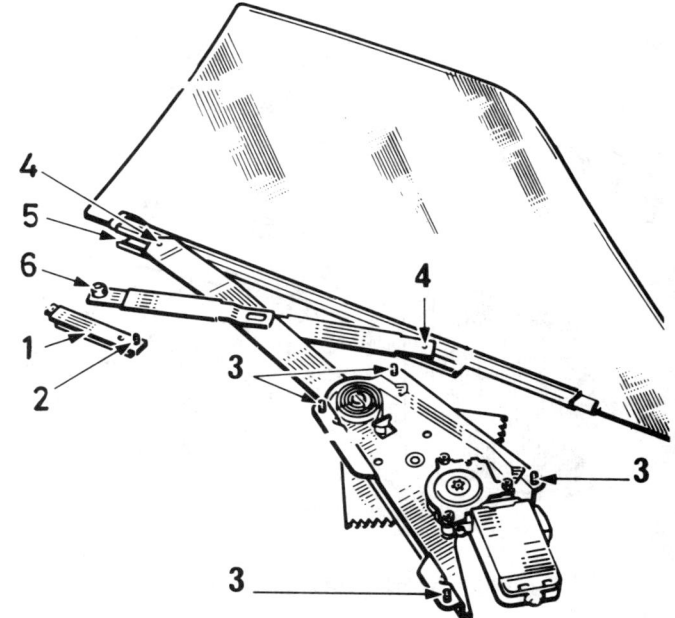

Fig. 12.53 Electrically-operated window winder

1 Bottom guide rail
2 Bottom guide rail mounting studs
3 Mounting studs
4 Roller
5 Glass channel
6 Roller

40 Clean the window only with water and detergent; never use solvents or abrasive compounds.
41 A broken grid wire can be repaired by using one of the silver conductive paints sold for the purpose.

Rear screen washer/wiper

42 Where these components are fitted, the control switch is located on the centre console or adjacent to the gear change lever.
43 The washer reservoir is located within the luggage compartment.
44 Jet adjustment and wiper motor removal and refitting operations are similar to those described for the windscreen wiper/washer in Chapter 9.

Rear fog lamp – installation

Models up to 1979
45 The neatest way to install a rear foglamp is to use the genuine Renault accessory.
46 Remove the right-hand rear lamp cluster base and substitute a new base (Part No. 77 01 021 578) which incorporates spade terminals and a reflector especially for the foglamp.
47 The foglamp wire should be routed under the luggage compartment mat, the sill kick strip and the mat near the wing inner panel.

48 Locate the switch, which should incorporate a warning ON lamp, in a convenient place.
49 Power to the switch should come from the headlamp dipped beam wire at the headlamp control switch. Connect the remaining switch terminal to the foglamp connecting wire.

Models 1980 on
50 On these later models, the foglamp wiring is already installed as part of the vehicle wiring harness.
51 A switch and rear lamp cluster base will, of course, still be required as described in earlier paragraphs of this Section.
52 On early 1980 models the foglamp connecting wire is installed only as far as a four-way junction box at the end of the harness near the accessories plate. This wire will have to be extended to connect the switch and the terminal on the headlamp beam control switch.

Auxiliary fog and driving lamps

53 Various designs of auxiliary fog and driving lamps are available

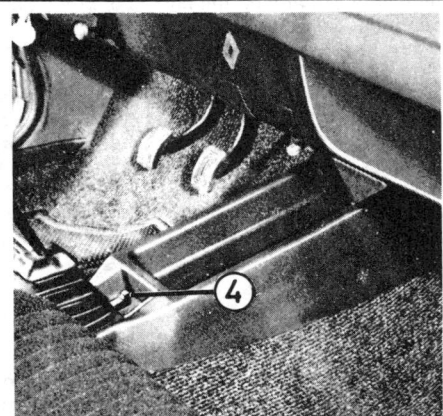

Fig. 12.54 Rear screen washer/wiper switch (4)

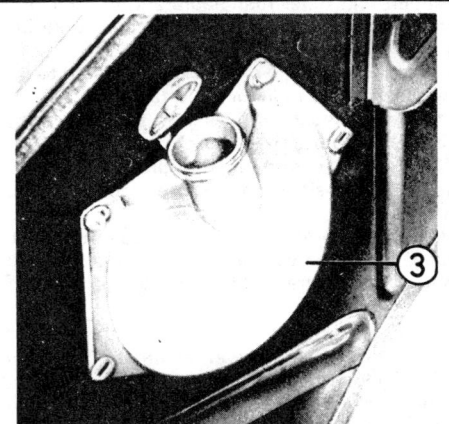

Fig. 12.55 Rear screen washer fluid reservoir (3)

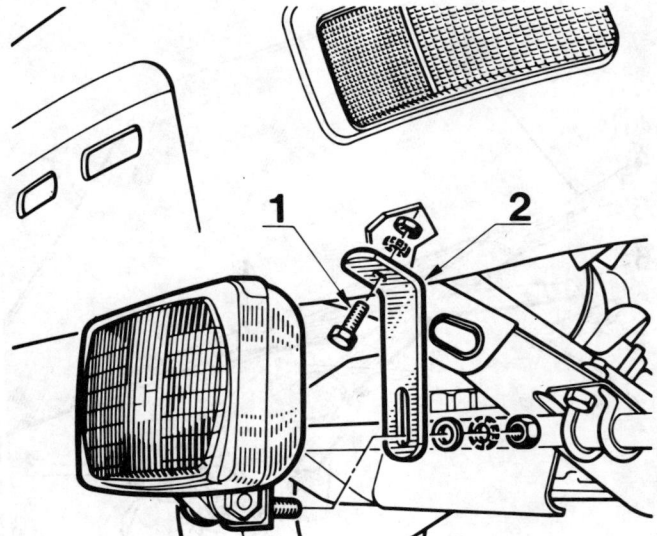

Fig. 12.56 Typical front fog lamp mounting

1 Bolt 2 Bracket

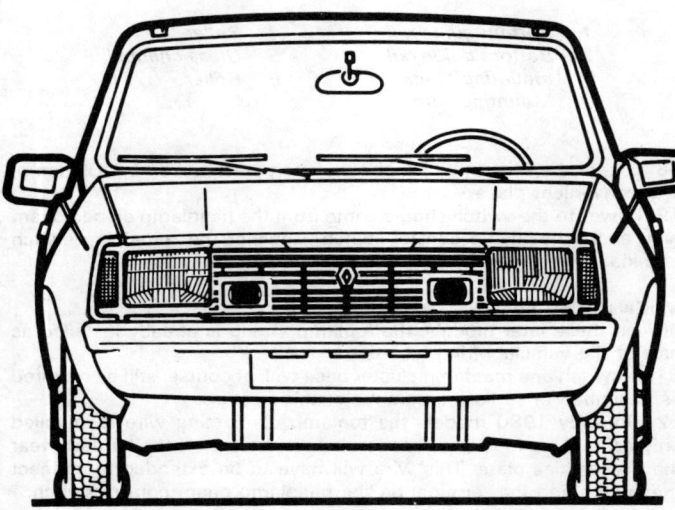

Fig. 12.57 Grille-mounted driving lamps

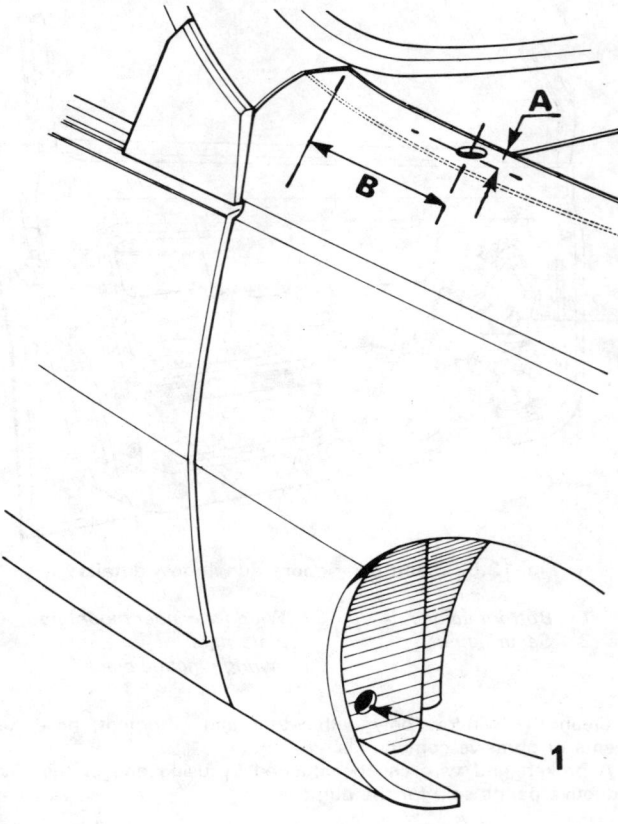

Fig. 12.58 Aerial fitting diagram (wing-mounted type)

A 18.0 mm (0.70 in) 1 Hole
B 140.0 mm (5.5 in)

from both Renault and other manufacturers. It is recommended that the lamps are of rectangular design not exceeding an overall fronted area of 185 x 100 mm (7.3 x 3.9 in).
54 The lamps can be mounted on brackets attached to the front lower panel/crossmember, or the radiator grille can be cut at the bottom two corners and the lamps mounted flush with the grille, supporting them on brackets attached to the front panel.
55 Always incorporate a relay in the circuit when fitting auxiliary lamps. Utilise a spare position on the fuse block or incorporate an in-line fuse holder.
56 The centres of the lamps must not be less than 610 mm (24 in) from the ground when the vehicle is normally laden. Adjust the light beams so that their tops are parallel with the ground.

Radio – installation
57 Provision is made at the lower centre of the fascia panel for installation of a standard size radio receiver.
58 Remove the blanking panel and offer the radio in, from the rear of the fascia panel. A pad of foam rubber should be placed at the rear of the receiver to act as an insulator. Fit spacers and locknuts as supplied for mounting purposes and push-on the tuning knobs.

Aerial – installation
59 The recommended location for the aerial is on the roof, just above the centre of the windscreen. Later models are already fitted with a blanking plug, but earlier versions will require a hole to be drilled in the following way.
60 Working inside the car, cut a hole in the headlining 34.0 mm (1.3 in) from the front edge of the lining at its centre point. Feel the location of the hole in the crossmember and cut the headlining to match it.
61 Drill the roof panel to accept the aerial base in line with the hole in the crossmember.
62 Scrape the paint from around the hole in the roof panel to ensure a good earth contact.
63 Route the aerial cable through the hole, then along the crossmember and down the left-hand windscreen pillar. It will ease the operation if the sun visor and the glove compartment are removed. The cable can then be gripped and pulled down using the door top hinge access hole.
64 Secure the aerial base with a locknut and shakeproof washer.
65 Route the aerial cable under the instrument panel, but keep it away from other wiring and instruments.
66 If for any reason the aerial must be wing mounted, the operations are similar to those just described, but drill the hole in accordance with the diagram.

Loudspeakers – installation

Single speaker
67 On Base and TL models, the most suitable position for a speaker is under the fascia panel attached to the engine compartment bulkhead. Custom designed components are available from Renault.
68 On GTL models 1979 on, the centre console is designed to accept an elliptical speaker.

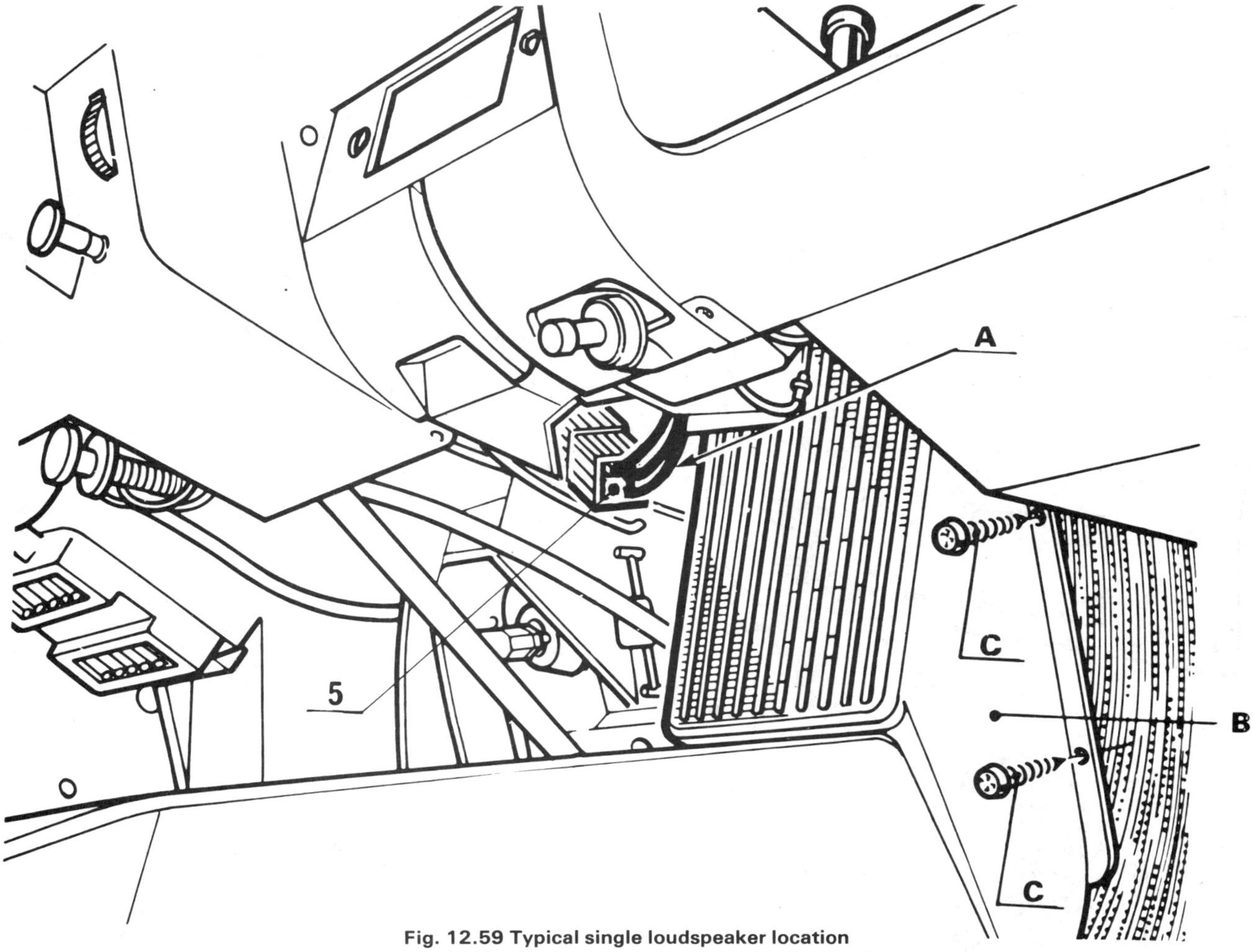

Fig. 12.59 Typical single loudspeaker location

A Wiring plug
B Loudspeaker casing
C Fixing screws
5 Spare location for power supply tapping

Fig. 12.60 Centre console and radio speaker grille (G)

Twin loudspeakers

69 On Base, TL and GTL models, the speakers can be mounted in the trim panels just forward of the door A pillars.

70 Alternatively, a speaker may be mounted in the door cavities, but this may require cutting the door trim panel on earlier models. Later versions are already equipped with door speaker grilles and wiring harness.

71 Route the speaker wires, on earlier models, through holes in the door edge which will have to be drilled and grommets fitted.

Four loudspeakers

72 In addition to front-mounted speakers, two speakers may be mounted on the rear parcels shelf.

73 The speakers may be of case-mounted type or recessed into the shelf. If the latter, then their grilles must not be more than 20.0 mm (0.79 in) thick or the shelf will not fold down properly.

74 Route the connecting wires neatly under the sill kick strip and fit a mixer so that the volume from the rear speakers can be adjusted. On vehicles with a removable parcel shelf use different terminals at the loudspeaker end, for instance one flat and one round, so that they can be reconnected in phase.

Radio wiring connections

75 Having installed the radio, speakers and aerial the wiring connections should now be made.

76 Plug the aerial cable into its socket in the radio and connect the loudspeaker wires.

Chapter 12 Supplement

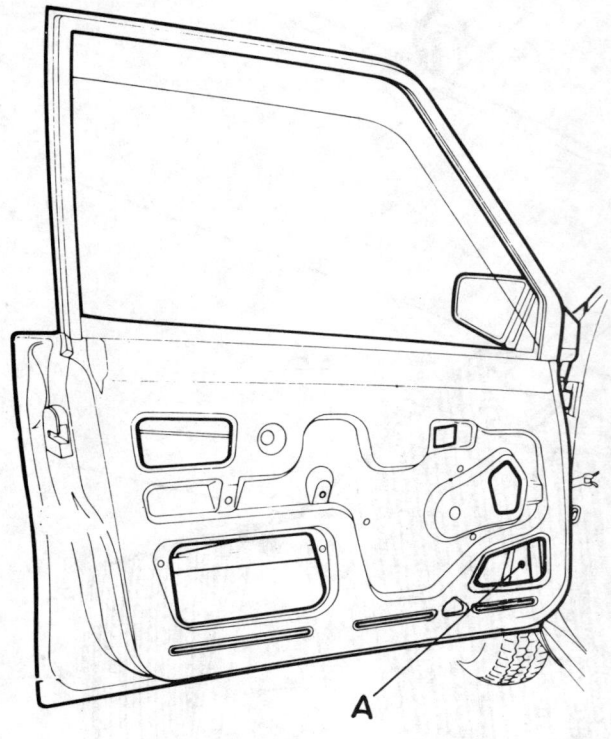

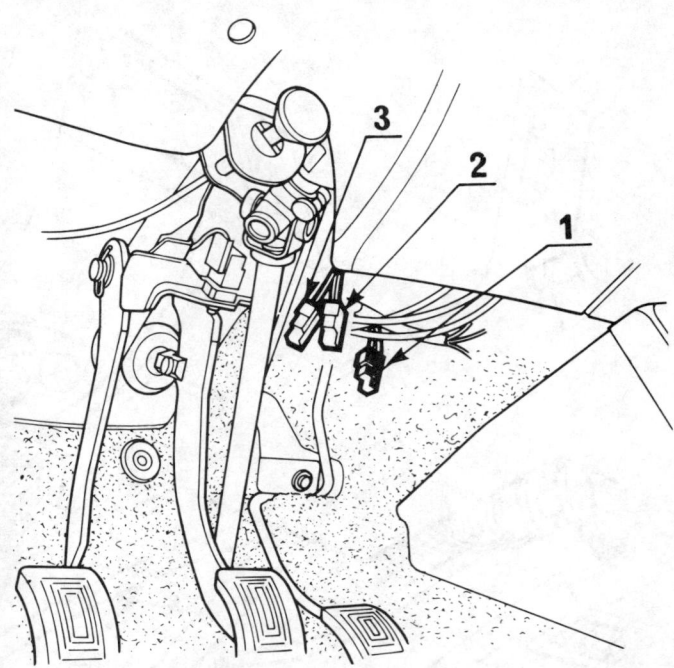

Fig. 12.62 Radio feed and earth pick-up point (TS 1979 on and GTL 1980 on)

1 Feed wire (+) and earth (black)
2 Loudspeaker
3 Loudspeaker

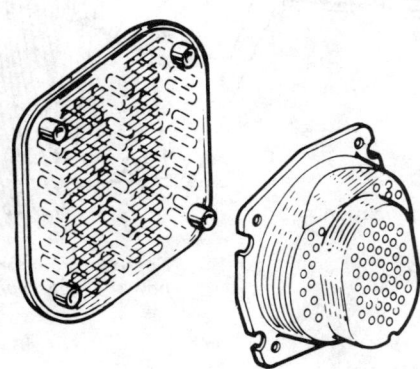

Fig. 12.61 Location of door-mounted loudspeaker A

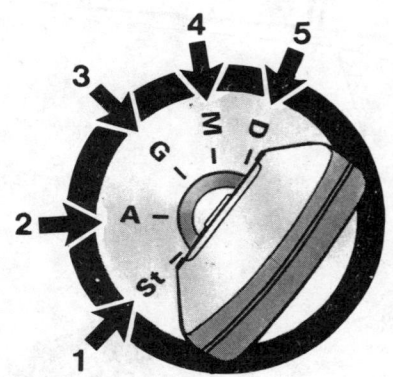

Fig. 12.63 Ignition switch key positions

ST Steering locked
A Accessories on
G Steering unlocked
M Ignition on
D Starter operation

77 Earth the radio casing direct to the engine bulkhead, using a braided type connection, to a clean metal surface.
78 Picking up the power supply depends on the year of the model concerned.

Base, TL and GTL to 1979
79 The feed may be taken from the spare hole (5) in the female component of the connecting plug A located under the instrument panel – see Fig. 12.59.

TS 1979 on and GTL 1980 on
80 On these models, the necessary feed and earth connections are to be found at the rear of the radio location. This is in the form of a black connector in which the female terminal is power (+) and the spade terminal is earth (–).
81 The radio may be operated when the ignition key is turned to either the A or M positions.

Radio – interference suppression
82 Having installed the radio, the first thing to do is to tune in to a weak station on the medium waveband and to trim the aerial using the small trim screw provided on the receiver. Once the signal is at its maximum and clearest, the aerial is correctly trimmed.
83 The ignition system is suppressed in production and it is possible that the radio will provide good reception without interference, and without the need for suppression. However, if a regular clicking is heard with the engine running, or a humming sound, carry out the following operations to eliminate the interference. All the operations described are unlikely to be required, but carry out the suppression in the sequence described until interference disappears or is at least reduced to an acceptable level.

Ignition coil (without ballast resistance)
84 Connect a condenser (60 mF + 10 000 pF) between the + terminal of the coil and the coil mounting bracket.

Ignition coil (with ballast resistance)
85 The condenser should be connected as just described, but it should be of 250 mF capacity.

Chapter 12 Supplement

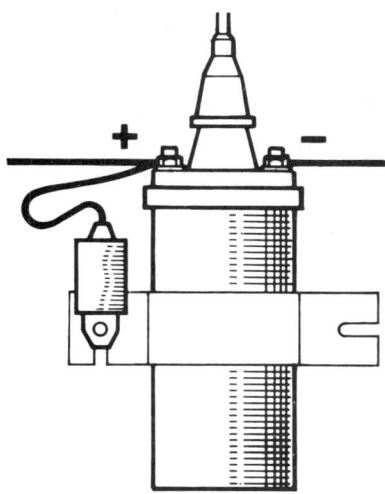

Fig. 12.64 Ignition coil suppressed

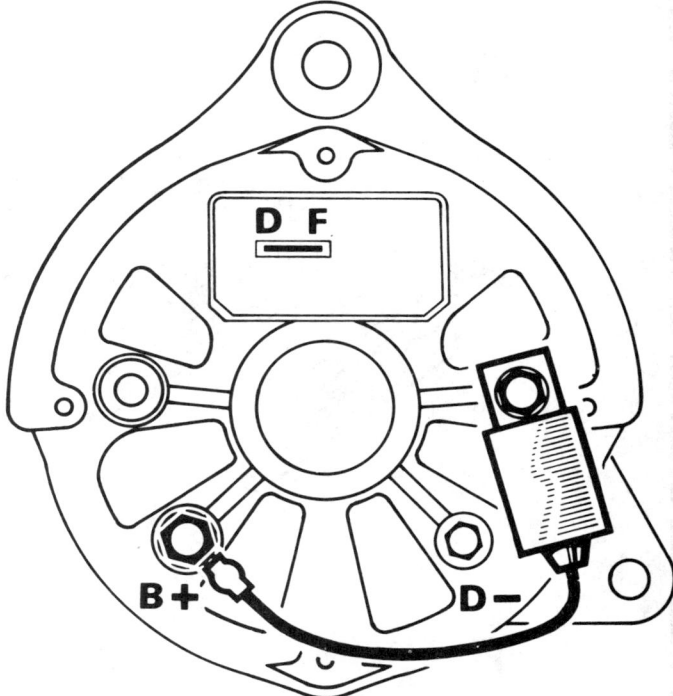

Fig. 12.65 Alternator suppressed

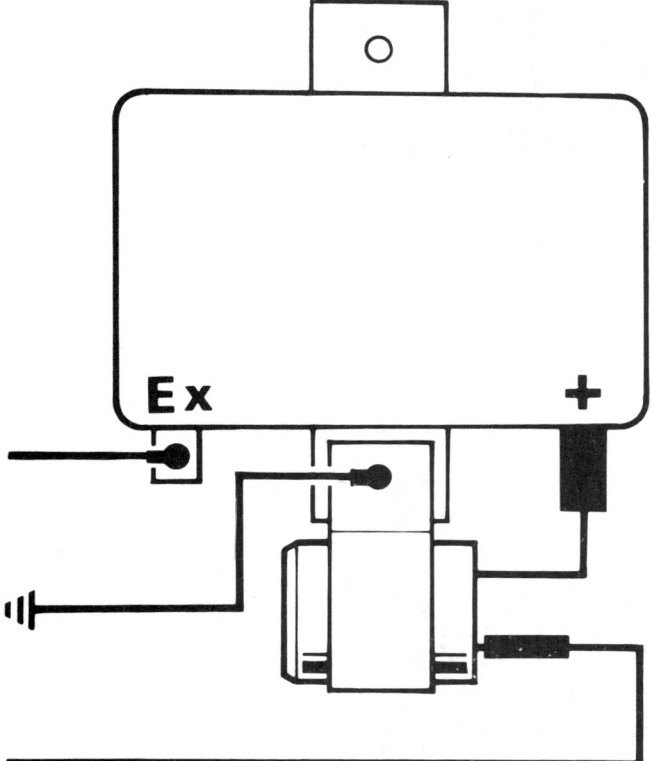

Fig. 12.66 Voltage regulator suppressed

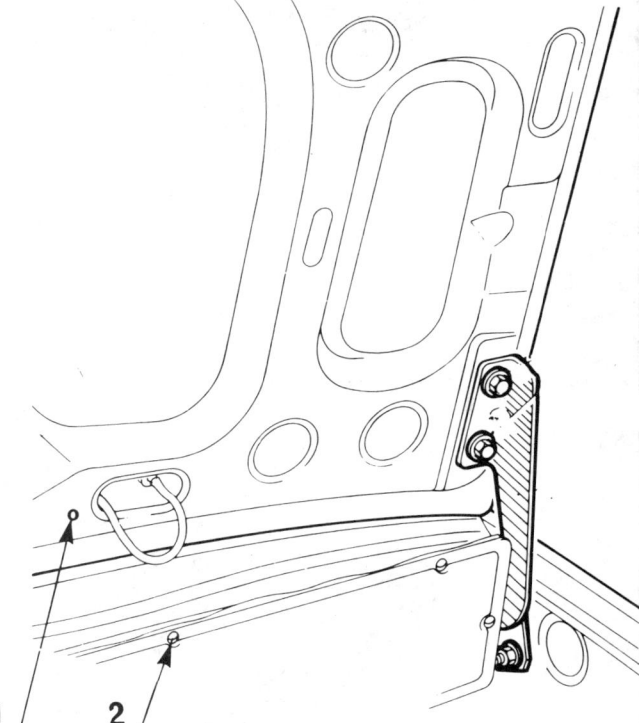

Fig. 12.67 Bonnet/bulkhead earth bonding points 1 and 2

Alternator
86 Connect a condenser (3 or 3.33 mF) between the B + terminal and the alternator earth screw.

Remotely mounted regulator
87 Connect a by-pass condenser (3 mF) between the + cable and the + terminal. Earth the condenser under the regulator mounting screw.

Earth bonding
88 Use a braided type earthing strap between the clutch cable sleeve stop on the bellhousing and the ignition coil mounting bracket bolt.
89 Bond the bonnet lid to one heater bulkhead mounting bolt.

General
90 Clicking from direction indicators, wiper motors and other electrical accessories is usually not severe enough to warrant the time, effort and money required to eliminate it, This also applies to any crackling which may occur when the brakes are applied and which is generated by the disc pads.

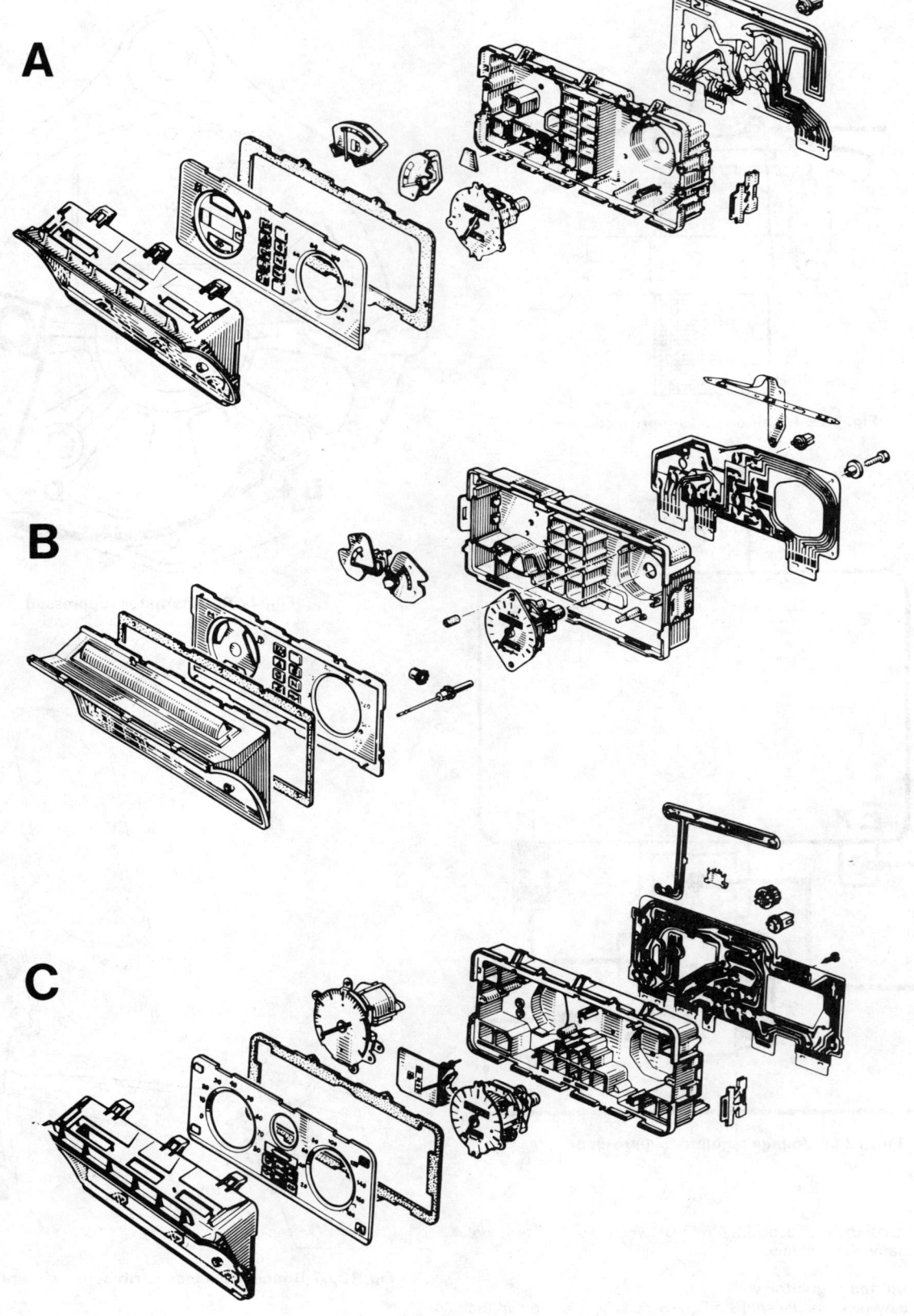

Fig. 12.68 Various types of instrument panel

A Jaeger   B Veglia   C Panel incorporating tachometer

Chapter 12 Supplement

Instruments and warning lights

91 Each warning light, tell-tale or gauge is supplied with current after the ignition/starter switch via the printed circuit board. Earth may be direct or variable.

Fuel gauge

92 This is in the form of a rheostat controlled by a lever and a float within the fuel tank. The current flow, variable with the level of fuel in the tank, reaches the gauge on the instrument panel where it is registered as a fuel level on the gauge.

Coolant temperature gauge

93 The current flow to the coolant temperature gauge is varied by a heat sensor, acting as a rheostat. The resulting current is registered on the gauge as the coolant temperature varies.

Tachometer

94 This is fed from the printed circuit board, supplying a direct feed after the ignition/starter switch, an earth, and a controlling earth taken from the ignition coil.

Battery charge indicator

95 This gauge is merely a volt meter. Its accuracy can be measured as follows.

The needle should be on the left-hand side of the central zone at a reading of 12.8 volts.
The needle should be in the centre at a reading of 13.5 volts.
The needle should be on the right-hand edge of the central zone at a reading of 15.6 volts.

General

96 Defective gauges are not usually repairable and should be replaced with new units.

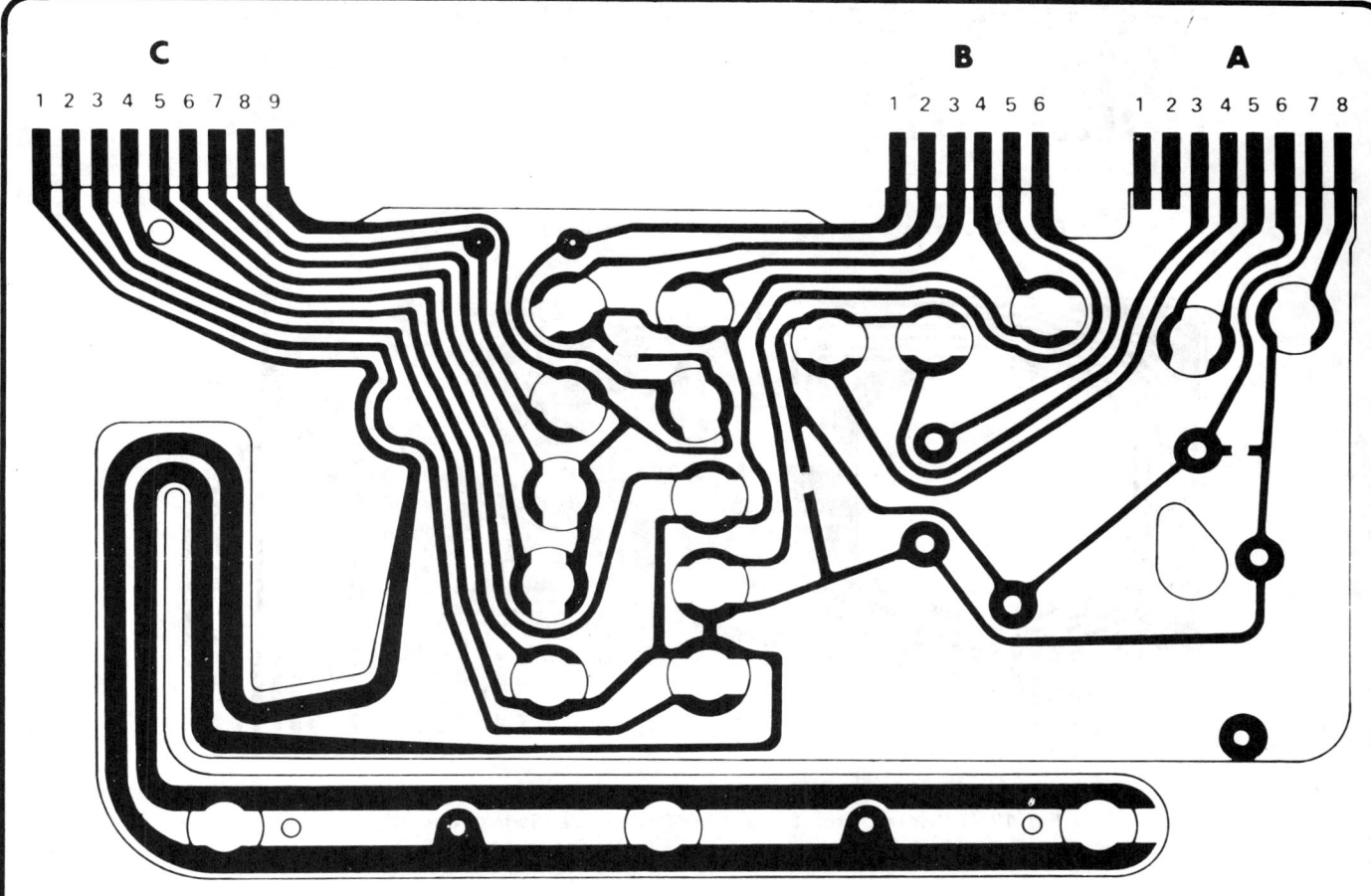

Fig. 12.69 Instrument panel printed circuit board, models up to 1979

Connector A
1   Not used
2   Not used
3   Not used
4   Not used
5   Not used
6   Not used
7   '+' after ignition/starter switch
8   Not used

Connector B
1   Coolant temperature
2   Not used
3   Not used
4   Direction indicators
5   Hazard warning lights system
6   Fuel gauge

Connector C
1   Lighting
2   Headlights
3   Earth
4   Not used
5   Rear screen demister
6   Brake warning light
7   Choke 'on' warning light
8   Oil pressure warning light
9   '+' after ignition/starter switch

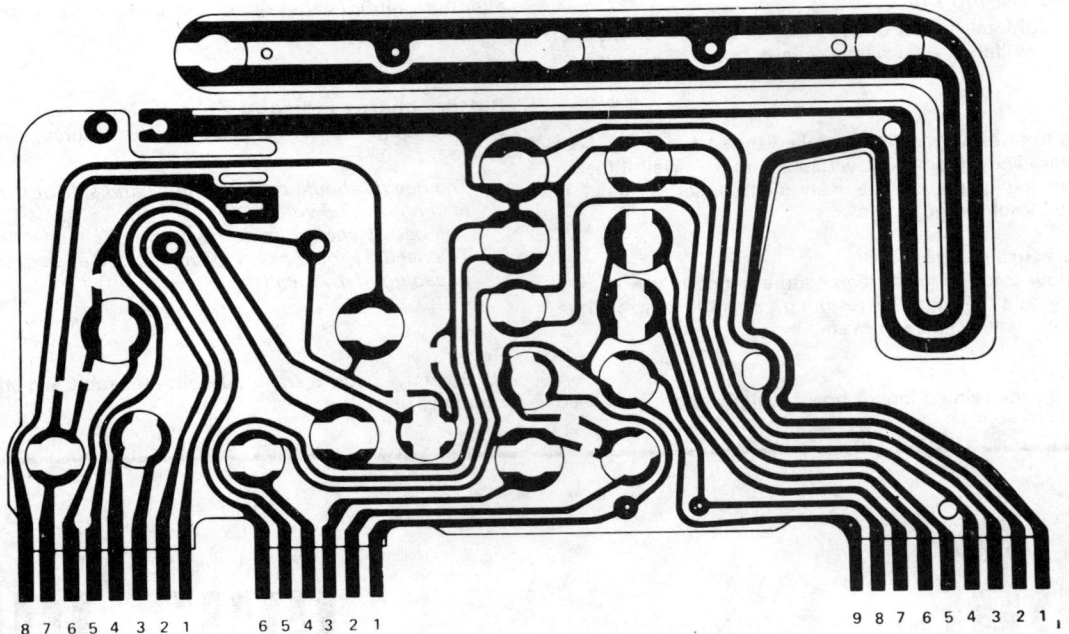

Fig. 12.70 Printed circuit broad for 1980 on Jaeger instrument panel

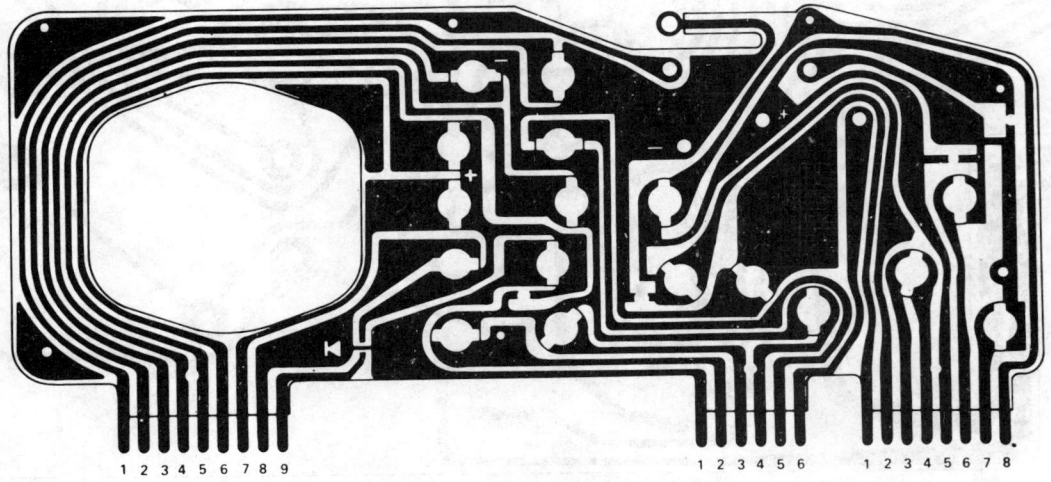

Fig. 12.71 Printed circuit board for 1980 on Veglia instrument panel

Key to Jaeger and Veglia instrument panel connections

Connector 30
1. Coolant temperature
2. Not used
3. Rear foglights
4. Direction indicators
5. Hazard warning lights
6. Fuel gauge

Connector 31
1. Not used
2. Not used
3. Not used
4. Not used
5. '+' after ignition/starter switch
6. Not used
7. Not used
8. Battery charge warning light

Connector 32
1. Instrument panel illumination
2. Headlights
3. Earth
4. Not used
5. Rear screen demister
6. Brake pressure drop warning light
7. Choke 'on' warning light
8. Oil pressure warning light
9. '+' after ignition/starter switch

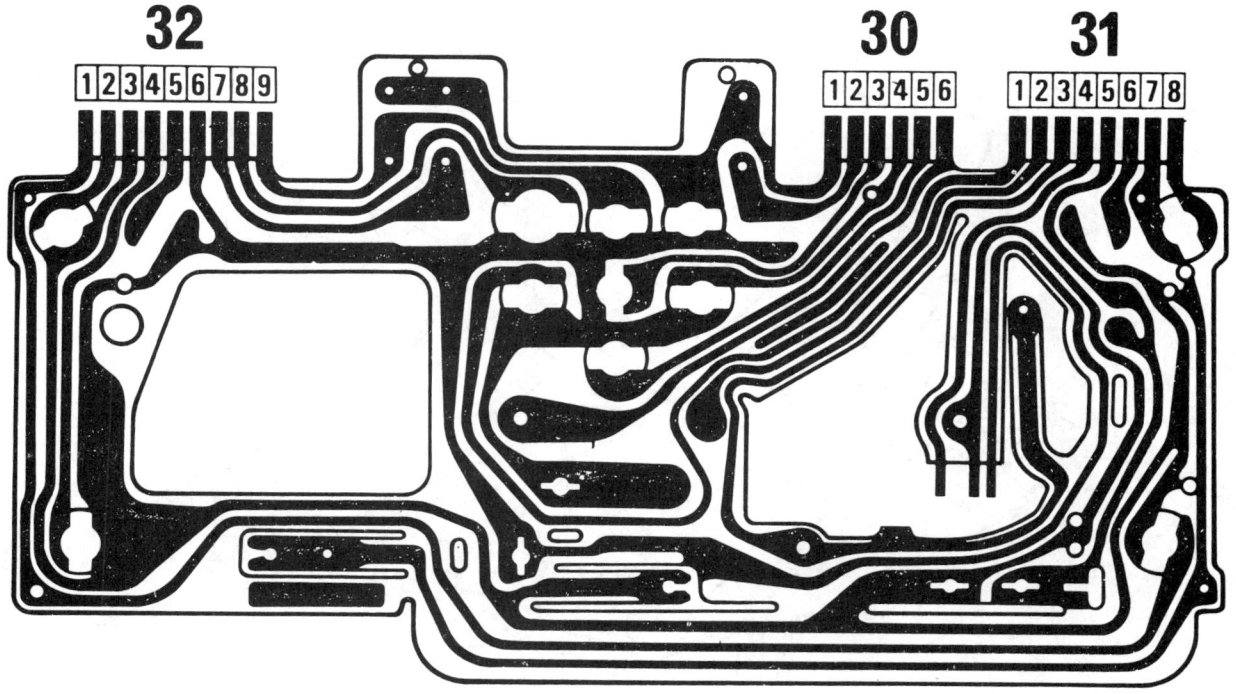

Fig. 12.72 Printed circuit board for instrument panel incorporating tachometer

Connector 32
1. Not used
2. Hazard warning lights tell-tale
3. Instrument panel earth
4. Headlight main beam warning light
5. Not used
6. '+' after ignition switch
7. Not used
8. Battery charge warning light
9. Brake fluid warning light

Connector 30
1. Oil pressure warning light
2. Not used
3. Choke 'on' warning light
4. Rear screen demister 'on' warning light
5. Direction indicators
6. Fuel gauge

Connector 31
1. Rear foglight 'on' warning light
2. Not used
3. Not used
4. Tachometer
5. Instrument panel lighting
6. Not used
7. Not used
8. Not used

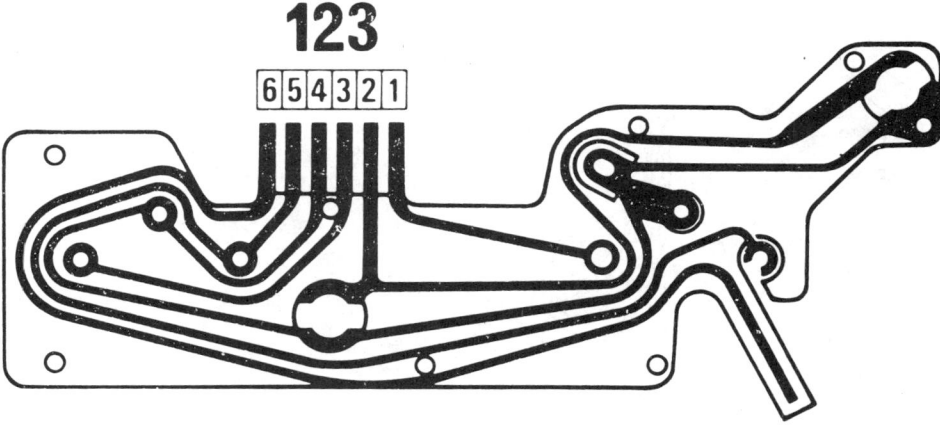

Fig. 12.73 Clock/coolant temperature gauge assembly

Connector 123
1. Not used
2. Illumination
3. Coolant temperature gauge
4. Earth
5. '+' ignition/starter switch
6. Feed to clock

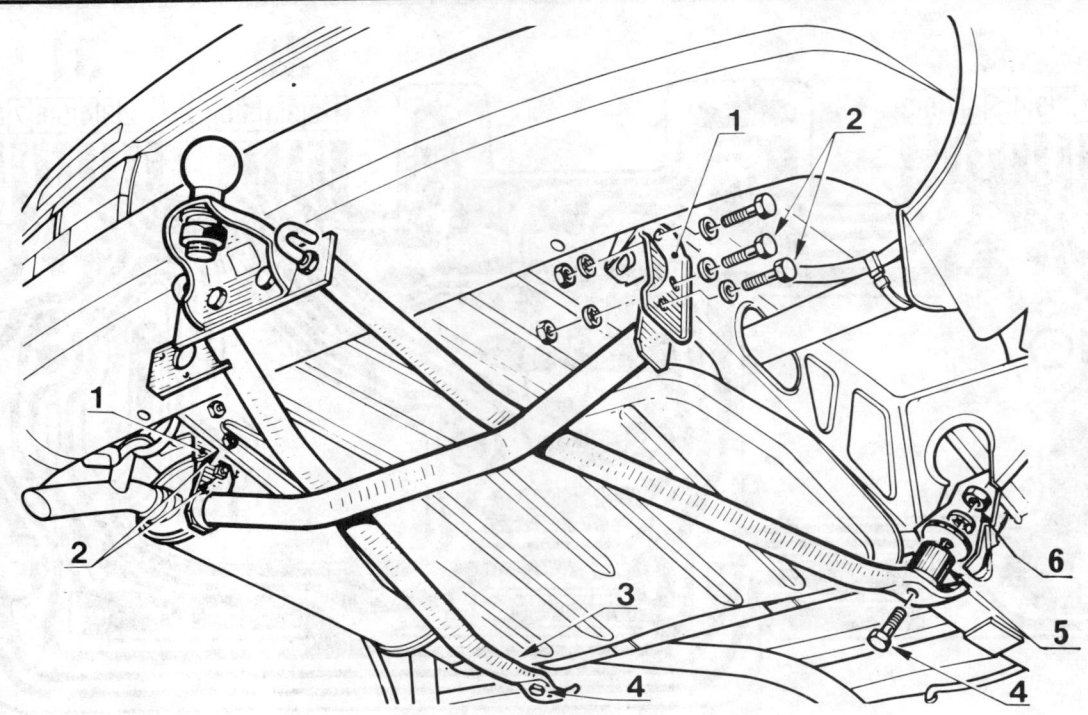

Fig. 12.74 Installing towing bracket

1 Plate
2 Bolts
3 Side struts
4 Bolts
5 Spacers
6 Nuts

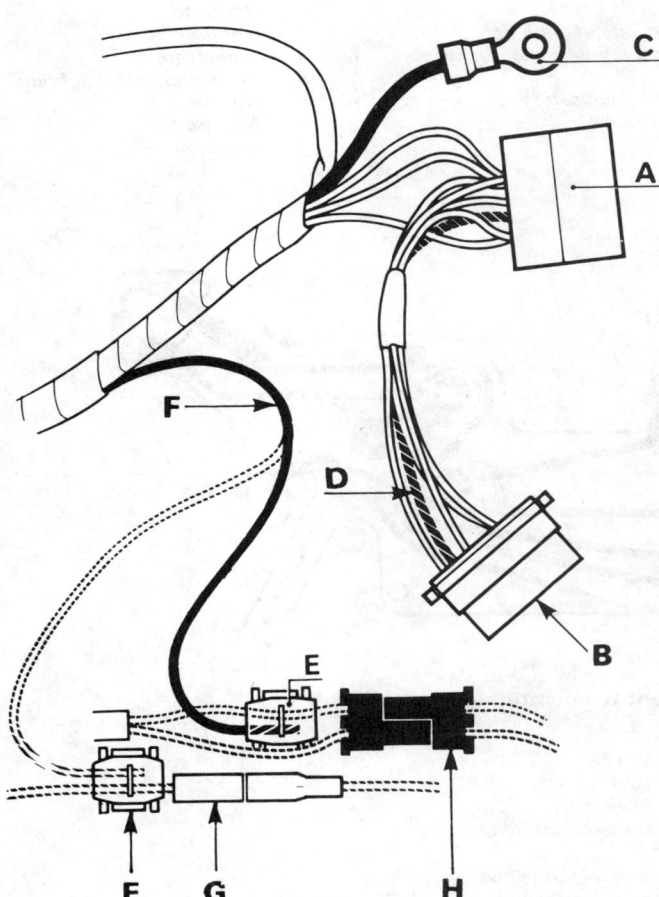

Fig. 12.75 Towing bracket harness

A Connecting plug
B Connector plug
C Earth wire
D Towing bracket harness wire (rear fog)
E Quick-fit connector
F Towing bracket harness wire
G Round plug and spocket
H Black connector

Chapter 12 Supplement

9 Bodywork

Towing bracket — installation

1 Towing brackets are available specifically to fit the Renault 14.
2 Offer up both plates (1) to the rear side members, but insert the bolts loosely at this stage (Fig. 12.74).
3 Offer up the towing bracket and connect it loosely to the plates.
4 At the front end of the bracket invert the bolts (4), spacers, washers and nuts.
5 Tighten all bolts and nuts evenly.
6 The wiring harness supplied with the towing bracket is designed to have its two connecting plugs connected in the following way:
Remove the back of the right-hand rear lamp cluster inside the luggage compartment. Pull the existing connector plug from the rear of the lamp and push on connector A of the towing bracket harness instead. Now join the removed connector to the connector B of the towing bracket harness (Fig. 12.75).
7 Connect the earth tag under the rear lamp cluster earth screw.
8 On vehicles produced up to 1979, connect wire (F) of the towing bracket harness, to the maroon coloured wire adjacent to the round plug and socket (G) using the quick-fit connector (E) supplied in the fitting kit.
9 On later models, connect wire (F) to the vehicle maroon coloured wire adjacent to the black connector (H) again using the quick-fit connector.
10 If a rear fog lamp is fitted to the vehicle, but the towing bracket wiring harness does not incorporate a wire for this lamp, run a wire (D) between the connectors (A) and (B). To do this, fit suitable tags to each end of the wire and insert them into the spare locations in each connector.
11 Route the new towing bracket wiring harness through the grommet used for the rear number plate lamp.
12 To fit a feed wire for the interior lamp on the trailer or caravan, run a heavy section wire under the sill kick plate and luggage compartment mat.
13 Incorporate an in-line fuse in the wire and then connect it to the feed screw at the back of the lighting switch after pulling off the bottom half casing.

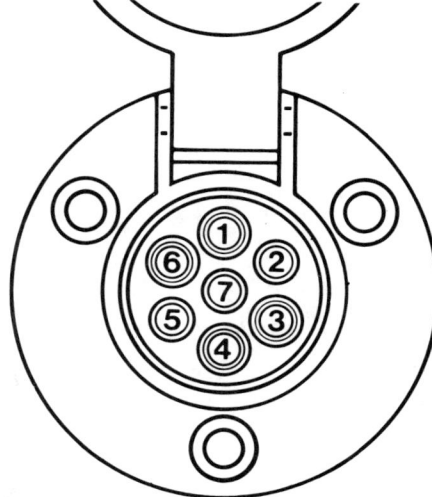

Fig 12.76 Towing bracket standard socket. For key see text

14 The standard socket wiring to DIN standards is as follows, using the diagram for plug identification (Fig. 12.76).

    1   LH rear direction indicator
    2   Interior lighting
    3   Earth
    4   RH rear direction indicator
    5   RH tail lamp and rear number plate
    6   Stop lamps
    7   LH tail lamp

15 It is recommended that a stabiliser or heavy duty shock absorbers are fitted to the rear of cars towing a trailer or caravan. These are available from Renault dealers and other sources and should be installed in accordance with the manufacturer's instructions.

See overleaf for wiring diagrams

Key to wiring diagrams

Colour code

B	Blue
Bc	White
Be	Beige
C	Clear
G	Grey
J	Yellow
M	Maroon
N	Black
Or	Orange
R	Red
S	Pink
V	Green
Vi	Violet

Wire identification code

Example: 162 N-5 74

162	Wire number
N	Wire colour
5	Wire diameter
74	Wire destination

Wire diameter code

No	Diameter (mm)
1	7/10
2	9/10
3	10/10
4	12/10
5	16/10
6	20/10
7	25/10
8	30/10
9	45/10
10	50/10
11	70/10
12	80/10

Harness code

A	Engine front
B	Rear
D	Cigar lighter
K	Starter
L	Interior light – door switches
P	Door Locks
Q	Tailgate
R	Engine

Component code

1	LH front sidelight and/or direction indicators		68	LH rear light assembly
2	RH front sidelight and/or direction indicators		69	RH rear light assembly
7	LH headlight main and dipped beams		70	Licence plate light
8	RH headlight main and dipped beams		71	Choke On warning light switch
9	LH horn		72	Reversing lights switch
10	RH horn		73	Rear light assemblies earth
12	Alternator		75	Heating-ventilating fan motor switch
13	LH earth		78	Rear screen wiper motor
14	RH earth		79	Rear screen washer pump
15	Starter		80	Junction block – front harness to engine harness
16	Battery		81	Junction block – front harness to rear harness
17	Engine cooling fan motor		82	Junction block – front harness to console harness
18	Ignition coil		85	Junction block – window winder harness
19	Distributor		93	Junction with LH wing light
20	Windscreen washer pump		94	Junction with RH wing light
21	Oil pressure switch		97	Bodyshell earth
22	Thermal switch on radiator		99	Dashboard earth
23	Thermal switch		103	Feed to accessories plate
26	Windscreen wiper		106	Rear foglights switch
27	Brake master cylinder		114	Windscreen wiper timer relay
28	Heating-ventilating fan motor		116	Junction – rear foglight
29	Instrument panel		117	Junction – front foglights
30	Connector No.1 – Instrument panel		121	Glovebox light
31	Connector No.2 – Instrument panel		123	Clock or connector
32	Connector No.3 – Instrument panel		127	Front centre speaker
34	Hazard warning lights switch		129	Front foglights switch
35	Rear screen demister switch		131	Electro-magnetic locks cut-out
36	Heating-ventilating fan motor rheostat		132	Electro-magnetic locks inertia switch
37	LH window winder switch		133	LH front door lock changeover switch
38	RH window winder switch		134	RH front door lock changeover switch
39	Earth junction plate under dashboard		135	LH front door electro-magnetic lock solenoid
40	LH door pillar switch		136	RH front door electro-magnetic lock solenoid
41	RH door pillar switch		137	LH front door electro-magnetic lock solenoid
42	LH window winder		138	RH rear door electro-magnetic lock solenoid
43	RH window winder		142	Wire junction – window winder harness
44	Accessories plate (fusebox)		144	Wire junction – interior light wiring
45	Junction block – front harness to accessories plate		146	Thermal switch
46	Junction block – front harness to accessories plate		150	LH front door loudspeaker
47	Junction block – front harness to accessories plate		151	RH front door loudspeaker
48	Junction box – front harness to accessories plate		152	Electro-magnetic door locks switch
49	Junction block – front harness to accessories plate		153	Car radio loudspeaker wires
51	Accessories plate feed		168	Wire junction – rear screen washer pump
52	Stoplights switch		171	Rear screen washer-wiper switch
53	Ignition/starter switch		180	Wire junction – LH door loudspeaker wiring
54	Heater control panel illumination		181	Wire junction – RH door loudspeaker wiring
55	Glovebox illumination		182	Tailgate RH counterbalance
56	Cigar lighter		183	Tailgate LH counterbalance
57	Radio feed		184	Luggage compartment light switch
58	Windscreen washer-wiper switch		185	Glove compartment light
59	Combination lighting switch		192	Tailgate earth
60	Direction indicators switch or connector		203	Wire junction – switch illumination to instrument panel
62	Interior light (LH)			
63	Interior light (RH)			
64	Handbrake On warning light			
65	Fuel gauge tank unit			
66	Rear screen demister			
67	Luggage compartment light			

Not all components are fitted to all models

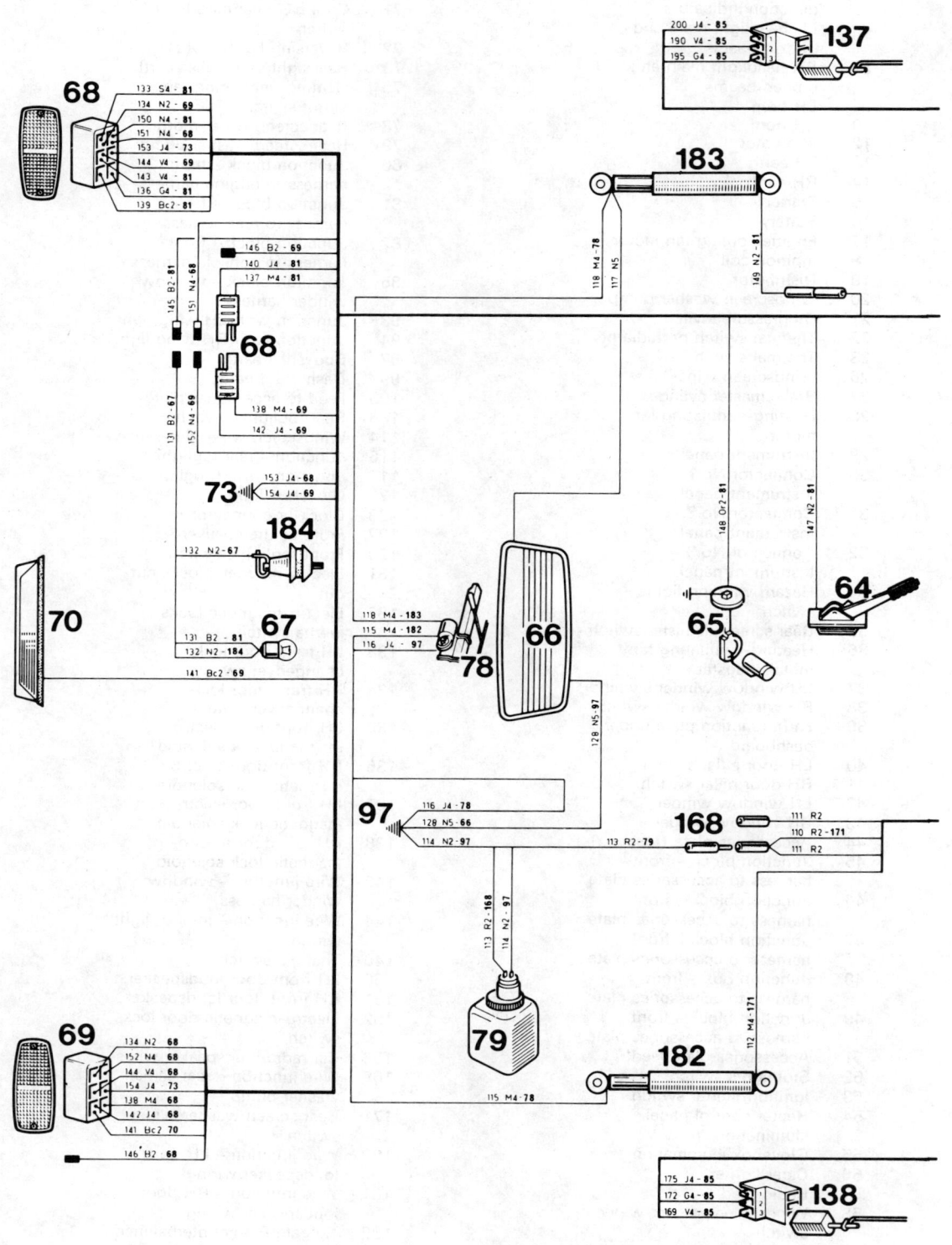

Fig. 12.77 Wiring diagram R1211 – 1979 on

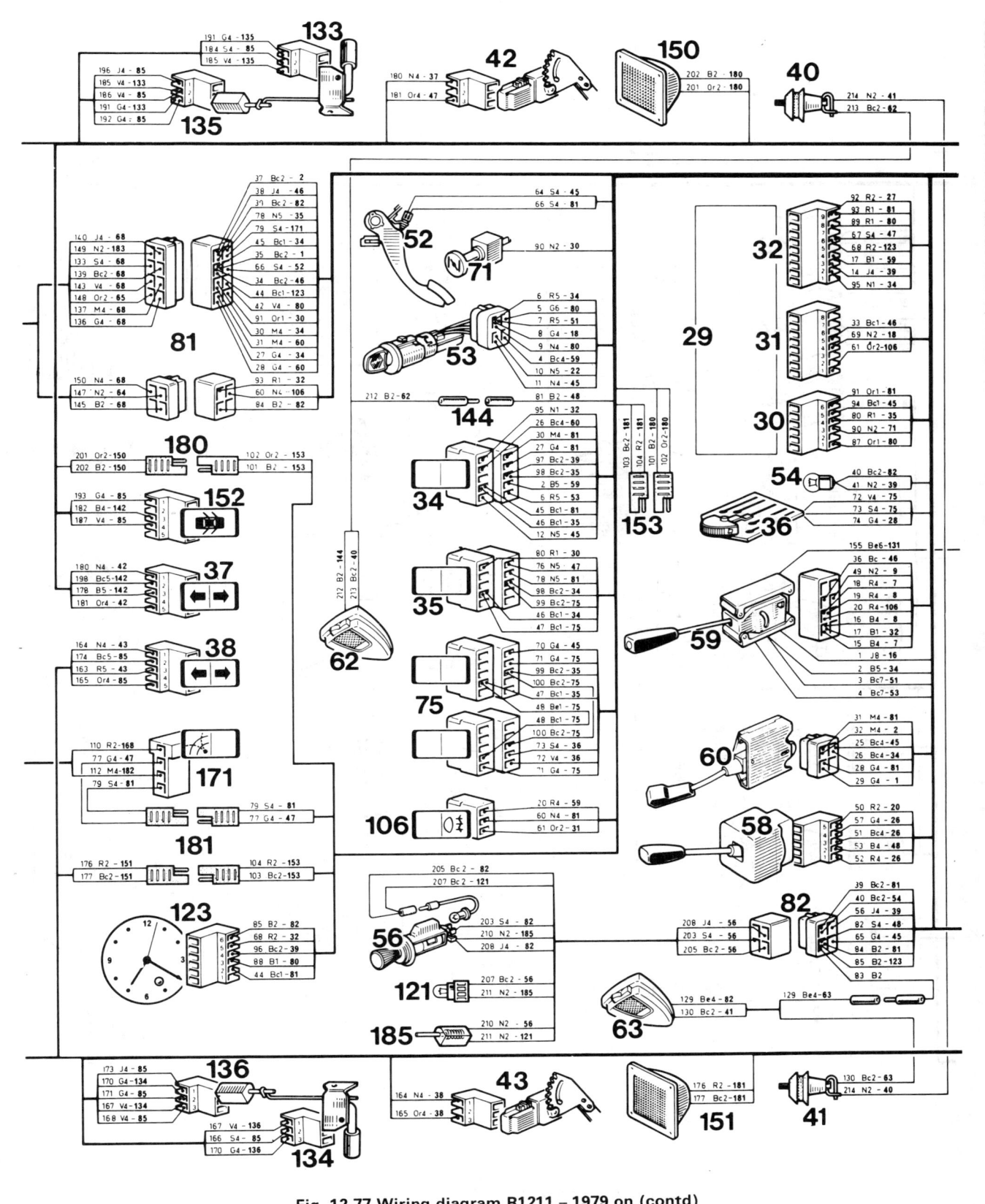

Fig. 12.77 Wiring diagram R1211 – 1979 on (contd)

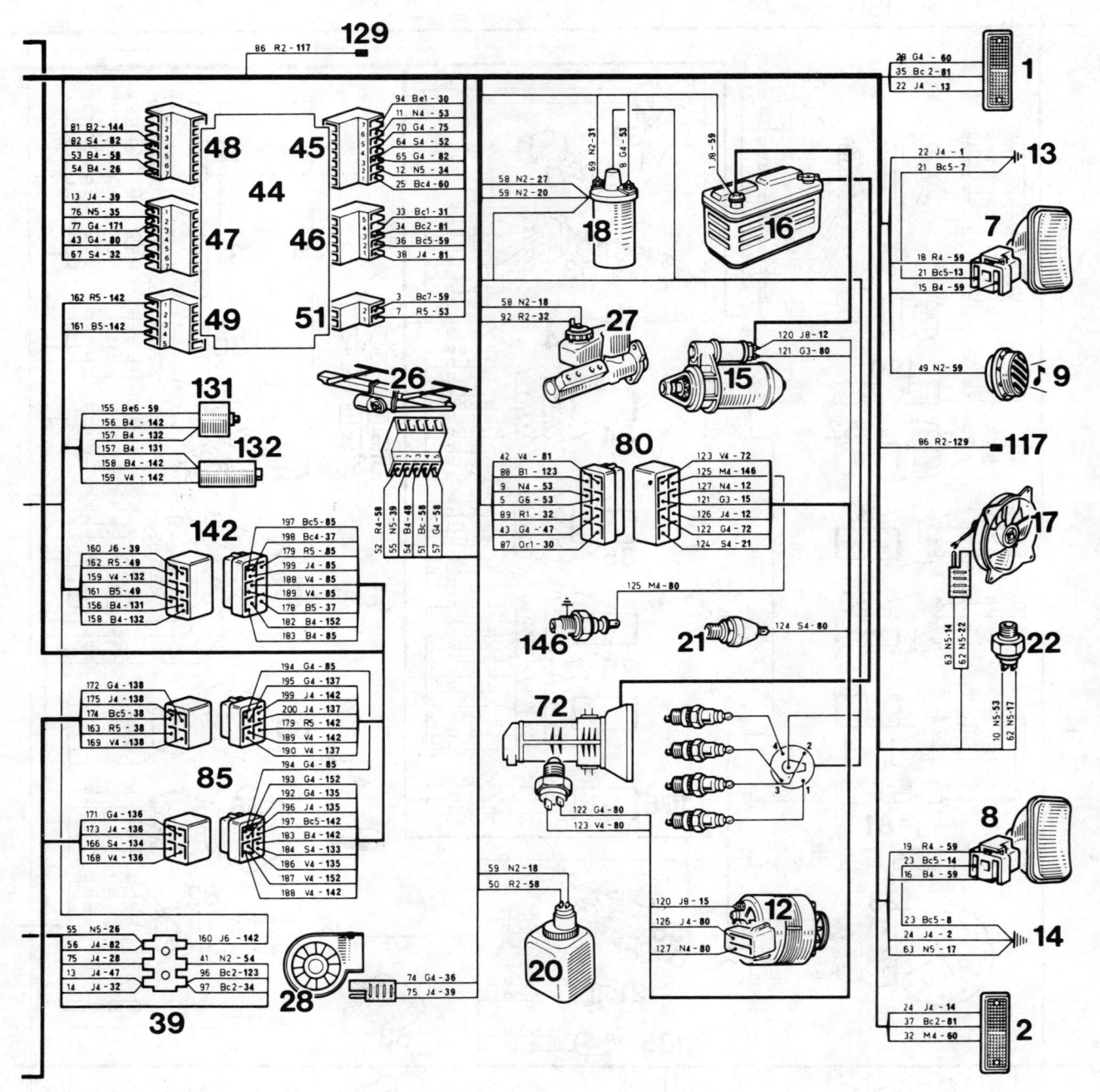

Fig. 12.77 Wiring diagram R1211 – 1979 on (contd)

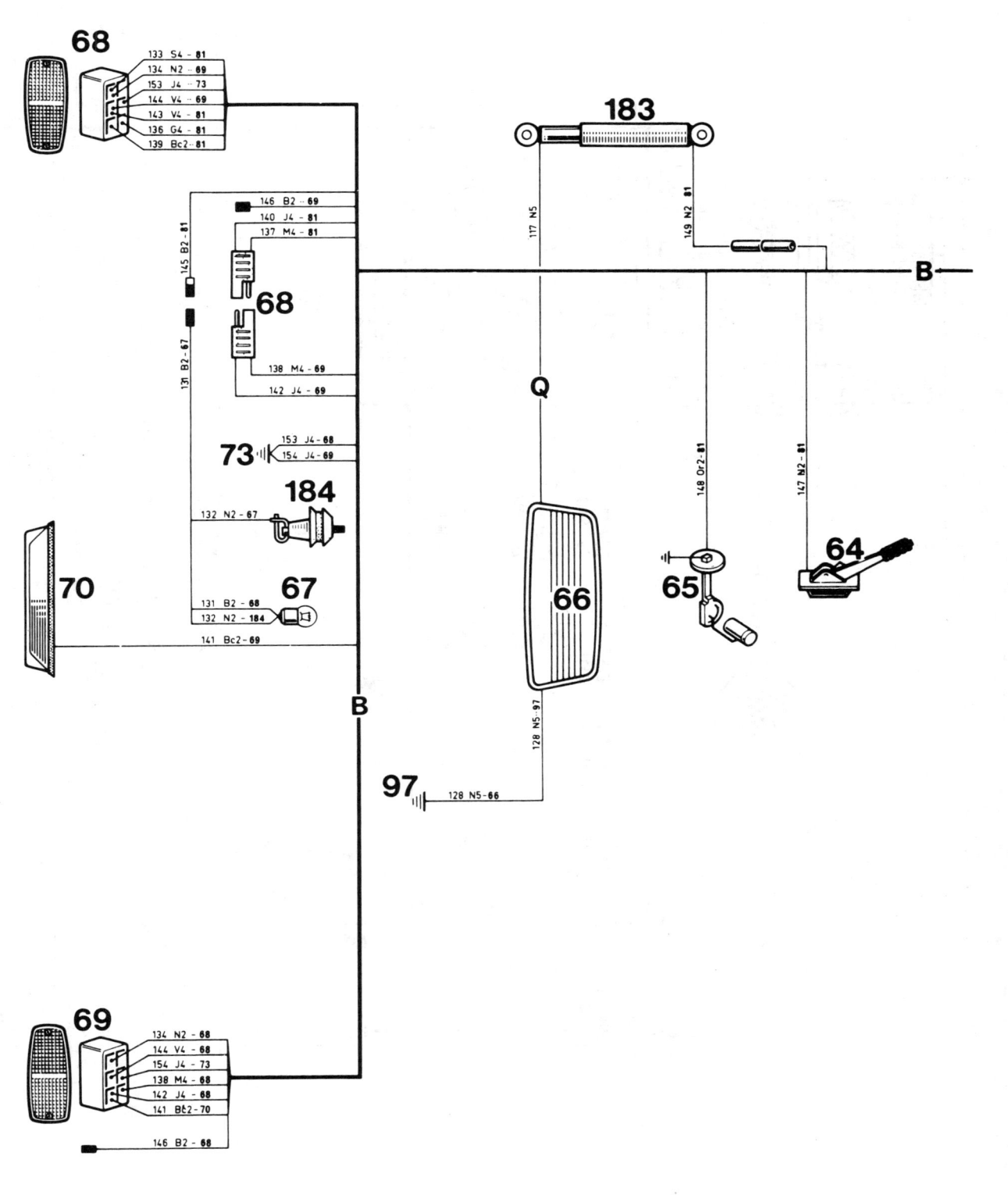

Fig. 12.78 Wiring diagram R1210 – 1980 on

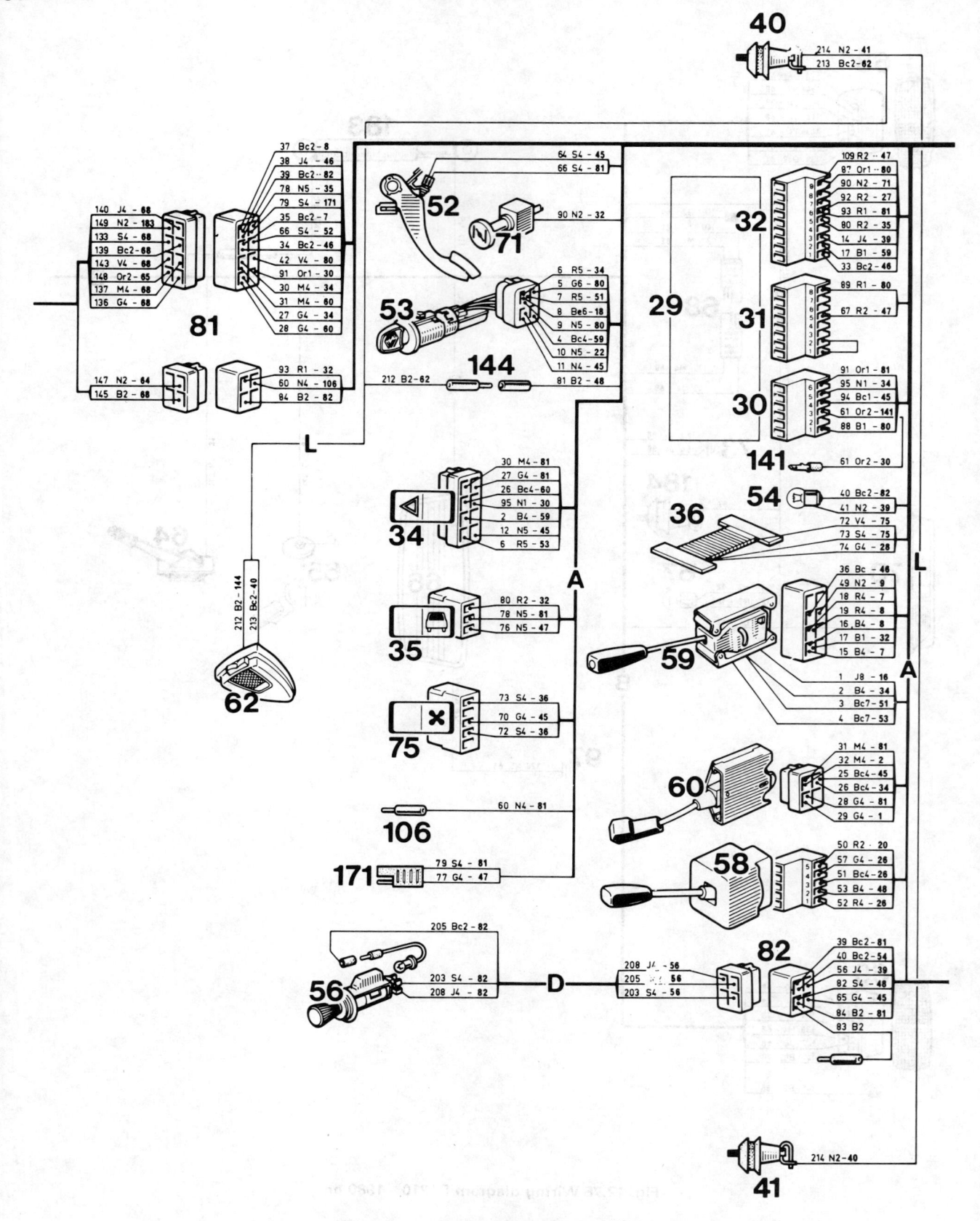

Fig. 12.78 Wiring diagram R1210 – 1980 on (contd)

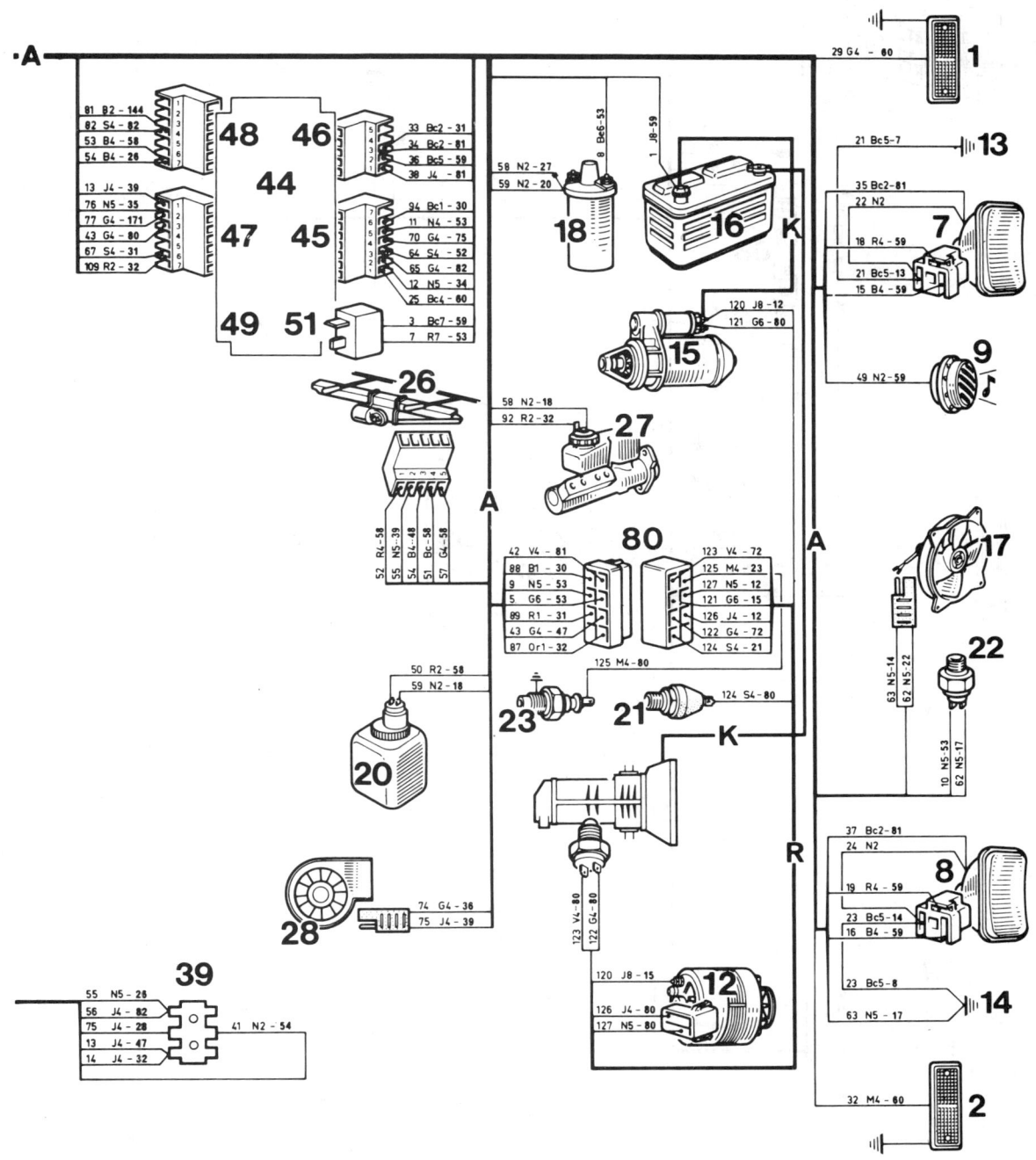

Fig. 12.78 Wiring diagram R1210 – 1980 on (contd)

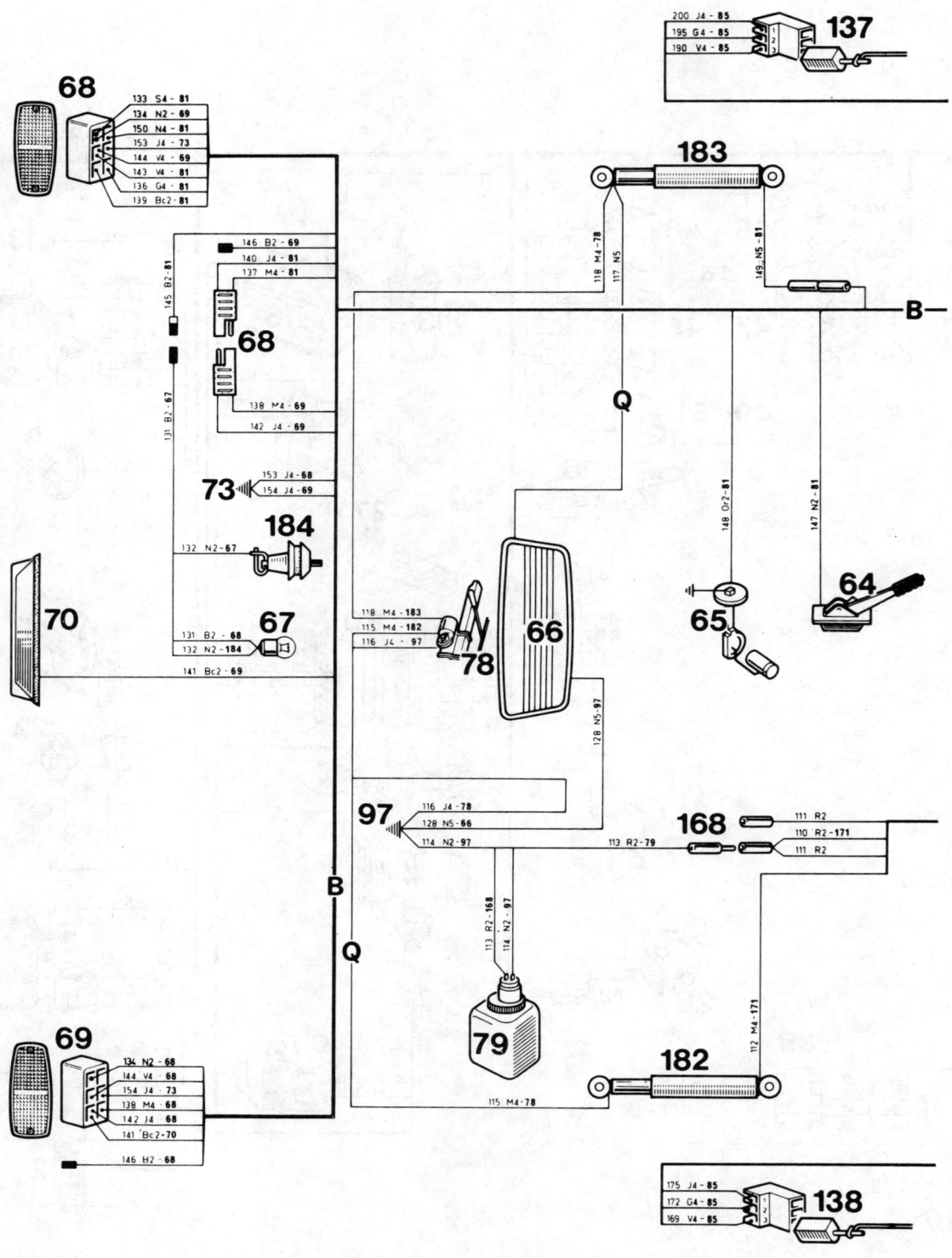

Fig. 12.79 Wiring diagram R1212 – 1980 on

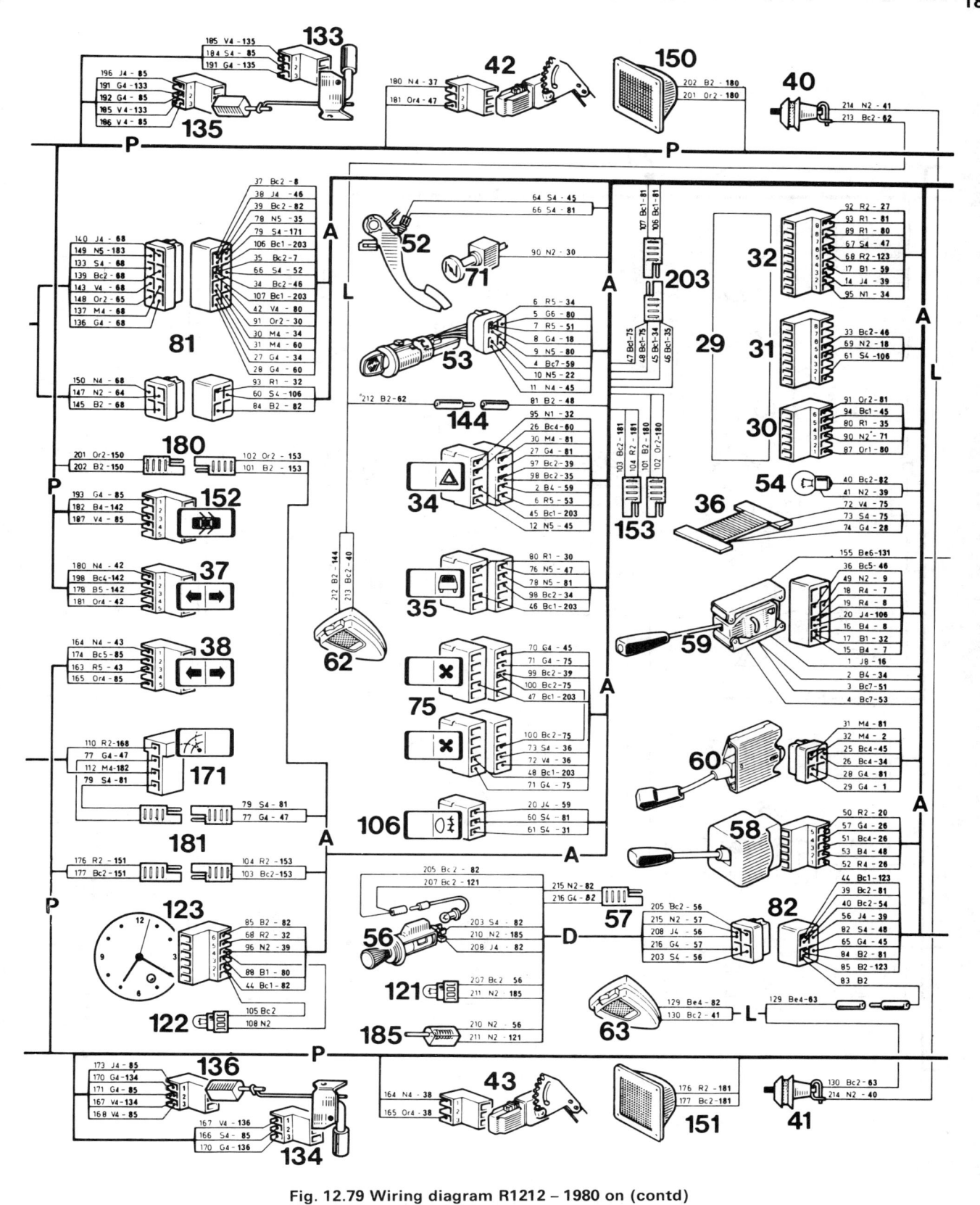

Fig. 12.79 Wiring diagram R1212 – 1980 on (contd)

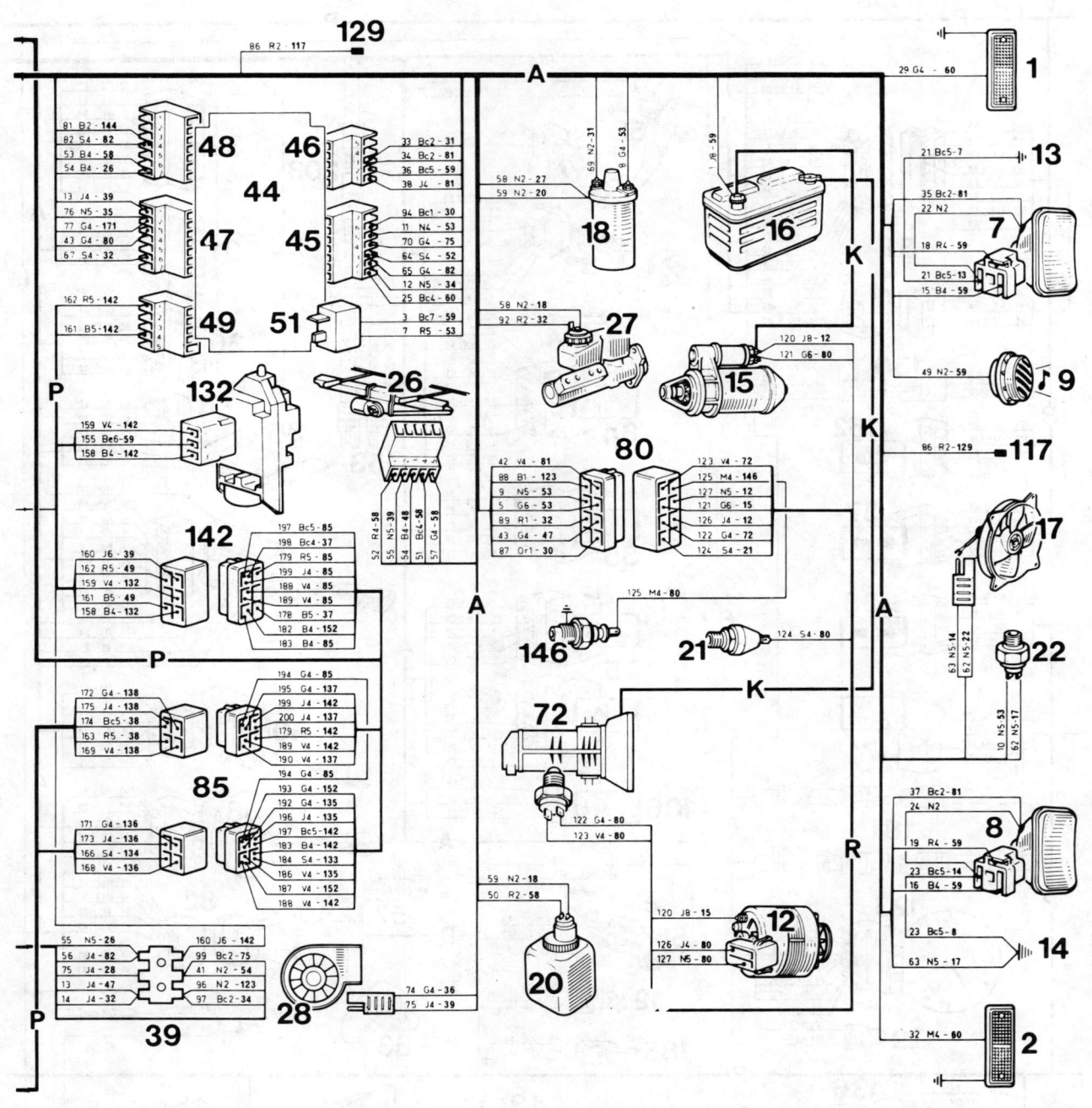

Fig. 12.79 Wiring diagram R1212 – 1980 on (contd)

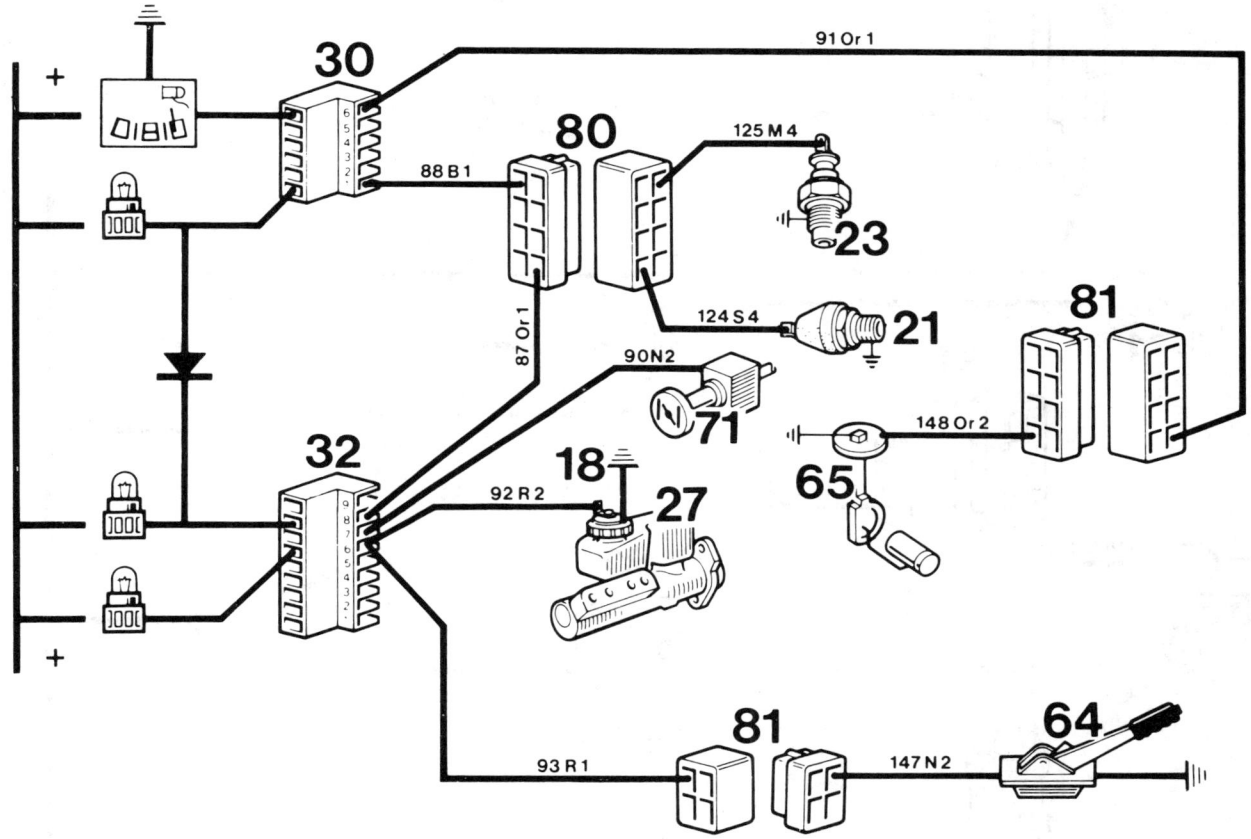

Fig. 12.80 Layout of gauge and switch connections for TL and GTL 1980 on

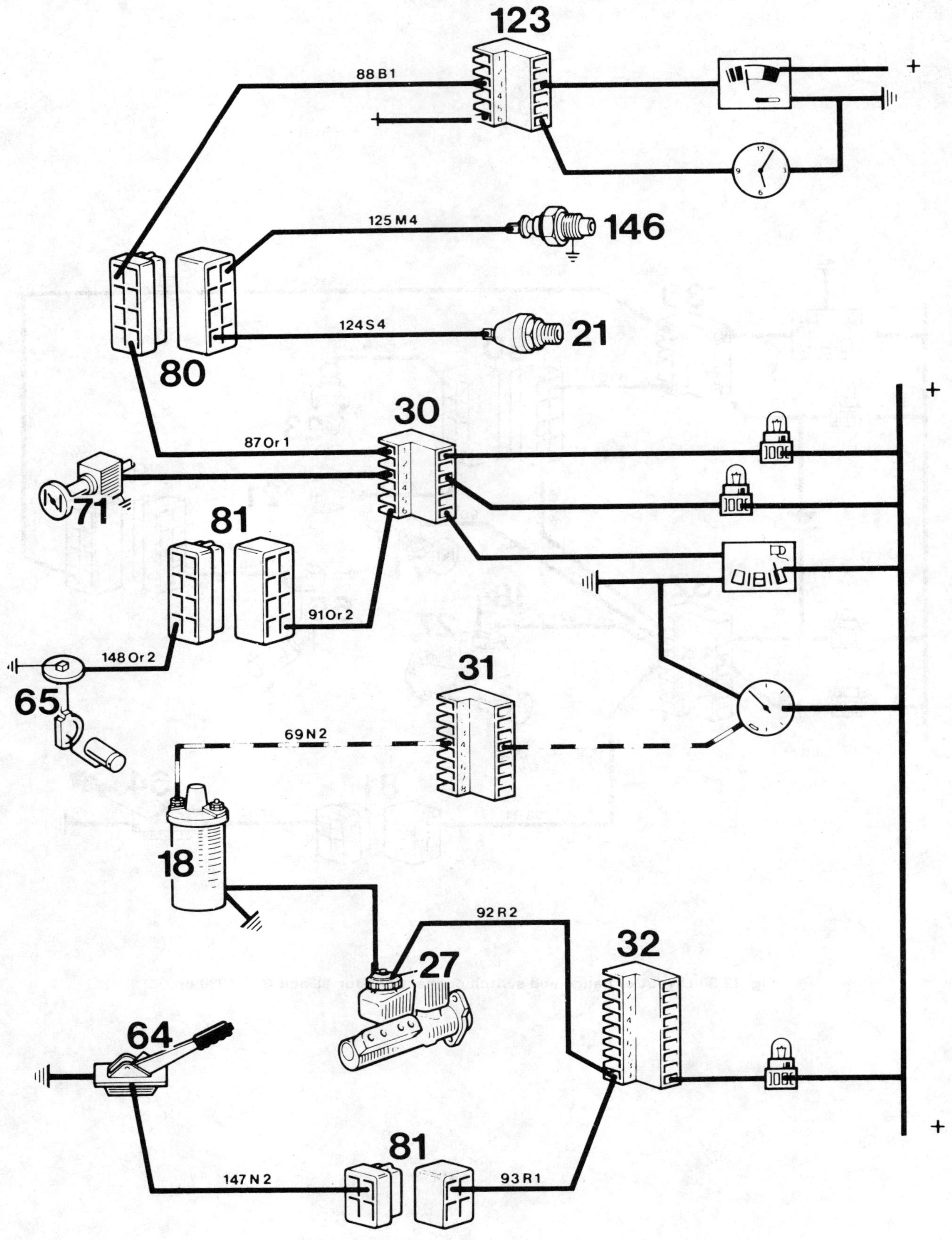

Fig. 12.81 Layout of gauge and switch connections for TS and LS models

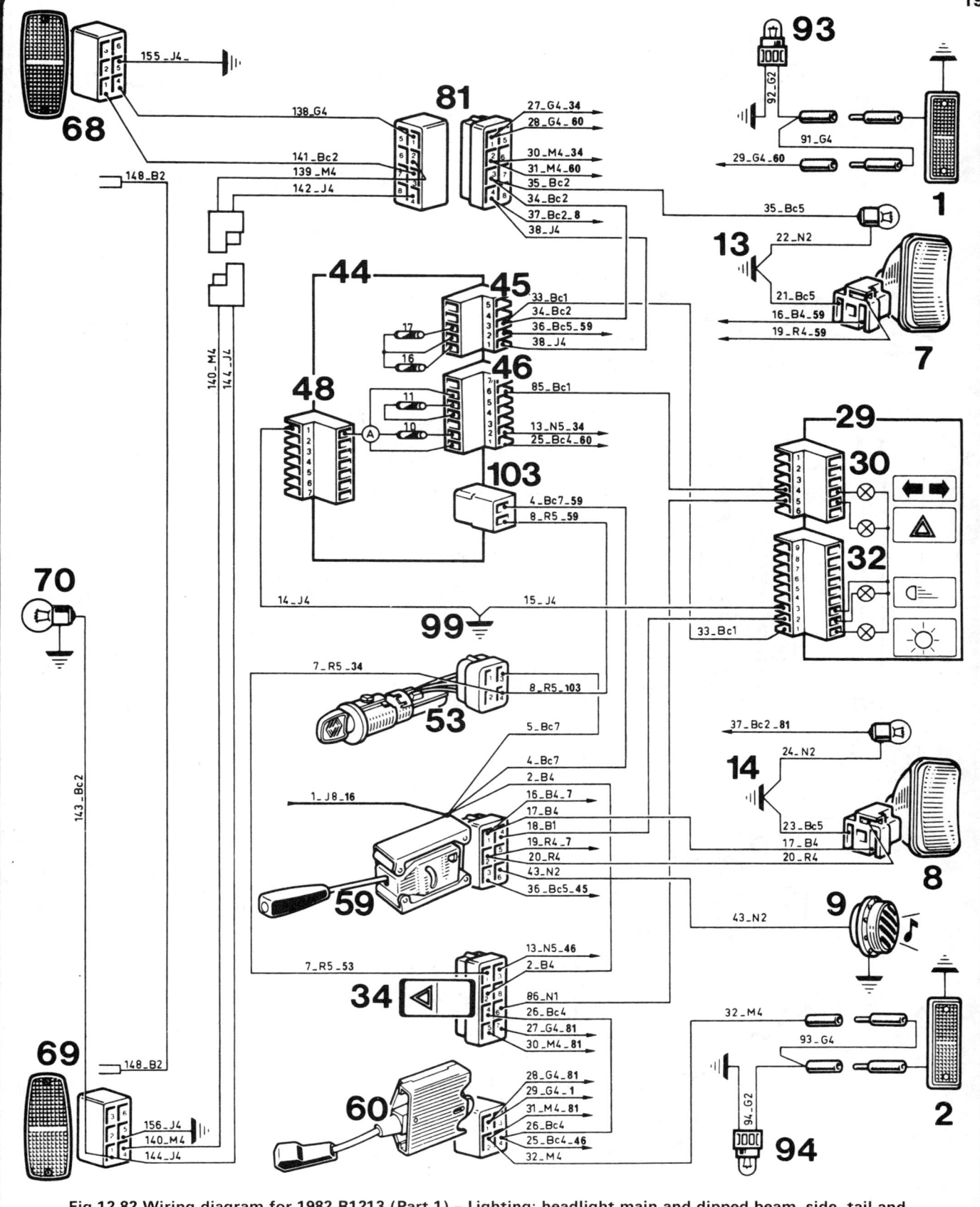

Fig 12.82 Wiring diagram for 1982 R1213 (Part 1) – Lighting; headlight main and dipped beam, side, tail and instrument panel

Letter A represents the flasher unit on the accessories plate 44

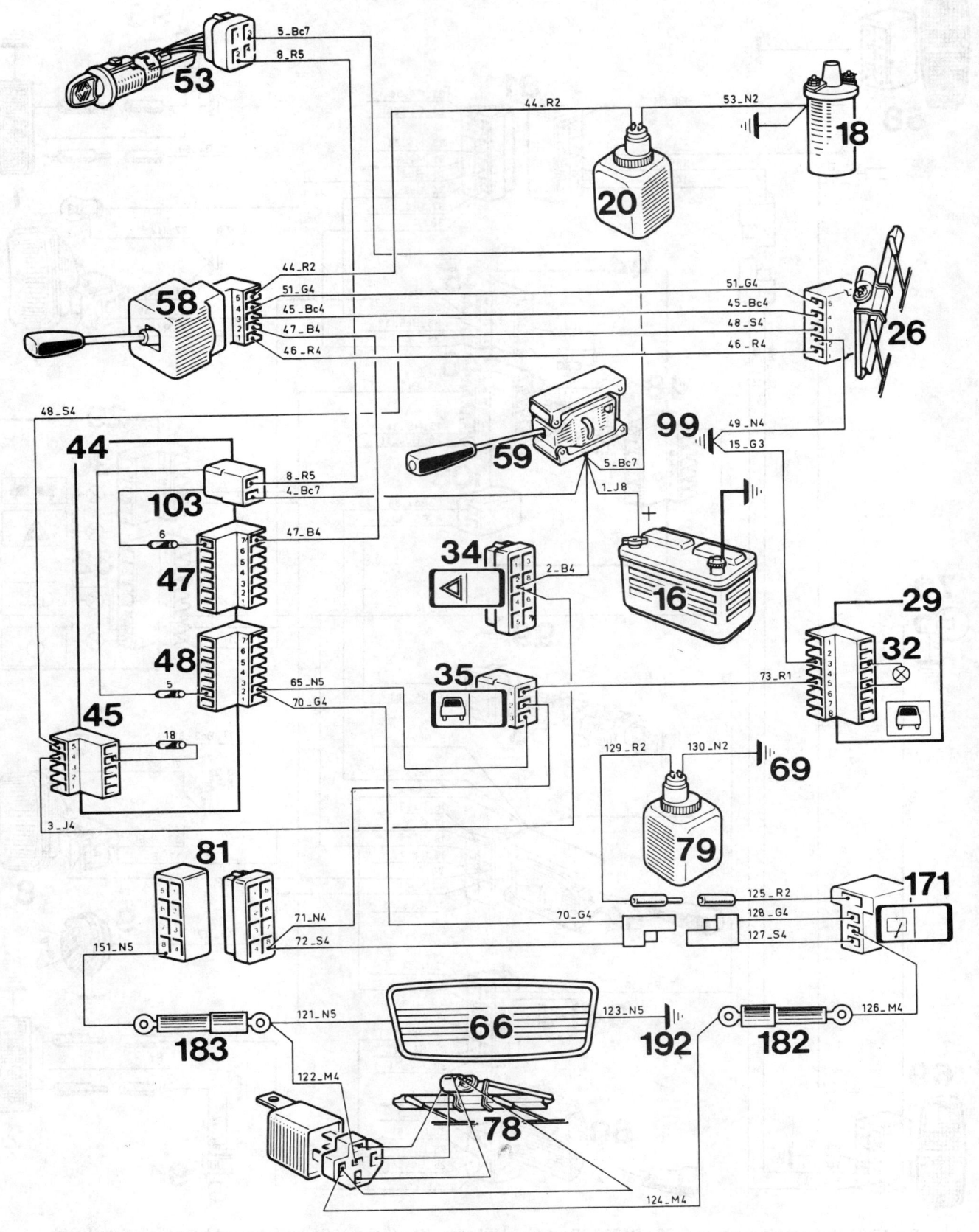

Fig. 12.82 Wiring diagram for 1982 R1213 (Part 2) – Windscreen wiper/washer, rear screen wiper/washer rear screen demister

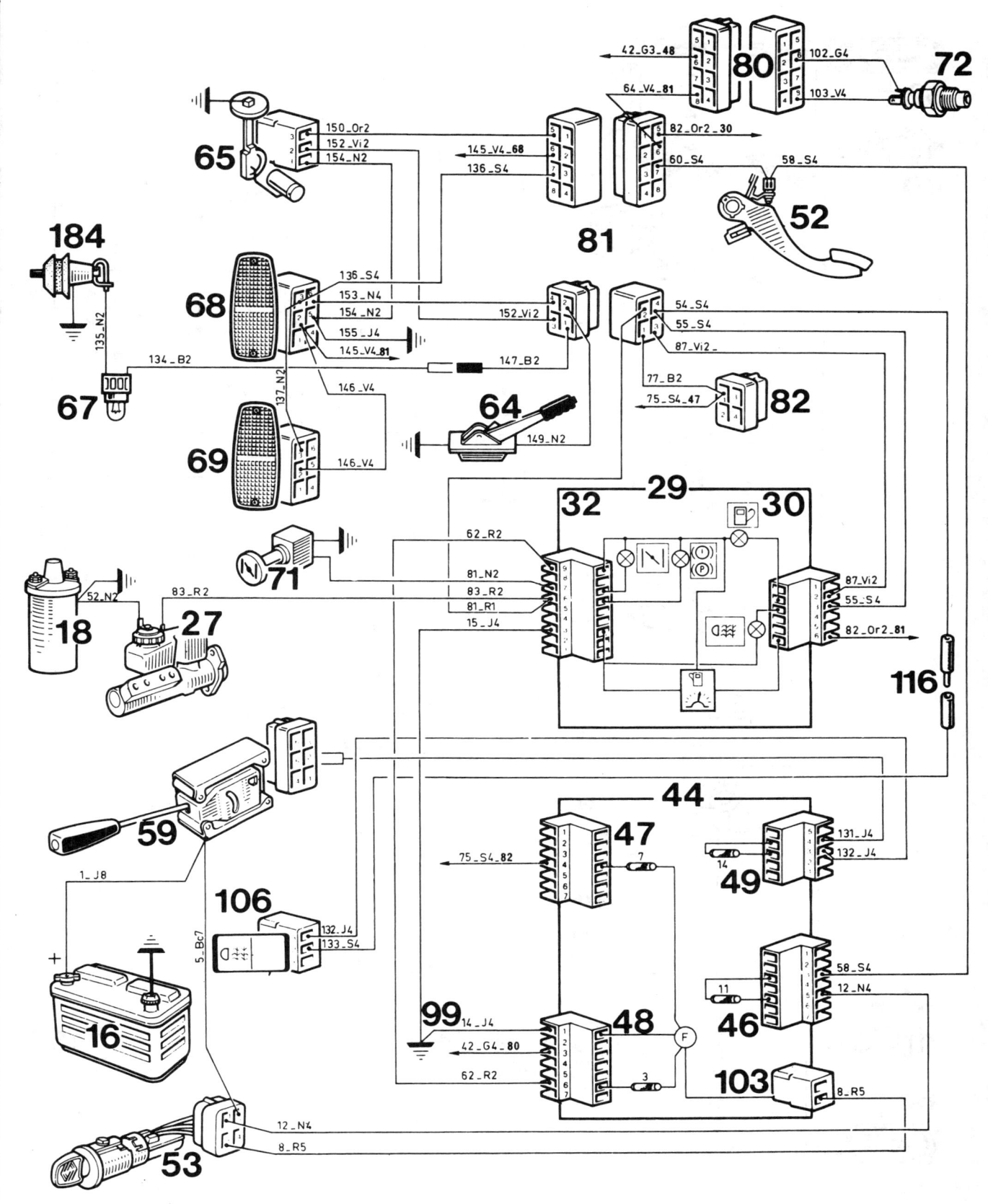

Fig. 12.82 Wiring diagram for 1982 R1213 (Part 3) – Fuel gauge and 'low fuel' warning light, handbrake, choke 'on' warning light switch, brake fluid warning light, rear foglight switch, luggage compartment light, stoplamps

Letter F represents the relay after the ignition switch on accessories plate 44

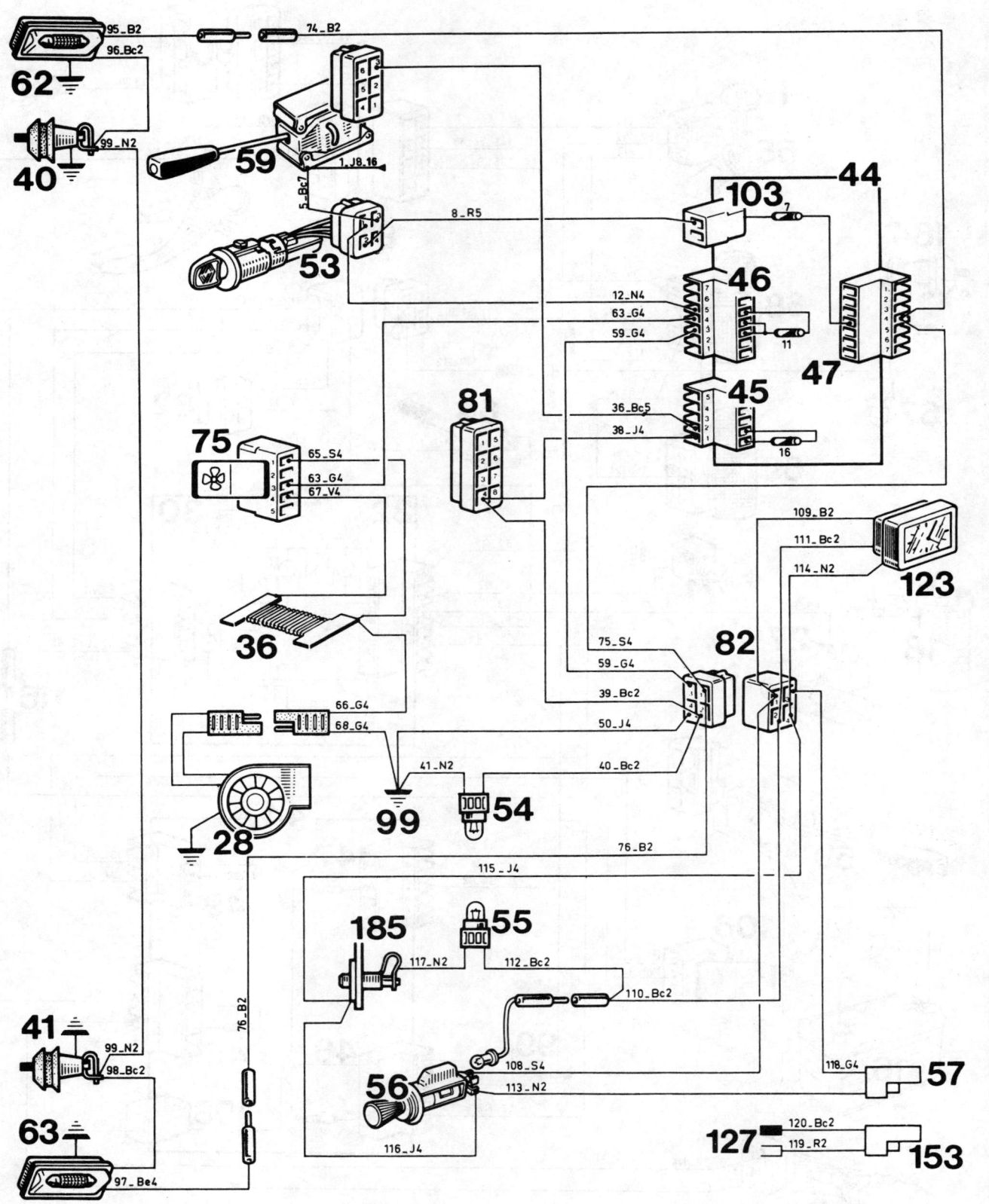

Fig. 12.82 Wiring diagram for 1982 R1213 (Part 4) – Lighting; interior lights, glove compartment, heater controls, clock, cigar lighter. Radio and loudspeaker, heater fan, cigar lighter

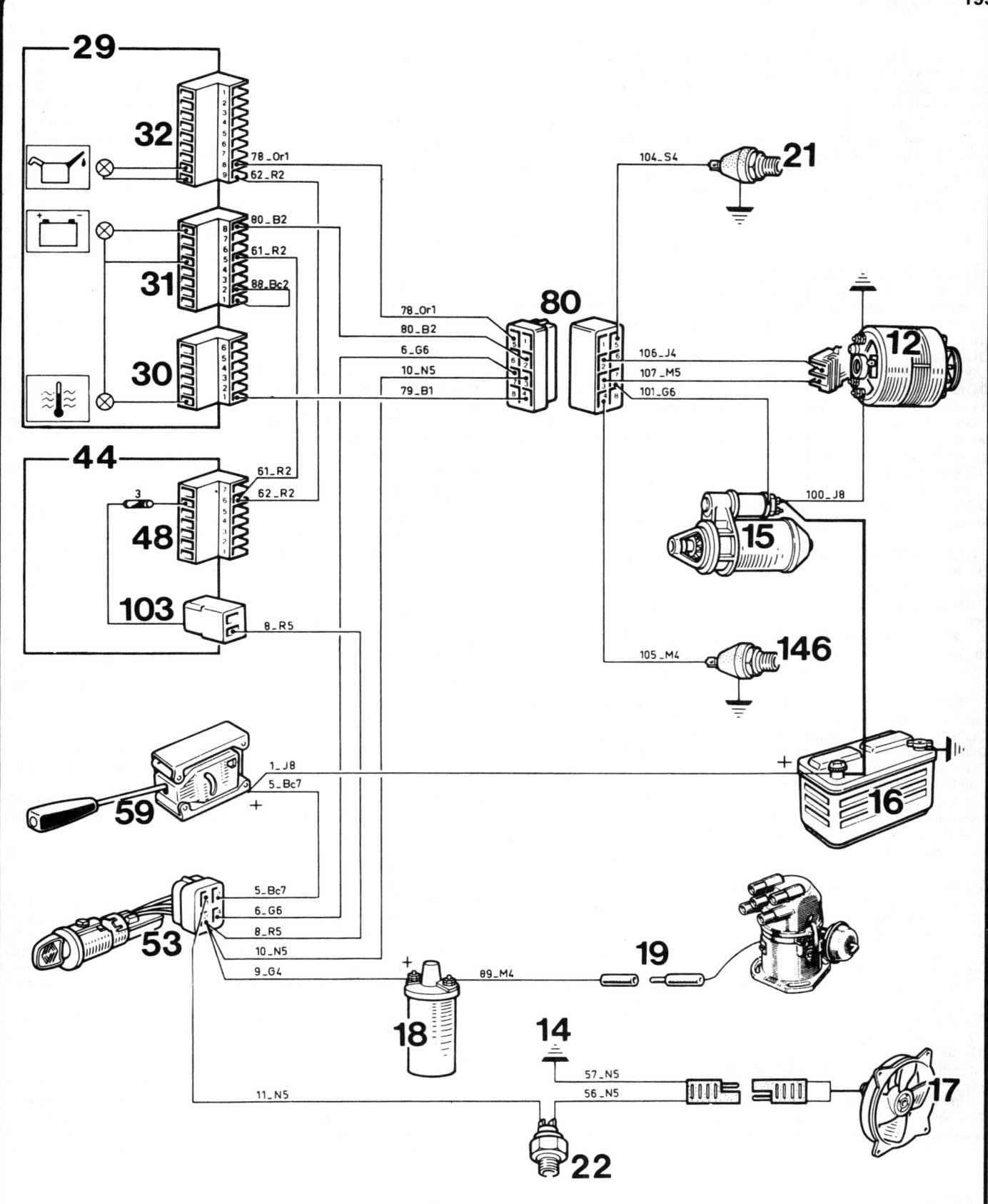

Fig. 12.82 Wiring diagram for 1982 R1213 (Part 5) – Charging circuit, starter circuit, ignition circuit, coolant temperature switch, engine cooling fan motor

Safety first!

Professional motor mechanics are trained in safe working procedures. However enthusiastic you may be about getting on with the job in hand, do take the time to ensure that your safety is not put at risk. A moment's lack of attention can result in an accident, as can failure to observe certain elementary precautions.

There will always be new ways of having accidents, and the following points do not pretend to be a comprehensive list of all dangers; they are intended rather to make you aware of the risks and to encourage a safety-conscious approach to all work you carry out on your vehicle.

Essential DOs and DON'Ts

DON'T rely on a single jack when working underneath the vehicle. Always use reliable additional means of support, such as axle stands, securely placed under a part of the vehicle that you know will not give way.

DON'T attempt to loosen or tighten high-torque nuts (e.g. wheel hub nuts) while the vehicle is on a jack; it may be pulled off.

DON'T start the engine without first ascertaining that the transmission is in neutral (or 'Park' where applicable) and the parking brake applied.

DON'T suddenly remove the filler cap from a hot cooling system – cover it with a cloth and release the pressure gradually first, or you may get scalded by escaping coolant.

DON'T attempt to drain oil until you are sure it has cooled sufficiently to avoid scalding you.

DON'T grasp any part of the engine, exhaust or catalytic converter without first ascertaining that it is sufficiently cool to avoid burning you.

DON'T allow brake fluid or antifreeze to contact vehicle paintwork.

DON'T syphon toxic liquids such as fuel, brake fluid or antifreeze by mouth, or allow them to remain on your skin.

DON'T inhale dust – it may be injurious to health (see *Asbestos* below).

DON'T allow any spilt oil or grease to remain on the floor – wipe it up straight away, before someone slips on it.

DON'T use ill-fitting spanners or other tools which may slip and cause injury.

DON'T attempt to lift a heavy component which may be beyond your capability – get assistance.

DON'T rush to finish a job, or take unverified short cuts.

DON'T allow children or animals in or around an unattended vehicle.

DO wear eye protection when using power tools such as drill, sander, bench grinder etc, and when working under the vehicle.

DO use a barrier cream on your hands prior to undertaking dirty jobs – it will protect your skin from infection as well as making the dirt easier to remove afterwards; but make sure your hands aren't left slippery. Note that long-term contact with used engine oil can be a health hazard.

DO keep loose clothing (cuffs, tie etc) and long hair well out of the way of moving mechanical parts.

DO remove rings, wristwatch etc, before working on the vehicle – especially the electrical system.

DO ensure that any lifting tackle used has a safe working load rating adequate for the job.

DO keep your work area tidy – it is only too easy to fall over articles left lying around.

DO get someone to check periodically that all is well, when working alone on the vehicle.

DO carry out work in a logical sequence and check that everything is correctly assembled and tightened afterwards.

DO remember that your vehicle's safety affects that of yourself and others. If in doubt on any point, get specialist advice.

IF, in spite of following these precautions, you are unfortunate enough to injure yourself, seek medical attention as soon as possible.

Asbestos

Certain friction, insulating, sealing, and other products – such as brake linings, brake bands, clutch linings, torque converters, gaskets, etc – contain asbestos. *Extreme care must be taken to avoid inhalation of dust from such products since it is hazardous to health.* If in doubt, assume that they *do* contain asbestos.

Fire

Remember at all times that petrol (gasoline) is highly flammable. Never smoke, or have any kind of naked flame around, when working on the vehicle. But the risk does not end there – a spark caused by an electrical short-circuit, by two metal surfaces contacting each other, by careless use of tools, or even by static electricity built up in your body under certain conditions, can ignite petrol vapour, which in a confined space is highly explosive.

Always disconnect the battery earth (ground) terminal before working on any part of the fuel or electrical system, and never risk spilling fuel on to a hot engine or exhaust.

It is recommended that a fire extinguisher of a type suitable for fuel and electrical fires is kept handy in the garage or workplace at all times. Never try to extinguish a fuel or electrical fire with water.

Note: *Any reference to a 'torch' appearing in this manual should always be taken to mean a hand-held battery-operated electric lamp or flashlight. It does NOT mean a welding/gas torch or blowlamp.*

Fumes

Certain fumes are highly toxic and can quickly cause unconsciousness and even death if inhaled to any extent. Petrol (gasoline) vapour comes into this category, as do the vapours from certain solvents such as trichloroethylene. Any draining or pouring of such volatile fluids should be done in a well ventilated area.

When using cleaning fluids and solvents, read the instructions carefully. Never use materials from unmarked containers – they may give off poisonous vapours.

Never run the engine of a motor vehicle in an enclosed space such as a garage. Exhaust fumes contain carbon monoxide which is extremely poisonous; if you need to run the engine, always do so in the open air or at least have the rear of the vehicle outside the workplace.

If you are fortunate enough to have the use of an inspection pit, never drain or pour petrol, and never run the engine, while the vehicle is standing over it; the fumes, being heavier than air, will concentrate in the pit with possibly lethal results.

The battery

Never cause a spark, or allow a naked light, near the vehicle's battery. It will normally be giving off a certain amount of hydrogen gas, which is highly explosive.

Always disconnect the battery earth (ground) terminal before working on the fuel or electrical systems.

If possible, loosen the filler plugs or cover when charging the battery from an external source. Do not charge at an excessive rate or the battery may burst.

Take care when topping up and when carrying the battery. The acid electrolyte, even when diluted, is very corrosive and should not be allowed to contact the eyes or skin.

If you ever need to prepare electrolyte yourself, always add the acid slowly to the water, and never the other way round. Protect against splashes by wearing rubber gloves and goggles.

When jump starting a car using a booster battery, for negative earth (ground) vehicles, connect the jump leads in the following sequence: First connect one jump lead between the positive (+) terminals of the two batteries. Then connect the other jump lead first to the negative (–) terminal of the booster battery, and then to a good earthing (ground) point on the vehicle to be started, at least 18 in (45 cm) from the battery if possible. Ensure that hands and jump leads are clear of any moving parts, and that the two vehicles do not touch. Disconnect the leads in the reverse order.

Mains electricity

When using an electric power tool, inspection light etc, which works from the mains, always ensure that the appliance is correctly connected to its plug and that, where necessary, it is properly earthed (grounded). Do not use such appliances in damp conditions and, again, beware of creating a spark or applying excessive heat in the vicinity of fuel or fuel vapour.

Ignition HT voltage

A severe electric shock can result from touching certain parts of the ignition system, such as the HT leads, when the engine is running or being cranked, particularly if components are damp or the insulation is defective. Where an electronic ignition system is fitted, the HT voltage is much higher and could prove fatal.

Fault diagnosis

Introduction

The vehicle owner who does his or her own maintenance according to the recommended schedules should not have to use this section of the manual very often. Modern component reliability is such that, provided those items subject to wear or deterioration are inspected or renewed at the specified intervals, sudden failure is comparatively rare. Faults do not usually just happen as a result of sudden failure, but develop over a period of time. Major mechanical failures in particular are usually preceded by characteristic symptoms over hundreds or even thousands of miles. Those components which do occasionally fail without warning are often small and easily carried in the vehicle.

With any fault finding, the first step is to decide where to begin investigations. Sometimes this is obvious, but on other occasions a little detective work will be necessary. The owner who makes half a dozen haphazard adjustments or replacements may be successful in curing a fault (or its symptoms), but he will be none the wiser if the fault recurs and he may well have spent more time and money than was necessary. A calm and logical approach will be found to be more satisfactory in the long run. Always take into account any warning signs or abnormalities that may have been noticed in the period preceding the fault – power loss, high or low gauge readings, unusual noises or smells, etc – and remember that failure of components such as fuses or spark plugs may only be pointers to some underlying fault.

The pages which follow here are intended to help in cases of failure to start or breakdown on the road. There is also a Fault Diagnosis Section at the end of each Chapter which should be consulted if the preliminary checks prove unfruitful. Whatever the fault, certain basic principles apply. These are as follows:

Verify the fault. This is simply a matter of being sure that you know what the symptoms are before starting work. This is particularly important if you are investigating a fault for someone else who may not have described it very accurately.

Don't overlook the obvious. For example, if the vehicle won't start, is there petrol in the tank? (Don't take anyone else's word on this particular point, and don't trust the fuel gauge either!) If an electrical fault is indicated, look for loose or broken wires before digging out the test gear.

Cure the disease, not the symptom. Substituting a flat battery with a fully charged one will get you off the hard shoulder, but if the underlying cause is not attended to, the new battery will go the same way. Similarly, changing oil-fouled spark plugs for a new set will get you moving again, but remember that the reason for the fouling (if it wasn't simply an incorrect grade of plug) will have to be established and corrected.

Don't take anything for granted. Particularly, don't forget that a 'new' component may itself be defective (especially if it's been rattling round in the boot for months), and don't leave components out of a fault diagnosis sequence just because they are new or recently fitted. When you do finally diagnose a difficult fault, you'll probably realise that all the evidence was there from the start.

Electrical faults

Electrical faults can be more puzzling than straightforward mechanical failures, but they are no less susceptible to logical analysis if the basic principles of operation are understood. Vehicle electrical wiring exists in extremely unfavourable conditions – heat, vibration and chemical attack – and the first things to look for are loose or corroded connections and broken or chafed wires, especially where the wires pass through holes in the bodywork or are subject to vibration.

All metal-bodied vehicles in current production have one pole of the battery 'earthed', ie connected to the vehicle bodywork, and in nearly all modern vehicles it is the negative (–) terminal. The various electrical components – motors, bulb holders etc – are also connected to earth, either by means of a lead or directly by their mountings. Electric current flows through the component and then back to the battery via the bodywork. If the component mounting is loose or corroded, or if a good path back to the battery is not available, the circuit will be incomplete and malfunction will result. The engine and/or gearbox are also earthed by means of flexible metal straps to the body or subframe; if these straps are loose or missing, starter motor, generator and ignition trouble may result.

Assuming the earth return to be satisfactory, electrical faults will be due either to component malfunction or to defects in the current supply. Individual components are dealt with in Chapter 9. If supply wires are broken or cracked internally this results in an open-circuit, and the easiest way to check for this is to bypass the suspect wire temporarily with a length of wire having a crocodile clip or suitable connector at each end. Alternatively, a 12V test lamp can be used to verify the presence of supply voltage at various points along the wire and the break can be thus isolated.

If a bare portion of a live wire touches the bodywork or other earthed metal part, the electricity will take the low-resistance path thus formed back to the battery: this is known as a short-circuit. Hopefully a short-circuit will blow a fuse, but otherwise it may cause burning of the insulation (and possibly further short-circuits) or even a fire. This is why it is inadvisable to bypass persistently blowing fuses with silver foil or wire.

Spares and tool kit

Most vehicles are supplied only with sufficient tools for wheel changing; the *Maintenance and minor repair* tool kit detailed in *Tools and working facilities*, with the addition of a hammer, is probably

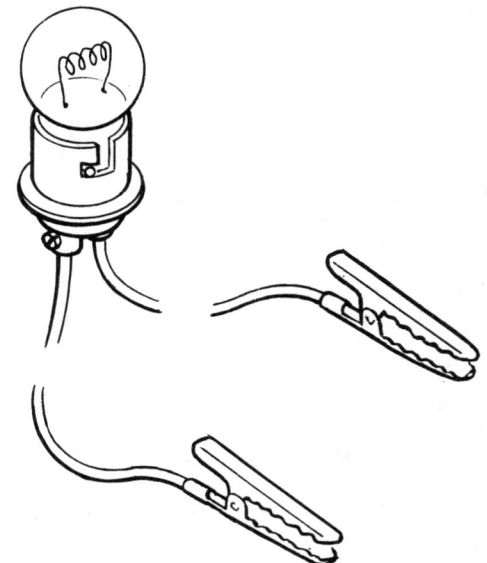

A simple test lamp is useful for investigating electrical faults

sufficient for those repairs that most motorists would consider attempting at the roadside. In addition a few items which can be fitted without too much trouble in the event of a breakdown should be carried. Experience and available space will modify the list below, but the following may save having to call on professional assistance:

Spark plugs, clean and correctly gapped
HT lead and plug cap – long enough to reach the plug furthest from the distributor
Distributor rotor, condenser and contact breaker points
Drivebelt(s) – emergency type may suffice
Spare fuses
Set of principal light bulbs
Tin of radiator sealer and hose bandage
Exhaust bandage
Roll of insulating tape
Length of soft iron wire
Length of electrical flex
Torch or inspection lamp (can double as test lamp)
Battery jump leads
Tow-rope
Ignition waterproofing aerosol
Litre of engine oil
Sealed can of hydraulic fluid
Emergency windscreen
'Jubilee' clips
Tube of filler paste

If spare fuel is carried, a can designed for the purpose should be used to minimise risks of leakage and collision damage. A first aid kit and a warning triangle, whilst not at present compulsory in the UK, are obviously sensible items to carry in addition to the above.

When touring abroad it may be advisable to carry additional spares which, even if you cannot fit them yourself, could save having to wait while parts are obtained. The items below may be worth considering:

Clutch and throttle cables
Cylinder head gasket
Dynamo or alternator brushes
Fuel pump repair kit
Tyre valve core

One of the motoring organisations will be able to advise on availability of fuel etc in foreign countries.

Engine will not start

Engine fails to turn when starter operated
Flat battery (recharge, use jump leads, or push start)
Battery terminals loose or corroded
Battery earth to body defective
Engine earth strap loose or broken
Starter motor (or solenoid) wiring loose or broken
Automatic transmission selector in wrong position, or inhibitor switch faulty
Ignition/starter switch faulty
Major mechanical failure (seizure)
Starter or solenoid internal fault (see Chapter 9)

Starter motor turns engine slowly
Partially discharged battery (recharge, use jump leads, or push start)
Battery terminals loose or corroded
Battery earth to body defective
Engine earth strap loose
Starter motor (or solenoid) wiring loose
Starter motor internal fault (see Chapter 9)

Starter motor spins without turning engine
Flat battery
Starter motor pinion sticking on sleeve
Flywheel gear teeth damaged or worn
Starter motor mounting bolts loose

Engine turns normally but fails to start
Damp or dirty HT leads and distributor cap (crank engine and check for spark)
Dirty or incorrectly gapped distributor points (if applicable)
No fuel in tank (check for delivery at carburettor)
Excessive choke (hot engine) or insufficient choke (cold engine)
Fouled or incorrectly gapped spark plugs (remove, clean and regap)
Other ignition system fault (see Chapter 4)
Other fuel system fault (see Chapter 3)
Poor compression (see Chapter 1)
Major mechanical failure (eg camshaft drive)

Engine fires but will not run
Insufficient choke (cold engine)
Air leaks at carburettor or inlet manifold
Fuel starvation (see Chapter 3)
Ballast resistor defective, or other ignition fault (see Chapter 4)

Engine cuts out and will not restart

Engine cuts out suddenly – ignition fault
Loose or disconnected LT wires
Wet HT leads or distributor cap (after traversing water splash)
Coil or condenser failure (check for spark)
Other ignition fault (see Chapter 4)

Engine misfires before cutting out – fuel fault
Fuel tank empty
Fuel pump defective or filter blocked (check for delivery)
Fuel tank filler vent blocked (suction will be evident on releasing cap)
Carburettor needle valve sticking
Carburettor jets blocked (fuel contaminated)
Other fuel system fault (see Chapter 3)

Engine cuts out – other causes
Serious overheating
Major mechanical failure (eg camshaft drive)

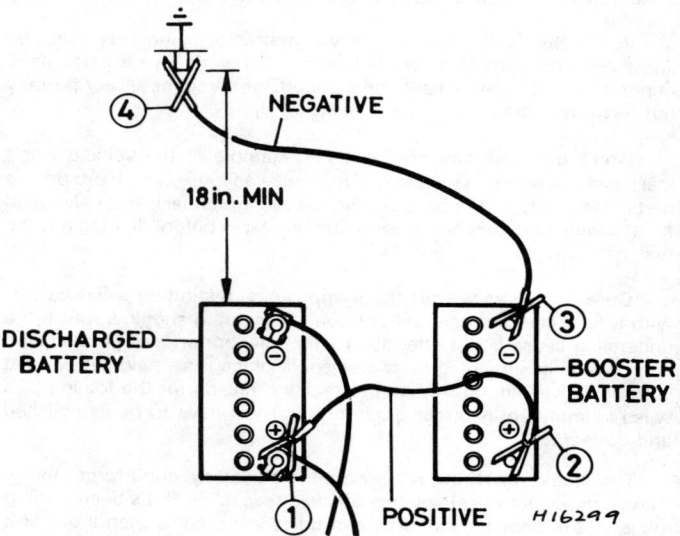

Jump start lead connections for negative earth – correct leads in order shown

Engine overheats

Ignition (no-charge) warning light illuminated
Slack or broken drivebelt – retension or renew (Chapter 2)

Fault diagnosis

Ignition warning light not illuminated
  Coolant loss due to internal or external leakage (see Chapter 2)
  Thermostat defective
  Low oil level
  Brakes binding
  Radiator clogged externally or internally
  Electric cooling fan not operating correctly
  Engine waterways clogged
  Ignition timing incorrect or automatic advance malfunctioning
  Mixture too weak

Note: *Do not add cold water to an overheated engine or damage may result*

Low engine oil pressure

Gauge reads low or warning light illuminated with engine running
  Oil level low or incorrect grade
  Defective gauge or sender unit
  Wire to sender unit earthed
  Engine overheating
  Oil filter clogged or bypass valve defective
  Oil pressure relief valve defective
  Oil pick-up strainer clogged
  Oil pump worn or mountings loose
  Worn main or big-end bearings

Note: *Low oil pressure in a high-mileage engine at tickover is not necessarily a cause for concern. Sudden pressure loss at speed is far more significant. In any event, check the gauge or warning light sender before condemning the engine.*

Engine noises

Pre-ignition (pinking) on acceleration
  Incorrect grade of fuel
  Ignition timing incorrect
  Distributor faulty or worn
  Worn or maladjusted carburettor
  Excessive carbon build-up in engine

Whistling or wheezing noises
  Leaking vacuum hose
  Leaking carburettor or manifold gasket
  Blowing head gasket

Tapping or rattling
  Incorrect valve clearances
  Worn valve gear
  Worn timing chain or belt
  Broken piston ring (ticking noise)

Knocking or thumping
  Unintentional mechanical contact (eg fan blades)
  Worn fanbelt
  Peripheral component fault (generator, water pump etc)
  Worn big-end bearings (regular heavy knocking, perhaps less under load)
  Worn main bearings (rumbling and knocking, perhaps worsening under load)
  Piston slap (most noticeable when cold)

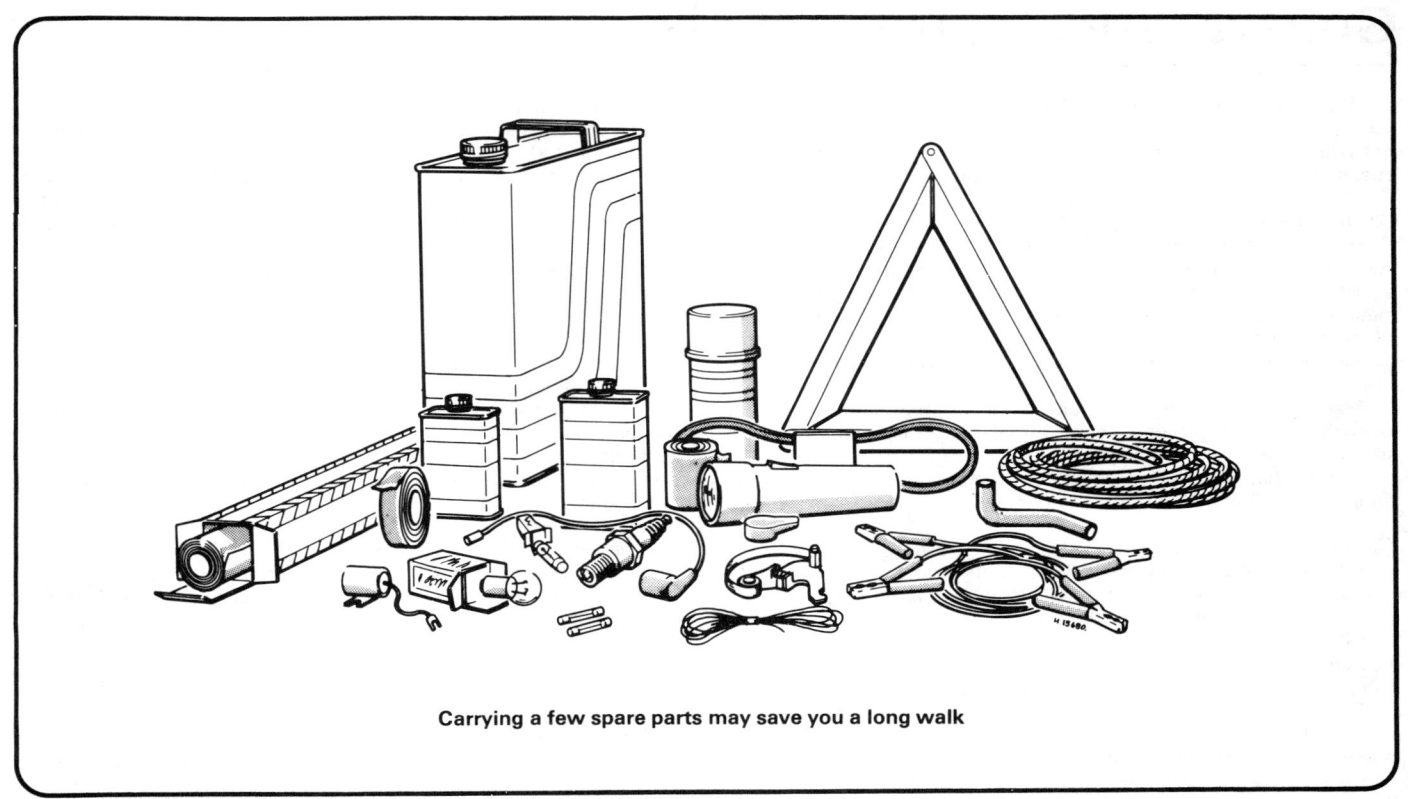

Carrying a few spare parts may save you a long walk

General repair procedures

Whenever servicing, repair or overhaul work is carried out on the car or its components, it is necessary to observe the following procedures and instructions. This will assist in carrying out the operation efficiently and to a professional standard of workmanship.

Joint mating faces and gaskets

Where a gasket is used between the mating faces of two components, ensure that it is renewed on reassembly, and fit it dry unless otherwise stated in the repair procedure. Make sure that the mating faces are clean and dry with all traces of old gasket removed. When cleaning a joint face, use a tool which is not likely to score or damage the face, and remove any burrs or nicks with an oilstone or fine file.

Make sure that tapped holes are cleaned with a pipe cleaner, and keep them free of jointing compound if this is being used unless specifically instructed otherwise.

Ensure that all orifices, channels or pipes are clear and blow through them, preferably using compressed air.

Oil seals

Whenever an oil seal is removed from its working location, either individually or as part of an assembly, it should be renewed.

The very fine sealing lip of the seal is easily damaged and will not seal if the surface it contacts is not completely clean and free from scratches, nicks or grooves. If the original sealing surface of the component cannot be restored, the component should be renewed.

Protect the lips of the seal from any surface which may damage them in the course of fitting. Use tape or a conical sleeve where possible. Lubricate the seal lips with oil before fitting and, on dual lipped seals, fill the space between the lips with grease.

Unless otherwise stated, oil seals must be fitted with their sealing lips toward the lubricant to be sealed.

Use a tubular drift or block of wood of the appropriate size to install the seal and, if the seal housing is shouldered, drive the seal down to the shoulder. If the seal housing is unshouldered, the seal should be fitted with its face flush with the housing top face.

Screw threads and fastenings

Always ensure that a blind tapped hole is completely free from oil, grease, water or other fluid before installing the bolt or stud. Failure to do this could cause the housing to crack due to the hydraulic action of the bolt or stud as it is screwed in.

When tightening a castellated nut to accept a split pin, tighten the nut to the specified torque, where applicable, and then tighten further to the next split pin hole. Never slacken the nut to align a split pin hole unless stated in the repair procedure.

When checking or retightening a nut or bolt to a specified torque setting, slacken the nut or bolt by a quarter of a turn, and then retighten to the specified setting.

Locknuts, locktabs and washers

Any fastening which will rotate against a component or housing in the course of tightening should always have a washer between it and the relevant component or housing.

Spring or split washers should always be renewed when they are used to lock a critical component such as a big-end bearing retaining nut or bolt.

Locktabs which are folded over to retain a nut or bolt should always be renewed.

Self-locking nuts can be reused in non-critical areas, providing resistance can be felt when the locking portion passes over the bolt or stud thread.

Split pins must always be replaced with new ones of the correct size for the hole.

Special tools

Some repair procedures in this manual entail the use of special tools such as a press, two or three-legged pullers, spring compressors etc. Wherever possible, suitable readily available alternatives to the manufacturer's special tools are described, and are shown in use. In some instances, where no alternative is possible, it has been necessary to resort to the use of a manufacturer's tool and this has been done for reasons of safety as well as the efficient completion of the repair operation. Unless you are highly skilled and have a thorough understanding of the procedure described, never attempt to bypass the use of any special tool when the procedure described specifies its use. Not only is there a very great risk of personal injury, but expensive damage could be caused to the components involved.

Conversion factors

Length (distance)
Inches (in)	X	25.4	= Millimetres (mm)	X 0.0394	= Inches (in)
Feet (ft)	X	0.305	= Metres (m)	X 3.281	= Feet (ft)
Miles	X	1.609	= Kilometres (km)	X 0.621	= Miles

Volume (capacity)
Cubic inches (cu in; in^3)	X	16.387	= Cubic centimetres (cc; cm^3)	X 0.061	= Cubic inches (cu in; in^3)
Imperial pints (Imp pt)	X	0.568	= Litres (l)	X 1.76	= Imperial pints (Imp pt)
Imperial quarts (Imp qt)	X	1.137	= Litres (l)	X 0.88	= Imperial quarts (Imp qt)
Imperial quarts (Imp qt)	X	1.201	= US quarts (US qt)	X 0.833	= Imperial quarts (Imp qt)
US quarts (US qt)	X	0.946	= Litres (l)	X 1.057	= US quarts (US qt)
Imperial gallons (Imp gal)	X	4.546	= Litres (l)	X 0.22	= Imperial gallons (Imp gal)
Imperial gallons (Imp gal)	X	1.201	= US gallons (US gal)	X 0.833	= Imperial gallons (Imp gal)
US gallons (US gal)	X	3.785	= Litres (l)	X 0.264	= US gallons (US gal)

Mass (weight)
Ounces (oz)	X	28.35	= Grams (g)	X 0.035	= Ounces (oz)
Pounds (lb)	X	0.454	= Kilograms (kg)	X 2.205	= Pounds (lb)

Force
Ounces-force (ozf; oz)	X	0.278	= Newtons (N)	X 3.6	= Ounces-force (ozf; oz)
Pounds-force (lbf; lb)	X	4.448	= Newtons (N)	X 0.225	= Pounds-force (lbf; lb)
Newtons (N)	X	0.1	= Kilograms-force (kgf; kg)	X 9.81	= Newtons (N)

Pressure
Pounds-force per square inch (psi; lbf/in^2; lb/in^2)	X	0.070	= Kilograms-force per square centimetre (kgf/cm^2; kg/cm^2)	X 14.223	= Pounds-force per square inch (psi; lbf/in^2; lb/in^2)
Pounds-force per square inch (psi; lbf/in^2; lb/in^2)	X	0.068	= Atmospheres (atm)	X 14.696	= Pounds-force per square inch (psi; lbf/in^2; lb/in^2)
Pounds-force per square inch (psi; lbf/in^2; lb/in^2)	X	0.069	= Bars	X 14.5	= Pounds-force per square inch (psi; lbf/in^2; lb/in^2)
Pounds-force per square inch (psi; lbf/in^2; lb/in^2)	X	6.895	= Kilopascals (kPa)	X 0.145	= Pounds-force per square inch (psi; lbf/in^2; lb/in^2)
Kilopascals (kPa)	X	0.01	= Kilograms-force per square centimetre (kgf/cm^2; kg/cm^2)	X 98.1	= Kilopascals (kPa)

Torque (moment of force)
Pounds-force inches (lbf in; lb in)	X	1.152	= Kilograms-force centimetre (kgf cm; kg cm)	X 0.868	= Pounds-force inches (lbf in; lb in)
Pounds-force inches (lbf in; lb in)	X	0.113	= Newton metres (Nm)	X 8.85	= Pounds-force inches (lbf in; lb in)
Pounds-force inches (lbf in; lb in)	X	0.083	= Pounds-force feet (lbf ft; lb ft)	X 12	= Pounds-force inches (lbf in; lb in)
Pounds-force feet (lbf ft; lb ft)	X	0.138	= Kilograms-force metres (kgf m; kg m)	X 7.233	= Pounds-force feet (lbf ft; lb ft)
Pounds-force feet (lbf ft; lb ft)	X	1.356	= Newton metres (Nm)	X 0.738	= Pounds-force feet (lbf ft; lb ft)
Newton metres (Nm)	X	0.102	= Kilograms-force metres (kgf m; kg m)	X 9.804	= Newton metres (Nm)

Power
Horsepower (hp)	X	745.7	= Watts (W)	X 0.0013	= Horsepower (hp)

Velocity (speed)
Miles per hour (miles/hr; mph)	X	1.609	= Kilometres per hour (km/hr; kph)	X 0.621	= Miles per hour (miles/hr; mph)

*Fuel consumption**
Miles per gallon, Imperial (mpg)	X	0.354	= Kilometres per litre (km/l)	X 2.825	= Miles per gallon, Imperial (mpg)
Miles per gallon, US (mpg)	X	0.425	= Kilometres per litre (km/l)	X 2.352	= Miles per gallon, US (mpg)

Temperature
Degrees Fahrenheit = (°C x 1.8) + 32   Degrees Celsius (Degrees Centigrade; °C) = (°F - 32) x 0.56

It is common practice to convert from miles per gallon (mpg) to litres/100 kilometres (l/100km), where mpg (Imperial) x l/100 km = 282 and mpg (US) x l/100 km = 235

Index

A

About this manual – 2
Acknowledgements – 2
Air filter element
   removal and installation – 53
Alternator
   brush holder renewal – 165
   description and precautions – 107
   drivebelt renewal and adjustment – 46
   integral regulator renewal – 164
   later models – 164
   new for old exchange – 165
   removal and refitting – 109
Antifreeze – 49
Anti-roll bar
   removal and refitting – 124

B

Battery
   charge indicator – 173
   charging – 106
   maintenance, removal and refitting – 105
Bearings
   big-end – 28
   hub – 91
   main – 24, 28, 33
Big-end bearings
   examination and renovation – 28
Bleeding – 95, 164
Body damage repair
   major – 138
   minor – 137
Bodywork and fittings – 136 *et seq*, 177
Bodywork and fittings
   bonnet – 139, 140
   bumpers – 146
   dashboard – 146
   doors – 138, 140, 141, 145, 166
   general description – 146
   grille panel – 146
   locks – 145, 166
   maintenance – 136, 137
   roof embellisher strips – 148
   specifications – 136
   tailgate – 145, 166, 167
   towing bracket – 177
   windows – 141, 146, 166
   wing panels – 139
   wiper arms and blades – 148
Bonnet
   release cable – 140
   removal and refitting – 139
Braking system – 93 *et seq*, 160 *et seq*
Braking system
   adjustments – 96
   bleeding – 95, 164
   brake shoes – 99, 161
   disc – 98, 163
   disc brake caliper – 97, 163
   disc pads – 96, 161
   drum brakes (rear) – 99, 161, 164
   fault diagnosis – 104
   general description – 95, 160
   handbrake – 104
   hydraulic pipes – 101
   limiter – 103
   master cylinder – 101, 160
   routine maintenance – 95
   servo unit – 101, 103
   specifications – 93
   wheel cylinder – 100, 164
Bulb renewal
   courtesy light – 113
   front combination – 113
   headlight – 113
   number plate – 113
   rear combination – 113
Bumper
   removal and refitting – 146
Buying spare parts – 5

C

Cables
   choke – 59
   clutch – 71
   heater – 51
   speedometer – 117
Caliper (disc brakes)
   removal, overhaul and refitting
      ATE – 97
      Bendix – 163
Camshaft
   examination and renovation – 29
   removal (with cylinder head) – 20
Carburettor *see*
   Solex 32-35 CICSA
   Solex 32 PBISA
   Solex 32 SHA
   Weber 32 IBR
Central door locking
   description – 165
   removal and refitting – 166
Choke cable
   removal and refitting – 59
Clutch – 70 *et seq*
Clutch
   actuating rod/fork and release bearing – 73
   adjustment – 70
   cable – 71
   fault diagnosis – 73
   general description – 70
   housing – 23, 41, 73
   inspection and renovation – 72
   installation – 72
   removal – 71

Index

specifications – 70
Coil (ignition) – 67
Coil springs (front)
  removal and refitting – 125
Combination lights
  front – 113
  rear – 113
Condenser
  testing, removal and refitting – 69
Contact breaker points
  adjustment – 63
  removal and refitting – 63
Conversion factors – 201
Cooling system – 45 *et seq*
Cooling system
  antifreeze – 49
  cooling fan – 49
  draining – 47
  drivebelt (water pump) – 46
  expansion bottle – 50
  fault diagnosis – 52
  filling – 47
  flushing – 47
  general description – 45
  heating system – 50, 51
  radiator – 49
  specifications – 45
  temperature gauge – 173
  thermal switch (cooling fan) – 50
  thermostat – 49
  water pump – 50
  water temperature sender unit – 50
Courtesy light bulb
  replacement – 113
Crankshaft
  examination and renovation – 28
  refitting – 33
  removal – 24
Cylinder head
  dismantling, inspection and renovation – 27
  installation – 35
  removal – 20
Cylinder liners
  examination and renovation – 28
  installation – 31

D

Damage repair (body)
  major – 138
  minor – 137
Dashboard
  removal and refitting – 146
Diagnostic socket (ignition) – 69
Differential – 89
  general – 78
  oil seal replacement – 86
Direction indicator switch
  removal and refitting – 114
Disc brake caliper
  removal, overhaul and refitting
    ATE – 97
    Bendix – 163
Disc
  examination, removal and refitting
    ATE – 98
    Bendix – 163
Disc pads
  inspection and refitting
    ATE – 96
    Bendix – 161
Distributor
  dismantling, inspection and reassembly – 66
  removal and installation – 66
  routine maintenance – 62

Doors
  locks – 145, 166
  rattles – 138
  removal, refitting and adjustment – 140
  trim and windows – 141
Driveshafts and hubs
  driveshaft joints – 89
  fault diagnosis – 135
  hub bearings – 91
  removal and refitting – 88
  specifications – 88
Driveshafts, hubs, wheels and tyres – 88 *et seq*
Driving lamps – 167
Drum brakes
  removal, inspection and refitting
    Bendix – 163
    Girling – 99

E

Electrical system – 105 *et seq*, 164 *et seq*
Electrical system
  alternator – 107, 109, 164, 165
  battery – 105, 106
  central door locking – 165, 166
  electric windows – 166
  fault diagnosis – 115, 117, 118
  foglamps – 167
  fuel tank sender unit – 116
  fuses – 111, 164
  general description – 105
  heated rear window – 166
  horn – 115
  instrument panel – 114
  instruments – 173
  interference suppression – 170
  lights – 113, 167
  radio – 168, 169
  rear wash/wipe – 167
  specifications – 105
  speedometer cable – 117
  starter motor – 109, 111
  switches – 59, 114, 116, 117
  warning lights – 173
  windscreen wipers and motor – 115, 116, 148, 167
  wiring diagrams – 119 to 121, 178 to 195
Engine – 14 *et seq*, 150 *et seq*
Engine
  big-end bearings – 28
  camshaft and rocker arms – 29
  clutch housing – 23, 41, 73
  crankcase – 24, 33
  crankshaft – 24, 28, 33
  cylinder head – 20, 27, 35
  cylinder liners – 28, 31
  dismantling (general) – 20
  examination and renovation (general) – 24
  fault diagnosis – 43, 44
  flywheel – 30, 39
  general description – 16, 150
  gudgeon pins – 29
  initial start-up after overhaul – 43
  installation – 42, 43, 73, 86
  lubrication system – 25
  main bearings – 24, 28, 33
  manifolds – 30
  mountings – 31
  oil filter – 25
  oil pressure switch – 117
  oil pump – 24, 30, 151
  operations possible with engine installed – 17
  operations requiring engine removal – 17
  pistons, piston rings and connecting rods – 24, 29, 31
  reassembly – 31, 41
  reconnection to transmission – 38, 86

removal – 17, 19
separation from transmission – 23
specifications – 14, 15, 149
timing chain and sprockets – 24, 29, 37, 150
timing cover – 23, 38, 150
transfer gears – 23, 29, 41
valve rocker clearances – 39
Exhaust system (general) – 59, 157
Expansion bottle (cooling system) – 50

F

Fan (cooling)
removal and refitting – 49
thermal switch – 50
Fan motor (heater) – 51
Fault diagnosis
braking system – 104
clutch – 73
cooling system – 52
driveshafts, hubs, wheels and tyres – 92, 135
electrical system – 117, 118
engine – 43, 44
fuel and exhaust systems – 61
general – 197 to 199
ignition system – 69
suspension and steering – 135
transmission – 86
windscreen wipers and drive motor – 115
Flywheel
examination and renovation – 30
installation – 39
Foglamps – 167
Fuel and exhaust systems – 53 *et seq*, 151 *et seq*
Fuel pump
removal, servicing and refitting – 59
Fuel system
air filter element – 53
choke cable – 59
fault diagnosis – 61
fuel gauge – 173
fuel pump – 59
fuel tank – 60, 116
general description – 53, 151
Solex 32-35 CICSA carburettor – 151, 156, 157
Solex 32 PBISA carburettor – 151 to 153
Solex 32 SHA carburettor – 53, 55, 57
specifications – 53, 150
Weber 32 IBR carburettor – 151, 153, 155
Fuel tank
gauge – 173
removal and refitting – 60
sender unit – 116
Fuses (general) – 111, 164

G

Gearbox *see* **Transmission**
Gearchange mechanism
removal and installation – 86, 158
General repair procedures – 200
Grille panel (front)
removal and refitting – 146
Gudgeon pin removal – 29

H

Handbrake – 104
Hazard warning switch – 117
Headlights
alignment – 113
bulb replacement – 113
removal – 113

Heated rear window – 166
Heating system
control cables – 51
fan motor – 51
general – 50
heater valve – 51
matrix – 51
Horn – 115
HT leads – 68
Hub bearings
removal and refitting
front wheels – 91
rear wheels – 91
Hydraulic fluid pipes
inspection and overhaul – 101
Hydraulic system bleeding – 95, 164

I

Ignition system – 62 *et seq*
Ignition system
condenser – 69
contact breaker points – 63
diagnostic socket and pick up – 69
distributor – 62, 66
fault diagnosis – 69
general description – 62
HT leads – 68
ignition coil – 67
ignition switch – 69
ignition timing – 67
routine maintenance – 62
spark plugs – 62, 68
specifications – 62, 150
Indicator switch
removal and refitting – 114
Instruments
general – 173
panel removal and refitting – 114
Interference suppression (radio) – 170, 171
Interior light bulb replacement – 113
Introduction – 2

J

Jacking – 8

L

Light (interior) bulb
replacement – 113
Lights (front)
auxiliary fog lamps – 167
driving lamps – 167
headlight alignment – 113
removal and bulb replacement – 113
Lights (rear)
fog lamp – 167
number plate lamp – 113
removal and bulb replacement – 113
Light switch
removal and replacement – 114
Limiter (braking system) – 103
Locks
electric – 165, 166
manual – 145
Loudspeakers installation – 168, 169
Lubricants and fluids – 10

M

MacPherson strut
removal and refitting – 125

Index

Main bearings
    examination and renovation – 28
    housing removal – 24
    reassembly – 33
Maintenance *see* **Routine maintenance**
Manifolds inspection – 30
Master cylinder
    removal, overhaul and refitting
        ATE – 101
        Bendix – 160
Matrix (heater) – 51

O

Oil filter
    removal and refitting – 25
Oil pressure switch
    removal and refitting – 117
Oil pump
    150 and X5J engines – 151
    examination and renovation – 30
    removal – 24

P

Pads (brakes)
    inspection and renewal
        ATE – 96
        Bendix – 161
Pistons, piston rings and connecting rods
    examination and renovation – 29
    installation – 31
    removal – 24

R

Radiator
    removal and refitting – 49
Radio
    installation – 168
    interference suppression – 170, 171
    loudspeakers – 168, 169
    wiring connections – 169, 170
Release bearing (clutch)
    removal, inspection and installation – 73
Reverse gear selection fault rectification – 158
Reversing light switch
    removal and refitting – 116
Rocker arms
    examination and renovation – 29
Rocker (valve) clearances – 39
Roof embellisher strips
    removal and refitting – 148
Routine maintenance
    bodywork and fittings – 136, 137
    braking system – 95, 101
    general (service intervals) – 11 to 13
    ignition system – 62
    suspension and steering – 123

S

Safety first! – 196
Service intervals (routine maintenance) – 11 to 13
Servo unit (brakes)
    general and maintenance – 101
    removal and installation – 103
Shock absorbers (rear)
    removal and refitting – 127
Shoes (brakes)
    Bendix – 161
    Girling – 99

Solex 32-35 CICSA carburettor
    adjustments – 156, 157
    description – 156
    general – 151
    idle adjustment – 151
Solex 32 PBISA carburettor
    adjustments – 152, 153
    description – 152
    general – 151
    idle adjustment – 151
Solex 32 SHA carburettor
    adjustments – 55
    description – 53
    dismantling, overhaul and reassembly – 57
    removal and installation – 57
Spare parts (buying) – 5
Spark plugs
    general – 68
    routine maintenance – 62
Specifications
    bodywork and fittings – 136
    braking system – 93
    clutch – 70
    cooling system – 45
    driveshafts, hubs, wheels and tyres – 88
    electrical system – 105
    engine – 14, 15, 149
    fuel system – 53, 150
    ignition system – 62, 150
    suspension and steering – 122
    transmission – 74, 150
Sprockets (timing)
    examination and renovation – 29
    installation – 37
Speedometer cable
    removal and refitting – 117
Starter motor
    dismantling and reassembly – 109
    drive pinion inspection and repair – 111
    removal and refitting – 109
    testing – 109
Steering
    arms – 133
    column intermediate shaft – 133
    rack – 133
    shaft lower coupling – 133
    wheel and column bushes – 131
Suspension and steering – 122 et seq
Suspension and steering
    alignment and geometry – 131
    anti-roll bar – 124
    coil springs – 125
    fault diagnosis – 135
    general description – 123
    MacPherson strut – 125
    routine maintenance – 123
    shock absorbers – 127
    specifications – 122
    steering – 131, 133
    suspension arm – 128
    testing – 123
    torsion bars – 127
Suspension arm
    removal and refitting – 128
Switch
    removal and installation
        direction indicator – 114
        hazard warning – 117
        ignition – 69
        light – 114
        oil pressure 117
        reversing light – 116
        thermal (cooling fan) – 50

Index

T

Tachometer – 173
Tailgate
    glass preparation and replacement – 146
    removal and refitting – 145
    wash/wipe system – 167
Temperature gauge – 173
Thermal switch (cooling fan)
    removal and refitting – 50
Thermostat
    removal, testing and refitting – 49
Timing chain and sprockets
    chain guide (150 engine) – 150
    examination and renovation – 29
    installation – 37
Timing cover
    installation – 38
    removal and installation (engine in car) – 23, 150
Timing (ignition) adjustment – 67
Tools and working facilities – 6, 7
Torsion bars
    removal, refitting and ride height adjustment – 127
Towing – 8, 177
Transfer gears
    inspection and renovation – 29
    refitting – 41
    removal – 23
Transmission – 74 *et seq*, 158 *et seq*
Transmission
    component inspection – 78
    'differential – 78, 86
    dismantling – 77
    fault diagnosis – 86
    gearchange mechanism – 86, 158
    general description – 75, 158
    installation – 42, 43, 73, 86
    primary shaft (5th speed gear) – 158
    reassembly – 79
    reconnection to engine – 38, 86
    removal – 17, 19, 75
    reverse fault (5-speed) – 158
    reversing light switch – 116
    secondary shaft (5th speed gear) – 158
    separation from engine – 23
    specifications – 74, 150
    speedometer cable – 117
Trim panel (door)
    removal and refitting – 141
Tyres *see* **Wheels and tyres**

V

Valve rocker clearances – 39
Vehicle identification numbers – 5

W

Water pump
    drivebelt – 46
    removal and refitting – 50
Water temperature sender unit
    removal and refitting – 50
Weber 32 IBR carburettor
    adjustments – 153, 155
    description – 153
    general – 151
    idle adjustment – 151
Wheel cylinder (rear brakes)
    removal, overhaul and refitting
        Bendix – 164
        Girling – 100
Wheels and tyres
    general – 92
    specifications – 88
Windows
    door (manual) – 14
    electric – 166
    tailgate – 146, 166
    windscreen – 146
Windscreen wash/wipe system
    fault diagnosis – 115
    general – 116
    removal and refitting – 115
    wiper arms and blades – 148
Wing panels
    removal and refitting – 139
Wiring diagrams – 119 to 121, 178 to 195
Working facilities – 6, 7

Printed by
J H Haynes & Co Ltd
Sparkford Nr Yeovil
Somerset BA22 7JJ England